Concepts and Strategies in Plant Sciences

Series Editor
Chittaranjan Kole, Raja Ramanna Fellow, Government of India, ICAR-National Institute for Plant Biotechnology, Pusa, Delhi, India

This book series highlights the spectacular advances in the concepts, techniques and tools in various areas of plant science. Individual volumes may cover topics like genome editing, phenotyping, molecular pharming, bioremediation, miRNA, fast-track breeding, crop evolution, IPR and farmers' rights, to name just a few. The books will demonstrate how advanced strategies in plant science can be utilized to develop and improve agriculture, ecology and the environment. The series will be of interest to students, scientists and professionals working in the fields of plant genetics, genomics, breeding, biotechnology, and in the related disciplines of plant production, improvement and protection.

Interested in editing a volume? Please contact Prof. Chittaranjan Kole, Series Editor, at ckoleorg@gmail.com

Shalini Tiwari
Jose L Gonzalez-Hernandez • Charu Lata
Editors

Decoding Non-Coding RNA

Enhancing Plant Abiotic Stress Tolerance

Springer

Editors
Shalini Tiwari
Agricultural and Biosystems Engineering
South Dakota State University
Brookings, SD, USA

Jose L Gonzalez-Hernandez
South Dakota State University
Brookings, SD, USA

Charu Lata
CSIR - National Institute of Science Communication and Policy Research (NIScPR)
New Delhi, Delhi, India

ISSN 2662-3188 ISSN 2662-3196 (electronic)
Concepts and Strategies in Plant Sciences
ISBN 978-981-96-9858-5 ISBN 978-981-96-9859-2 (eBook)
https://doi.org/10.1007/978-981-96-9859-2

This Springer imprint is published by the registered company Springer Nature Singapore Pte Ltd.
The registered company address is: 152 Beach Road, #21-01/04 Gateway East, Singapore 189721, Singapore

Preface

In the face of accelerating climate change and increasing global food demands, the need to understand and enhance plant resilience to abiotic stresses has never been more urgent. For decades, research in plant genomics focused predominantly on protein-coding regions, overlooking vast stretches of the genome once dismissed as "junk DNA." However, it is now evident that non-coding regions—including long non-coding RNAs (lncRNAs), small RNAs (miRNAs, siRNAs), introns, intergenic regions, and cis-regulatory elements—are central to gene regulation, stress signaling, and genome plasticity in plants.

This book, *Decoding Non-coding RNA: Enhancing Plant Abiotic Stress Tolerance*, compiles cutting-edge insights into the structural, functional, and evolutionary aspects of non-coding regions and their roles in plant adaptation to abiotic stressors such as drought, salinity, heat, and cold. Bringing together leading scientists from across the globe, this volume provides a comprehensive overview of how non-coding elements orchestrate complex regulatory networks that enable plants to perceive, respond to, and survive environmental challenges.

The opening chapters introduce readers to the diversity and functional relevance of non-coding regions, laying a foundation for understanding their regulatory potential. Subsequent chapters delve into the mechanisms through which non-coding RNAs and regulatory DNA elements influence abiotic stress responses, including transcriptional and epigenetic regulation, mobile genetic elements, and hormone-responsive signaling pathways. Special emphasis is placed on genome-wide profiling technologies, CRISPR-based functional analyses, and bioinformatics resources that are transforming our ability to explore and harness non-coding regions in crop improvement.

This book also addresses the evolutionary dynamics of stress-responsive regulatory sequences, highlighting their adaptive significance across plant lineages. The final chapters explore real-world applications in agriculture, with examples of how non-coding elements have been successfully utilized to enhance stress tolerance in crops and conclude with a critical reflection on the challenges and future directions of this rapidly evolving field.

We are grateful to the distinguished contributors whose expertise and dedication have shaped this volume. Their collective work bridges molecular biology, genomics, biotechnology, and systems biology, offering a forward-looking perspective on the untapped potential of non-coding regions. We hope this book will serve as a valuable reference for plant scientists, geneticists, breeders, and students aiming to develop climate-resilient crops and advance the frontiers of plant molecular biology.

We extend our sincere thanks to all chapter authors for their invaluable contributions and to our peer reviewers for their critical insights and constructive feedback. We are especially grateful to South Dakota State University and CSIR—National Institute of Science Communication and Policy Research for providing a supportive academic environment during the development of this volume. Additional thanks go to the editorial and production teams for their meticulous assistance in coordinating and compiling the book. This work was made possible in part through the collaboration and shared vision of researchers committed to advancing sustainable agriculture through scientific innovation.

Brookings, SD, USA Shalini Tiwari
Brookings, SD, USA Jose L. Gonzalez-Hernandez
New Delhi, Delhi, India Charu Lata

Contents

Editors and Contributors

About the Editors

Shalini Tiwari is a Post-Doctoral Research Associate at the South Dakota State University, USA. Her research interest is in the area of plant stress signaling and plant-microbe interaction. She has worked on how interactions between plants and PGPR (Plant Growth-Promoting Rhizobacteria) can enhance resilience to abiotic stresses. One of her notable achievements lies in elucidating these mechanisms by employing cutting-edge techniques such as transcriptomics and comprehensive profiling of non-coding RNAs, including long non-coding RNAs (lncRNA) and microRNAs (miRNA), particularly under stressful conditions. She actively contributes to the scientific community by advancing her understanding of environmental microbiology, plant biology, stress responses, and symbiotic relationships.

Jose L. Gonzalez Hernandez is working as a Professor at the South Dakota State University, USA. His research area is in the field of Plant Genetics and Genomics. His research focuses on unraveling the genetic basis of complex plant traits essential for agricultural sustainability and food security. He also utilizes state-of-the-art Next-Generation Sequencing (NGS) technologies to conduct comprehensive analyses of genomic regions associated with traits such as yield, disease resistance, and stress tolerance. His commitment to advancing scientific understanding and practical applications in plant genetics and genomics is crucial for addressing global challenges and fostering innovation in agricultural biotechnology.

Charu Lata is a Principal Scientist at the CSIR—National Institute of Science Communication and Policy Research, New Delhi. Her research interest has been in the field of crop improvements, especially abiotic stress responses and tolerance in crop plants. She is primarily focusing in the research areas pertaining to the elucidation of mechanisms regulating abiotic stress tolerance in legumes and cereal crops through transcriptomics, small RNA profiling, and genome editing. Her research underscores the critical role of genomics and molecular biology in addressing global agricultural challenges and advancing crop improvement strategies.

Contributors

Samreen Abbas Department of Biotechnology, Institute of Engineering and Technology, Lucknow, Uttar Pradesh, India

Bibek Adhikari Department of Biology/Microbiology, South Dakota State University, Brookings, SD, USA

Navjeet Ahalawat Department of Bioinformatics and Computational Biology, College of Biotechnology, CCS Haryana Agricultural University, Hisar, Haryana, India

T. G. Ajith Department of Plant Science, Central University of Kerala, Kasaragod, Kerala, India

P. P. Amritha Department of Plant Science, Central University of Kerala, Kasaragod, Kerala, India

Mehar H. Asif CSIR-National Botanical Research Institute, Lucknow, Uttar Pradesh, India

Asmatullah Dr Panjwani Center for Molecular Medicine and Drug Research, International Center of Chemical and Biological Sciences, University of Karachi, Karachi, Pakistan

Neethu Asokan Department of Life Sciences, Sri Sathya Sai University for Human Excellence, Kalaburagi, Karnataka, India

Marya Bibi Department of Agronomy, Horticulture & Plant Science, South Dakota State University, Brookings, SD, USA

Nisha Boora Department of Bioinformatics and Computational Biology, College of Biotechnology, CCS Haryana Agricultural University, Hisar, Haryana, India

Priyanka Chauhan School of Sciences, P P Savani University, Surat, Gujarat, India

Academy of Scientific and Innovative Research (AcSIR), Ghaziabad, Uttar Pradesh, India

Vrushali Desai School of Sciences, P P Savani University, Surat, Gujarat, India

Shah Fahad Department of Biology/Microbiology, South Dakota State University, Brookings, SD, USA

Nikhil LNUSirmast Faiz Dr Panjwani Center for Molecular Medicine and Drug Research, International Center of Chemical and Biological Sciences, University of Karachi, Karachi, Pakistan

Department of Agronomy, Horticulture & Plant Science, South Dakota State University, Brookings, SD, USA

Dhananjay Gotarkar Department of Fisheries, Wildlife and Conservation Biology, University of Minnesota, Minneapolis, MN, USA

Sanchita Gupta CSIR-National Botanical Research Institute, Lucknow, Uttar Pradesh, India

S. Jeevan Department of Life Sciences, Sri Sathya Sai University for Human Excellence, Kalaburagi, Karnataka, India

Nitish Joshi Department of Agronomy, Horticulture & Plant Science, South Dakota State University, Brookings, SD, USA

Bhawna Kalra Faculty of Agricultural Sciences, Department of Genetics and Plant Breeding, SGT University, Gurugram, Haryana, India

Imran Khan Center for Life Science Ventures, Cornell University, Ithaca, NY, USA

Pradeep Kumar Agricultural and Biosystems Engineering, South Dakota State University, Brookings, SD, USA

A. K. Mall ICAR-Indian Institute of Sugarcane Research, Lucknow, Uttar Pradesh, India

A. T. Meharban Department of Plant Science, Central University of Kerala, Kasaragod, Kerala, India

Bhartendu Nath Mishra Department of Biotechnology, Institute of Engineering and Technology, Lucknow, Uttar Pradesh, India

Department of Biotechnology, Institute of Engineering and Technology, Lucknow, Uttar Pradesh, India

Manoj Kumar Mishra Department of Biotechnology, S.R. Institute of Management & Technology, Lucknow, Uttar Pradesh, India

Parneeta Mishra Molecular Biology and Biotechnology Division, CSIR-National Botanical Research Institute, Lucknow, Uttar Pradesh, India

Prabhatam Mishra CSIR-National Botanical Research Institute, Lucknow, Uttar Pradesh, India

Priyamvada Mishra CSIR-National Botanical Research Institute, Lucknow, Uttar Pradesh, India

Varucha Misra ICAR-Indian Institute of Sugarcane Research, Lucknow, Uttar Pradesh, India

Taufiq Nawaz Department of Biology/Microbiology, South Dakota State University, Brookings, SD, USA

Pei Jia Ng Department of Biochemistry and Molecular Biology, Oklahoma State University, Stillwater, OK, USA

Pranesh Prabhu Department of Biotechnology, Institute of Engineering and Technology, Lucknow, Uttar Pradesh, India

Sribash Roy Department of Plant Sciences, School of Life Sciences, University of Hyderabad, Hyderabad, Telangana, India

Muhammad Sarmad Huazhong Agriculture University, Wuhan, Hubei Province, People's Republic of China

Shah Saud College of Life Science, Linyi University, Linyi, Shandong, China

Parul Saxena A*STAR, Agency for Science, Technology and Research, Singapore, Singapore

Jasmine M. Shah Department of Plant Science, Central University of Kerala, Kasaragod, Kerala, India

Saniya Shakir Department of Biotechnology, Institute of Engineering and Technology, Lucknow, Uttar Pradesh, India

Santosh Kumar Singh Department of Life Sciences, Sri Sathya Sai University for Human Excellence, Kalaburagi, Karnataka, India

Shalini Tiwari Agricultural and Biosystems Engineering, South Dakota State University, Brookings, SD, USA

Ashutosh Tripathi Academy of Scientific and Innovative Research (AcSIR), Ghaziabad, Uttar Pradesh, India

Divison of Microbial Technology, Lucknow, Uttar Pradesh, India

Mohammad Hassan Ukasha Department of Public Health, Cornell University, Ithaca, NY, USA

Neha Verma Department of Biotechnology, S.R. Institute of Management & Technology, Lucknow, Uttar Pradesh, India

Pankaj Kumar Verma Department of Plant and Soil Sciences, Texas Tech University, Lubbock, TX, USA

Pratibha Verma Academy of Scientific and Innovative Research (AcSIR), Ghaziabad, Uttar Pradesh, India

Divison of Microbial Technology,, Lucknow, Uttar Pradesh, India

Shikha Verma French Associates Institute for Agriculture and Biotechnology of Drylands, Ben-Gurion University of the Negev, Be'er Sheva, Israel

Vikash Kumar Yadav School of Biological Sciences and Biotechnology, Goa University, Taleigao, Goa, India

Chapter 1
Introduction to Noncoding Regions

Saniya Shakir, Bhartendu Nath Mishra, Manoj Kumar Mishra, and Neha Verma

Abstract Noncoding regions of the genome, once deemed "junk DNA," have emerged as critical players in genomic architecture and cellular function. This chapter delves into the definition, variations, and historical perspectives of noncoding regions, charting their discovery and evolving significance from early hypotheses to modern genomics. Highlighting their role in gene regulation, genome stability, RNA splicing, and evolutionary processes, we discuss their functional diversity and disease associations. Special emphasis is placed on the unique characteristics of noncoding DNA in plant genomes, including introns, intergenic regions, regulatory elements, and noncoding RNAs. By exploring their interactions, epigenetic modifications, and evolutionary adaptations, this chapter underscores the integral role of noncoding regions in shaping biological complexity and adaptability, inviting a reevaluation of their contribution to genomic and functiona l innovation.

Keywords Gene regulation · Genome stability · Junk DNA · Noncoding region · RNA splicing

1.1 Introduction

1.1.1 Definition of Noncoding Regions

Noncoding regions, also known as noncoding DNA, are segments of the genome that do not directly encode for functional proteins. These regions constitute a significant portion of the genome and play crucial roles in a wide range of cellular

S. Shakir · B. N. Mishra
Department of Biotechnology, Institute of Engineering and Technology, Lucknow, Uttar Pradesh, India

M. K. Mishra · N. Verma (✉)
Department of Biotechnology, S.R. Institute of Management & Technology, Lucknow, Uttar Pradesh, India

S. Tiwari et al. (eds.), *Decoding Non-Coding RNA*, Concepts and Strategies in Plant Sciences, https://doi.org/10.1007/978-981-96-9859-2_1

processes, including gene expression regulation, chromatin organization, and genome stability, although they do not directly specify the amino acid sequence of proteins (Santosh et al. 2015). Noncoding DNA encompasses a diverse array of elements, including introns, intergenic regions, regulatory sequences, and noncoding RNA genes, which collectively contribute to the complexity and dynamic regulation of the genome. These noncoding regions play essential roles in shaping the transcriptional and posttranscriptional landscape of the cell, as well as in mediating epigenetic modifications that modulate gene activity and genome function (Szymański and Barciszewski 2009).

In plant genomes, the distinction between coding and noncoding regions is fundamental to understanding genetic function and regulation (Poliseno et al. 2024). Coding regions, known as exons, are sequences that directly encode proteins, which are crucial for a plant's structural, functional, and regulatory processes within cells. In contrast, noncoding regions, or introns, do not contribute to protein synthesis; instead, they play vital roles in regulating gene expression and maintaining genomic integrity (Fig. 1.1). While exons collectively form the exome, which represents the genetic information responsible for phenotypic traits, introns serve more complex roles that can influence the efficiency of protein production and the overall expression of genes. Interestingly, a significant portion of plant genomes is composed of noncoding DNA, with estimates suggesting that only about 2% of the genome encodes proteins, highlighting the extensive regulatory and structural function of noncoding sequences (Zhao et al. 2022). This phenomenon is exemplified by the bladderwort plant, *Utricularia gibba*, which possesses a remarkably efficient genome with only 3% noncoding DNA, indicating a streamlined approach to genetic information that minimizes unnecessary sequences while maximizing functional output (Ibarra-Laclette et al. 2013). Ultimately, both coding and noncoding regions

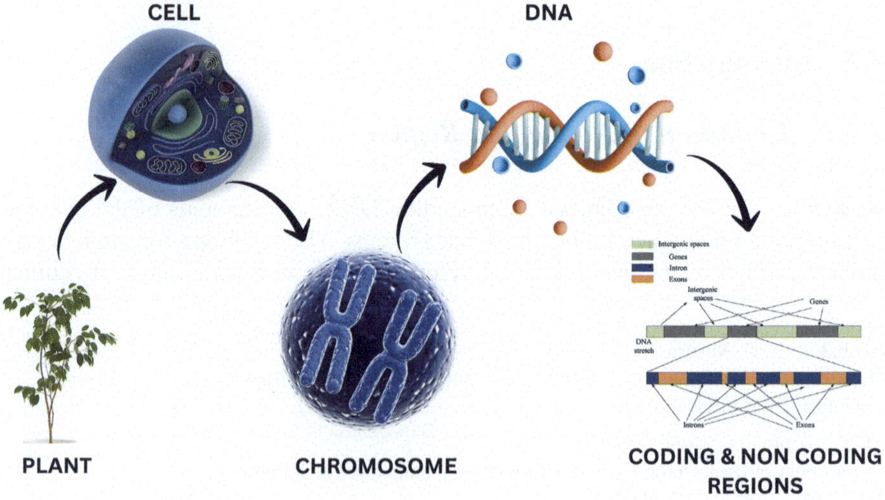

Fig. 1.1 Diagrammatic representation of coding and noncoding regions in a plant genome

are essential, yet they serve distinct purposes within the genomic landscape, shaping the overall biology and adaptability of plants.

1.1.2 Variations in Coding and Noncoding Regions

The variations in coding and noncoding regions among different plant species reveal intriguing patterns influenced by genomic characteristics (Fig. 1.2) (Gullotta et al. 2023). For instance, Arabidopsis species, which are often studied for their genetic properties, may present an anomaly in the context of noncoding regions, as they contain relatively low levels of repetitive DNA compared to other plant genomes. This suggests that while many plant species experience significant changes in genome size due to the accumulation of repetitive sequences, Arabidopsis may maintain a more stable genomic structure. Additionally, a comparative study between *Arabidopsis lyrata* and *Arabidopsis thaliana* showed no notable differences in intergenic indel patterns and lengths, hinting at similarities in the proportions of noncoding regions despite their differences in overall genome size. Such findings underscore the complexity of noncoding DNA, which, while generally perceived as lacking in function, may possess variable degrees of functionality across species. The intergenic regions selected for analysis were specifically chosen for

Fig. 1.2 Graph showing the proportion of noncoding DNA across major crop genomes

their low gene density and apparent lack of function, which raises questions about the selective pressures acting on these regions in comparison to coding DNA. Furthermore, there is little evidence of selection on noncoding regions adjacent to coding sequences in the studied plant species, indicating that these areas may not be as evolutionarily constrained as once thought (He et al. 2012). Overall, the variability in noncoding and coding region proportions among plant species suggests a rich tapestry of genomic evolution, driven by factors such as repetitive DNA accumulation and selective pressures that vary across different environments and species.

1.1.3 Analysis of Genomic Regions

Analyzing and comparing genomic regions efficiently requires a diverse array of methodologies, each tailored to meet specific research objectives and constraints. Chromosomal Microarray Analysis (CMA) is a widely used technique that offers a cost-effective and reliable means to detect chromosomal abnormalities. However, it lacks the depth and resolution provided by Next Generation Sequencing (NGS), which is celebrated for its ability to generate extensive datasets with high accuracy. NGS enables the precise identification of complex genomic variations, including single-nucleotide polymorphisms (SNPs), insertions, deletions, and structural changes. Despite its advantages, NGS has limitations such as high costs and the requirement for specialized expertise, which can restrict its accessibility (Chen et al. 2023). Selecting the appropriate microarray for a given study poses additional challenges due to the variety of available functionalities, emphasizing that genomic analysis must be customized for each research scenario. Often, combining CMA and NGS provides a comprehensive approach by utilizing the unique strengths of both methodologies. Tools like AsmVar and Assemblytics further enhance genomic analysis by facilitating the detection and characterization of structural variations, including inversions, translocations, and large insertions or deletions. These advanced computational tools are particularly useful for comparing genomic assemblies against reference sequences (Domínguez-Rosas et al. 2023). Ultimately, the choice of methodology is dictated by the specific requirements of the study, with sensitivity and precision metrics guiding the evaluation of each approach's effectiveness in analyzing the relevant genomic regions.

1.1.4 Historical Perspective

The historical perspective of noncoding DNA highlights the shift from viewing it as "junk" to recognizing its vital roles in genome function and regulation (Lakhotia 2018). The understanding of noncoding DNA has evolved significantly over the years:

1.1.4.1 Early Discoveries (1940s–1950s)

The foundation of molecular biology was laid with the discovery of the double helical structure of DNA by Watson and Crick (1953). This period also saw the development of the one gene-one enzyme hypothesis and the demonstration of DNA as the genetic material. These discoveries emphasized the importance of coding regions of DNA, which were believed to be the primary functional components of the genome.

1.1.4.2 Central Dogma and "Junk DNA" (1960s–1980s)

Francis Crick proposed the Central Dogma of molecular biology, which described the flow of genetic information from DNA to RNA to proteins. During this time, large portions of the genome were considered "junk DNA" because they did not code for proteins and their functions were not understood. The concept of "selfish DNA" also emerged, suggesting that noncoding DNA sequences were merely parasitic elements without significant function (Finnegan 1971).

1.1.4.3 Reevaluation and Discovery of Noncoding RNAs (1990s–2000s)

The discovery of various noncoding RNAs (ncRNAs) such as microRNAs (miRNAs), small interfering RNAs (siRNAs), and long noncoding RNAs (lncRNAs) challenged the notion of "junk DNA." Research revealed that these ncRNAs play critical roles in gene regulation, chromatin remodeling, and other cellular processes (Ricker and Kaji 1992). The sequencing of different eukaryotic genomes, including the human genome, led to a renewed interest in the significance of noncoding RNAs.

1.1.4.4 Modern Understanding (2000s–Present)

Advances in high-throughput sequencing technologies and bioinformatics have led to the identification of numerous noncoding DNA elements and their functions. Studies have shown that noncoding DNA is involved in complex regulatory networks that control gene expression, development, and disease (Santosh et al. 2015). The concept of "junk DNA" has been largely debunked, and there is now a greater appreciation for the diverse roles that noncoding DNA plays in maintaining genome stability and function.

1.2 Significance of Noncoding Regions

Noncoding regions have emerged as crucial and versatile components of plant genomes, with extensive evidence demonstrating their critical roles in diverse aspects of plant biology (Zhang and Chen 2017). These regions play pivotal roles in regulating gene expression, shaping genome organization, and orchestrating various developmental processes and stress response pathways in plants (Fig. 1.3).

Recent advances in functional genomics, particularly through transcriptome sequencing, have shed light on the diverse functions of noncoding RNAs and other noncoding sequences in the plant kingdom (Chand et al. 2021). The role of the non-coding region includes:

1.2.1 Gene Regulation

Noncoding DNA is instrumental in the precise regulation of gene expression, orchestrating complex molecular processes that dictate the timing, location, and levels of gene activity. These regions harbor specific sequences functioning as regulatory elements essential for the coordinated control of transcription and downstream genetic activities (Gil and Ulitsky 2020). Key regulatory components include:

- *Promoters*: Promoters are fundamental in recruiting the transcriptional machinery, including RNA polymerase and transcription factors, to initiate transcription. They are typically located upstream of the gene, within close proximity to the transcription start site, and contain conserved motifs such as the TATA box and initiator elements that facilitate the precise assembly of the transcriptional complex.

Fig. 1.3 Significance of noncoding regions

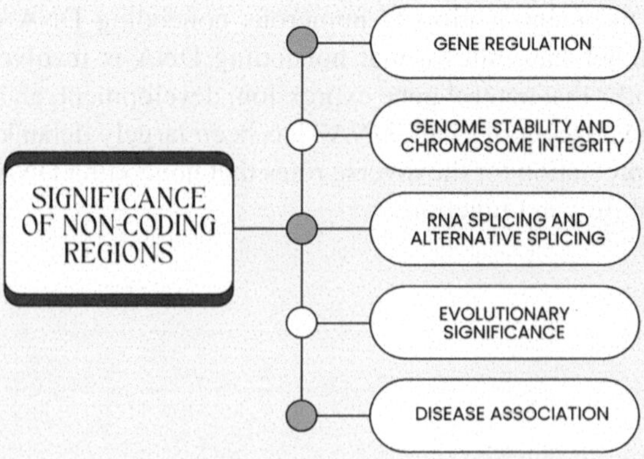

- *Enhancers*: Enhancers amplify gene expression by serving as binding sites for activator proteins, such as transcription factors and coactivators. These elements can act over long distances, looping through three-dimensional chromatin architecture to interact with target promoters, ensuring spatial and temporal specificity of gene activation. Enhancers are often highly tissue-specific and responsive to environmental cues.
- *Silencers*: Silencers repress transcription by recruiting repressor proteins that inhibit the assembly of the transcriptional machinery or modify chromatin to a repressive state. These elements can function bidirectionally and are often involved in fine-tuning gene expression to prevent aberrant activation.
- *Insulators*: Insulators perform dual functions in gene regulation. Enhancer-blocker insulators prevent enhancers from inappropriately activating nontarget promoters, while barrier insulators maintain active chromatin states by blocking the spread of heterochromatin into euchromatic regions.

1.2.2 Genome Stability and Chromosome Integrity

Noncoding DNA plays a critical role in maintaining the structural integrity and stability of the genome, ensuring proper chromosome function and segregation (di Fagagna 2014). Key structural elements include:

- *Telomeres*: Comprising repetitive noncoding sequences such as TTAGGG in humans, telomeres cap the ends of chromosomes, preventing exonucleolytic degradation and end-to-end fusions. Telomeric DNA also interacts with shelterin protein complexes, safeguarding chromosomal termini during cell division and playing a role in cellular aging and replicative senescence.
- *Centromeres*: Centromeres are essential noncoding regions that facilitate chromosome alignment and segregation during mitosis and meiosis. They serve as platforms for kinetochore assembly, a protein complex that anchors spindle microtubules, ensuring accurate distribution of chromosomes to daughter cells.
- *Heterochromatin*: Heterochromatic regions, enriched in repetitive noncoding sequences, contribute to chromosome stability by protecting against transposon activity, maintaining genome integrity, and regulating gene expression through position-effect variegation. These regions are also implicated in nuclear organization.

1.2.3 RNA Splicing and Alternative Splicing

Noncoding DNA is deeply involved in RNA processing mechanisms, particularly splicing, which removes introns from pre-mRNA and joins exons to form mature transcripts. Intronic sequences often harbor splice sites and branch points critical for the splicing machinery (Chen and Narry Kim 2024). Alternative splicing, regulated by splicing enhancers and silencers within noncoding regions, allows a single gene to produce multiple protein isoforms, vastly expanding the proteome and enabling organisms to adapt to diverse functional demands.

1.2.4 Evolutionary Significance

Noncoding DNA has been a driving force in the evolution of complex life forms. Regulatory elements within noncoding regions have facilitated the emergence of novel gene expression patterns, underpinning phenotypic diversity and adaptability. These regions act as reservoirs of genetic variation, with mutations in noncoding sequences often serving as substrates for evolutionary selection. Comparative genomic studies have revealed that changes in noncoding DNA contribute significantly to species-specific traits and developmental innovations. The evolution and complexity of plant genomes have been a subject of extensive and ongoing research, capturing the attention of the scientific community in recent years. Deciphering the intricate mechanisms underlying the dynamic interplay between coding and noncoding regions within plant genomes has emerged as a critical focus area, shedding light on the remarkable diversity and adaptability of these essential organisms. By unraveling the hidden roles and significance of noncoding regions, researchers are poised to unveil the profound influence these enigmatic genomic elements exert on the very fabric of plant genome evolution and complexity. The rapid advancements in deep sequencing technologies have revolutionized the field, enabling researchers to delve deeper into the expansive noncoding regions of plant genomes (Cao et al. 2024). These noncoding RNAs, once considered as mere by-products of transcriptional activities, are now recognized as crucial bioactive molecules that contribute to the remarkable adaptability and resilience of plant species. The complexity of higher-level organisms, including plants, appears to be driven by a more sophisticated regulatory architecture, heavily reliant on the functional integration of noncoding RNAs rather than an exclusive expansion of protein-coding genes. Plants have evolved a myriad of intricate molecular mechanisms to navigate various biotic and abiotic stresses encountered throughout their evolutionary journey. These sophisticated regulatory networks, often relying on the dynamic interplay between coding and noncoding regions within plant genomes, have conferred upon plants the remarkable capacity to thrive in a wide range of environmental conditions. By harnessing the power of noncoding RNAs and other regulatory elements, plants have developed multifaceted strategies to perceive, respond, and ultimately overcome the challenges posed by their ever-changing surroundings, enabling them to maintain their remarkable diversity and adaptability as a testament to their evolutionary resilience.

Mounting evidence has shed light on the pivotal roles of diverse noncoding RNA species, including long noncoding RNAs, microRNAs, and small interfering RNAs, in orchestrating a myriad of essential plant developmental processes and stress response pathways (Chou et al. 2024). These noncoding RNAs have emerged as crucial regulatory players, exerting profound influence on gene expression, chromatin remodeling, and posttranscriptional modifications, thus contributing to the remarkable adaptability and resilience of plant genomes in the face of ever-changing environmental challenges.

1.2.5 *Disease Association*

Aberrations in noncoding DNA are increasingly recognized as key contributors to a wide range of diseases. Mutations in regulatory elements, such as enhancers and promoters, can lead to dysregulated gene expression, manifesting in developmental disorders, cancer, and other pathologies. For instance, enhancer hijacking in oncogenesis enables tumor cells to exploit regulatory sequences for abnormal growth (Bhatti et al. 2021). Advances in genome editing and high-throughput sequencing are shedding light on the intricate roles of noncoding DNA in disease mechanisms, paving the way for targeted therapeutic strategies that address these regulatory disruptions (Nemeth et al. 2024).

As our understanding of the regulatory landscape within plant genomes continues to evolve, the significance of these noncoding regions in governing the complexity and adaptability of plants is becoming increasingly apparent. Noncoding regions are now recognized as essential players in the intricate molecular mechanisms that have evolved in plants, allowing them to adapt and thrive in the face of various biotic and abiotic stresses (Yadav et al. 2024). These regions contribute to the plants' genome and phenotypic diversity, with only a small fraction of the transcribed RNAs actually coding for proteins. The diverse roles of noncoding RNAs, including miRNAs, lncRNAs, and other classes, in regulating developmental processes and stress responses in plants have been the focus of extensive research.

1.3 Types of Noncoding DNA in Plant Genomes

Plant genomes contain a diverse array of noncoding DNA elements that play vital roles in genome organization, regulation, and adaptation (Table 1.1). These include:

1.3.1 *Introns*

Introns, the noncoding regions within genes that are removed during the process of splicing, play a crucial role in various aspects of gene expression and regulation in plant genomes (Fahad et al. 2024). These intronic sequences are not simply inactive spacers between coding regions; they are now recognized as important regulatory

Table 1.1 Summary table of noncoding DNA and RNA types along with their functions

S. no.	Type of noncoding DNA	Description	Functions	Example	References
1	Introns	Noncoding sequences within genes removed during RNA splicing	Regulate gene expression, influence alternative splicing, and serve as sources for regulatory RNAs like siRNAs	Introns in AtGRP7 gene regulating circadian rhythm in Arabidopsis	Staiger et al. (2003)
2	Group I introns	Self-splicing introns found in organellar and nuclear genomes	Catalyze their own excision and sometimes encode proteins like maturases	rRNA introns in *Tetrahymena thermophila*	Csank et al. (1990)
3	Group II introns	Self-splicing introns found in organelles and some bacterial genomes	Serve as evolutionary precursors to spliceosomal introns and encode reverse transcriptase	Introns in nad1 gene in plant mitochondria	Dombrovska and Qiu (2004)
4	Spliceosomal introns	Introns spliced out by the spliceosome in eukaryotic genomes	Increase proteomic diversity through alternative splicing.	Introns in rubisco small subunit genes	Nagarajan and Gill (2018)
5	tRNA introns and archaeal BHB introns	Short introns in tRNA and bulge-helix-bulge (BHB) motifs in archaeal RNA	Facilitate proper RNA folding, splicing, and stability	Introns in archaeal tRNA genes	Yoshihisa (2014)
6	Intergenic regions	DNA sequences between genes that do not code for proteins	Contain regulatory elements (enhancers, silencers), contribute to chromatin organization, and mediate long-range interactions	Enhancers in Sh1 gene regulating maize seed shattering	Yoon et al. (2014)
7	Regulatory elements	Include promoters, enhancers, silencers, and insulators	Control the timing, location, and level of gene expression in response to internal and external signals	Promoters in heat-shock genes	Narberhaus (1999)

(continued)

Table 1.1 (continued)

S. no.	Type of noncoding DNA	Description	Functions	Example	References
8	Promoters	DNA sequences where transcription machinery binds to initiate transcription	Regulate transcription initiation and gene expression	TATA box in Arabidopsis ADH1 gene	Paul and Ferl (1998)
9	Enhancers	DNA elements that boost transcription levels of target genes	Enhance gene expression over long distances and in a tissue-specific manner	Enhancers in AGAMOUS floral identity gene in Arabidopsis	Yang et al. (2011)
10	Silencers	DNA elements that repress transcription	Inhibit gene expression to prevent aberrant activation	Silencers in ferritin genes in plants	Tanabe et al. (2019)
11	Insulators	DNA elements that block interaction between enhancers and nontarget promoters	Maintain proper transcriptional domains and chromatin structure	Insulators in gypsy retrotransposons in drosophila	Kukushkina et al. (2020)
12	Repetitive DNA	Sequences repeated multiple times, including transposable elements and tandem repeats	Impact genome structure, regulate gene expression, and contribute to genetic diversity and evolution	Copia-like LTR elements in maize	SanMiguel and Vitte (2009)
13	Noncoding RNAs (ncRNAs)	RNA molecules not translated into proteins, including miRNAs, siRNAs, and lncRNAs	Regulate gene expression at transcriptional and posttranscriptional levels, stress responses, and developmental processes	miR156 regulating flowering in Arabidopsis	Zheng et al. (2019)
14	miRNAs	Small RNAs that regulate mRNA stability and translation	Control gene expression by targeting mRNAs for degradation or translational inhibition	miR399 in phosphate homeostasis	Tian et al. (2018)
15	siRNAs	Small RNAs derived from double-stranded RNA precursors	Involved in gene silencing through RNA interference (RNAi) pathways	siRNAs in virus defense in tobacco	Liu et al. (2021)

(continued)

Table 1.1 (continued)

S. no.	Type of noncoding DNA	Description	Functions	Example	References
16	lncRNAs	Long noncoding RNAs that modulate gene expression and chromatin states	Act as molecular scaffolds, decoys, or guides for chromatin modifiers	COOLAIR regulating vernalization in Arabidopsis	Zaretskaya et al. (2023)
17	Telomeres	Repeated sequences at chromosome ends	Protect chromosome ends from degradation and prevent fusion during cell division	Telomeric repeats in Arabidopsis	Schrumpfova et al. (2014)
18	Centromeres	Noncoding DNA at the constriction point of chromosomes	Facilitate accurate chromosome segregation during cell division	Centromeric repeats in rice	Bao et al. (2006)
19	Heterochromatin	Densely packed noncoding regions rich in repetitive elements.	Maintain genome stability, suppress transposon activity, and regulate gene expressio	Pericentromeric heterochromatin in maize	Thiemann et al. (2014)

elements that contribute to the complexity and fine-tuning of gene expression in plants. Introns can harbor regulatory sequences, such as enhancers, silencers, and splicing regulatory elements, that influence transcription, mRNA processing, and posttranscriptional regulation (Ariel et al. 2015). Additionally, introns may contain noncoding RNAs, like snoRNAs and miRNAs, which can further modulate gene expression through diverse mechanisms. The variable lengths and sequences of introns also play a role in alternative splicing, allowing for the generation of multiple mRNA isoforms from a single gene and increasing the regulatory capacity and proteomic diversity within plant cells. These intronic sequences are not merely inactive spacers, but rather important regulatory elements that contribute to the complexity and fine-tuning of gene expression in plants. Overall, the multifaceted functions of introns demonstrate their crucial significance in the intricate regulatory networks governing plant growth, development, and stress responses (Galli et al. 2020). Based on tertiary structure and mechanism of their excision, introns are classified into four main categories.

1.3.1.1 Group I Introns

They are small RNAs ranging in size from 250 to 500 nucleotides (Alonso and Mondragón 2021). These introns are categorized as primary examples of catalytic RNAs and are formally recognized as group I ribozymes, due to their ability to execute RNA cleavage and ligation as part of their splicing process. Some of them exhibit a highly complex organization, presenting functional genes and other sequences, establishing deep relationships with the genomes of their hosts (Nielsen and Johansen 2009). In addition to their self-splicing ability, group I introns can encode specific DNA homing endonucleases (HEs) that make them mobile elements within genomes through double-strand DNA breaks (DSB) and homologous recombination (Megarioti and Kouvelis 2020).

Group I Introns occur in all life domains. In bacteria, they are primarily found in rRNA and tRNA genes but are less common in coding genes. In some protists, plants, and fungi, these introns can be found in nuclear rRNA genes as well as in rRNA, tRNA, and protein-coding genes of organelles, such as fungal mitochondria and plant mitochondria and chloroplasts (Rogers 2019). The branches of the tree of life known to date as exhibiting a notable abundance of group I introns include fungi, plants, as well as red and green algae. These specific groups account for approximately 90% of all the identified group I Introns. They are absent in metazoans, except for a few nonbilateral basal lineages and five shark species (Corsaro and Venditti 2020).

1.3.1.2 Group II Introns

They are mobile genetic elements that are found in bacterial and organellar genomes and are thought to be ancestors of spliceosomal introns and retrotransposons in eukaryotes (Garcia and Virginia Sanchez-Puerta 2024). They consist of a catalytically active intron RNA ("ribozyme") and an intron-encoded protein (IEP), whose combined activities enable intron proliferation within genomes. The group II intron RNA catalyzes its own splicing via transesterification reactions that are the same as those of spliceosomal introns, yielding spliced exons and an excised intron lariat RNA. The IEP is a multifunctional reverse transcriptase (RT), which is related to non-LTR-retrotransposon RTs and assists splicing by stabilizing the catalytically active RNA structure. It then remains bound to the excised intron RNA in an RNP that invades DNA sites. DNA invasion occurs by a remarkable mechanism in which the intron RNA uses its ribozyme activity to reverse splice directly into a DNA strand, after which it is reverse transcribed back into DNA by the IEP. Cycles of RNA splicing and reverse splicing enable the introns to proliferate to new DNA sites, while minimally impairing gene expression.

The characteristics of group II introns, including their splicing and mobility mechanisms, active-site structure, and naturally occurring variants that are split into two or more functionally reassociating segments, suggest an evolutionary scenario for the origin of introns, the spliceosome, and retrotransposons in eukaryotes. An evolutionary relationship between group II and spliceosomal introns appears increasingly plausible in light of newly obtained structural information, and indeed, a recent hypothesis asserts that group II intron invasion was the major driving force for the emergence of eukaryotes (Martin and Koonin 2006).

1.3.1.3 Spliceosome

The spliceosome is a dynamic and complex molecular machinery essential for RNA splicing. It consists of small nuclear ribonucleoproteins (snRNPs) and numerous associated proteins that collaboratively execute the removal of introns and ligation of exons (Wahl et al. 2009). Key components of the spliceosome include:

- *snRNPs*: These complexes, such as U1, U2, U4/U6, and U5, recognize and bind to conserved splice sites at intron-exon junctions. U1 snRNP binds to the 5′ splice site, while U2 snRNP interacts with the branch point sequence, facilitating the recruitment of additional components (Cordin et al. 2012).
- *Splice Sites and Branch Point*: Conserved sequences within pre-mRNA, including the 5′ and 3′ splice sites and the branch point adenine, are crucial for spliceosome assembly and catalytic activity.
- *Catalytic Core Formation*: During splicing, the spliceosome undergoes significant structural rearrangements to form a catalytic core that mediates two sequential transesterification reactions. The first reaction cleaves the 5′ splice site, creating a lariat structure at the branch point. The second reaction joins the upstream exon with the downstream exon, excising the intron as a lariat (Yogi et al. 2025).

The dynamic nature of the spliceosome enables precise and flexible splicing, ensuring the production of functional mRNA transcripts. Dysregulation or mutations in spliceosomal components are linked to various diseases, including cancers and neurodegenerative disorders, highlighting their critical role in maintaining cellular homeostasis.

1.3.1.4 tRNA Introns and Archaeal BHB Introns

tRNA introns are short, noncoding sequences embedded within the anticodon loop of precursor tRNA molecules. These introns are excised by a highly conserved splicing mechanism that involves endonucleases specific to tRNA introns. The removal of these introns is critical for proper tRNA maturation, as the resulting conformational changes in the tRNA structure ensure accurate translation during protein synthesis (Hu et al. 2025). Archaeal introns, often referred to as BHB (bulge-helix-bulge) introns, are unique noncoding sequences found in the tRNA, rRNA, and mRNA of archaea. These introns adopt a characteristic secondary structure composed of two bulges flanking a central helix. The splicing of BHB introns is mediated by endonucleases that recognize this structure, followed by a ligation step facilitated by RNA repair enzymes. The conservation of BHB introns across diverse archaeal species underscores their functional significance, which may include regulatory roles and contributions to RNA stability. Additionally, these introns provide insights into the evolutionary origins of splicing mechanisms, bridging the gap between self-splicing introns and the complex spliceosome (Yuan et al. 2023).

1.3.2 *Intergenic Regions*

Intergenic regions, often underestimated in their significance, play a vital role in the structural organization of the genome. These regions, which consist of large stretches of DNA separating genes in eukaryotic genomes, are not merely filler space; rather, they are integral to the regulatory landscape of the genome. For instance, promoters located within these intergenic areas are crucial for initiating the transcription of adjacent genes, thereby influencing gene expression and activity (Peela 2025). Additionally, intergenic regions house enhancers and silencers that fine-tune gene expression, demonstrating their regulatory complexity and importance in maintaining proper gene function. Interestingly, regulatory elements within these regions can be located far from the genes they control, indicating a sophisticated level of organization where intergenic distance and genome architecture are nonrandomly influenced by the regulatory information present in noncoding DNA. This relationship is particularly evident in species with compact genomes, where the species-specific loss of nonfunctional DNA has unveiled a rich landscape of regulatory information that shapes genomic structure (Nelson et al. 2004). Furthermore, the arrangement of genes and their respective intergenic regions—often divergently arranged on different strands of the double-stranded replicative form (ds RF)—is pivotal for the overall architecture of the genome, impacting the functionality of open reading frames (ORFs) and the initiation of processes such as rolling circle replication (RCR) (Ali et al. 2025). Thus, intergenic regions are essential for the proper structural organization of the genome, acting as critical hubs for regulation and gene expression.

1.3.2.1 Functions of Intergenic Regions

Intergenic regions play a crucial role in the regulation of neighboring gene expression through various mechanisms that enhance co-expression and ensure proper transcriptional dynamics. Notably, intergenic regions located within approximately 400 bp of adjacent genes have been shown to significantly increase the likelihood of co-expression, suggesting a close regulatory relationship between these genes (Chatterjee et al. 2025). This heightened co-expression can be attributed to shared chromatin domains that facilitate regulatory interactions, rather than direct transcriptional influences, particularly beyond the 400 bp threshold, where co-expression levels become independent of intergenic distance. Furthermore, intergenic regions can contain transcriptional regulatory sites that allow for divergent transcription, enabling the sharing of promoter regions and enhancing the expression correlation of neighboring genes (Zhang et al. 2024b). This bi-directional transcription not only supports a positive feedback loop for gene expression but also contributes to the co-expression patterns observed in genes with related functions, indicating a functional relationship that intergenic regions help maintain (Chen et al. 2010). Moreover, the presence of enhancer elements and insulator sites within these regions further underscores their regulatory significance, as these features help modulate the precise expression patterns of genes by preventing unwanted cross talk between enhancers and target genes (Nelson et al. 2004). Thus, intergenic regions are not merely passive sequences but active participants in shaping the regulatory landscape of the genome, impacting both gene expression and overall genomic architecture.

1.3.2.2 Variations in Intergenic Regions

Variations in intergenic regions not only contribute to gene co-expression but also play a pivotal role in maintaining genomic stability and integrity. For instance, specific variants in these regions can influence the deposition of epigenetic marks, which are crucial for regulating gene expression and maintaining chromatin structure (Ignatieva and Matrosova 2021). Furthermore, the identification of disease-associated loci within intergenic regions underscores their potential regulatory functions, suggesting that these noncoding sequences can affect not only the expression of nearby genes but also the overall stability of the genome. Despite this significance, the incomplete annotation of the human transcriptome creates challenges in fully understanding how variations in these regions correlate with genomic stability (Akcesme et al. 2025). Additionally, single-nucleotide variations can disrupt essential interactions between proteins and DNA or RNA, which may hinder the binding of crucial regulatory elements such as promoters and enhancers. This disruption can lead to misregulation of gene expression and ultimately impact cellular function. Moreover, single-nucleotide variations in intergenic regions can induce large conformational changes in the DNA structure, thereby affecting the chromatin state and, subsequently, the accessibility of genetic material for transcription and replication (Bartonicek et al. 2017). Collectively, these variations in intergenic regions highlight their importance in genomic architecture and integrity, revealing a complex interplay that extends beyond mere gene regulation.

1.3.3 Regulatory Elements

The primary regulatory elements involved in gene expression include promoters, enhancers, and silencers, which are critical for modulating the transcriptional activity of genes. These regulatory elements are specific DNA sequences that can be found both upstream and downstream of core promoters, and they play a vital role in controlling the rate at which genes are transcribed (de Boer and Taipale 2024). Promoters are essential as they help RNA polymerase recognize the start sites of transcription, with specific sequences typically comprising 35 and 10 bases in length that are crucial for binding (Weake and Workman 2010). Enhancers function by binding to transcription factors, which can either enhance or suppress gene expression, depending on whether they are activators or repressors. The interaction between these regulatory elements and transcription factors is central to the regulation of gene expression, as it determines the levels of transcription that occur in response to various cellular signals (Göös et al. 2022). In eukaryotes, this regulation is further complicated by the need for the transcriptional machinery to be recruited to specific gene promoters, highlighting the intricate interplay between regulatory elements and the components involved in transcription initiation (Mattick et al. 2023). Consequently, disruptions in these regulatory systems can lead to significant pathologies, including various cancers and metabolic disorders, underscoring the importance of understanding how these elements interact within the cellular environment (Jaenisch and Bird 2003).

1.3.3.1 Function of Regulatory Elements

Enhancers, silencers, and insulators serve as key players in the intricate regulation of gene expression, each fulfilling unique yet interconnected roles. These regulatory DNA elements are embedded within metazoan genomes and are crucial during developmental processes, coordinating the modulation of gene expression in a synergistic manner (Pang et al. 2023). Enhancers are known to act as transcription units that enhance the proximity of target promoters to transcriptional hotspots, thus facilitating increased transcriptional activity (Segert et al. 2021). In contrast, silencers act as the repressive counterparts, reducing transcription from target promoters and thus preventing gene expression during critical phases such as differentiation and cell cycle progression. Interestingly, recent studies have shown that silencers can exhibit bifunctionality; depending on the cellular context, they can also act as enhancers, demonstrating their versatile nature in gene regulation (Kolovos et al. 2012). Although enhancers have been extensively studied, the significance of silencers is often overlooked, despite their enrichment for expression quantitative trait loci (eQTLs) and disease-associated variants, which underscores their importance in developmental biology and disease contexts. Moreover, while no common "silencer chromatin signature" has yet been identified, it is evident that silencers might exist in various subclasses, each employing distinct mechanisms for their repressive actions (Capitanchik et al. 2024). Collectively, these regulatory elements exemplify the complex interplay of genetic control, highlighting how enhancers, silencers, and insulators collaboratively shape the landscape of gene expression.

1.3.3.2 Role of Epigenetic Modifications

Epigenetic modifications play a critical role in the regulation of gene expression by providing a dynamic layer of control that operates independently of the DNA sequence itself (Siddique and Ansari 2024). These modifications, which include changes in DNA methylation and histone acetylation patterns, are crucial in determining when and how genes are expressed in response to various internal and external stimuli. For instance, aberrant epigenetic modifications can disrupt normal cellular function, leading to chronic health conditions such as cancer and neurological disorders (Moosavi and Ardekani 2016). A deeper understanding of how these modifications can be altered offers significant potential for genetic engineering applications, particularly in the realm of targeted therapies for cancer. Researchers are increasingly focusing on how specific alterations in epigenetic patterns can influence oncogenes and tumor suppressor genes, which are vital in tumorigenesis (Lu et al. 2020). Moreover, the interplay between metabolic changes and epigenetic regulation further underscores the potential for manipulating these processes; for example, the synthesis of oncometabolites can influence the activity of essential epigenetic enzymes, thereby altering gene expression patterns related to cancer (Wu et al. 2023). The use of epi-drugs, which can modify epigenetic markers, demonstrates promising outcomes in enhancing antitumoral effects, highlighting the therapeutic implications of targeting epigenetic modifications in cancer treatments (Pagiatakis et al. 2021). Ultimately, the heritable and reversible nature of epigenetic modifications presents a compelling avenue for advancing genetic engineering strategies aimed at improving health outcomes and combating disease.

1.3.4 Noncoding RNAs (ncRNAs)

Noncoding RNAs (ncRNAs) encompass a diverse group of RNA molecules that play crucial roles in regulating gene expression through various mechanisms (Table 1.1). Among these, microRNAs (miRNAs), long noncoding RNAs (lncRNAs), and PIWI-interacting RNAs (piRNAs) are significant contributors to the intricate regulatory networks within cells. MiRNAs are known for their ability to target messenger RNAs (mRNAs) to inhibit their expression, effectively controlling gene expression at the posttranscriptional level (Moafian et al. 2021). In contrast, lncRNAs can function as competing endogenous RNAs (ceRNAs), interacting with miRNAs to regulate one another through shared binding sites, leading to the decay of targeted lncRNAs and impacting overall gene regulation. Additionally, certain lncRNAs have the potential to encode small nucleolar RNAs (snoRNAs) and miRNAs, further enriching the complexity of gene regulatory mechanisms (Slack and Chinnaiyan 2019). Moreover, lncRNAs also contribute to epigenetic regulation, influencing chromatin structure through histone modifications and DNA methylation, thus impacting gene expression at the transcriptional level (López-Jiménez and Andrés-León 2021). The interplay between these ncRNAs showcases a dynamic cross talk that not only modulates the levels of protein-coding mRNAs but also participates in RNA-dependent protein-protein and protein-DNA interactions, which are pivotal for cellular processes such as proliferation and stress responses (Zhang et al. 2019). Overall, the interactions among ncRNAs underscore their essential roles in maintaining the delicate balance of gene expression regulation within biological systems (Matsui and Corey 2017).

1.3.4.1 Interaction of ncRNAs with Transcription Factors

The intricate interplay between ncRNAs and transcription factors is pivotal in modulating gene activity, demonstrating the multifaceted roles these molecules play in gene regulation. Noncoding RNAs serve as essential scaffolds, facilitating the docking of various proteins involved in gene regulation, which allows for a coordinated response to cellular signals (Kugel and Goodrich 2012). For instance, ncRNAs can directly influence the assembly of transcription factor complexes, thereby regulating the transcription process more effectively. Additionally, ncRNAs can mimic functional DNA elements, which can have profound effects on transcription by altering the recruitment of transcriptional activators and coregulators, thus impacting overall gene expression levels (Statello et al. 2021). Specific examples illustrate this functionality, such as the lncRNA DEANR1, which aids in the recruitment of SMAD2/3 to the FOXA2 promoter, facilitating gene activation. Moreover, ncRNAs also interact with chromatin remodeler proteins to regulate chromatin accessibility, further emphasizing their influence on the transcription machinery and the overall gene expression landscape (Li 2023). Furthermore, the role of lncRNAs is not strictly defined, as they can act both as transcriptional activators and repressors depending on their interactions with various protein partners, showcasing their versatile regulatory capabilities. As research in this field expands, it is evident that ncRNAs hold significant promise in elucidating the complexities of gene regulation through their diverse interactions with transcription factors and other regulatory proteins (Shukla and Datta 2024).

1.3.4.2 Degradation and Stability of ncRNAs

The intricate relationship between the stability and degradation of noncoding RNAs (ncRNAs) plays a pivotal role in gene regulation, as evidenced by the complex structure of genome networks. The stability of ncRNAs can significantly influence gene expression, since these molecules can modulate mRNA stability and activity through various mechanisms. For instance, long noncoding RNAs (lncRNAs) have been shown to interact with target mRNAs and RNA-binding proteins, effectively preventing their degradation and thus maintaining their expression levels (Das et al. 2024). This interaction is critical, as it allows lncRNAs to act as regulators of gene expression, ensuring that specific mRNAs remain stable and functional within the cellular environment. Furthermore, lncRNAs also influence mRNA stability through epitranscriptomic modifications, adding another layer of complexity to their regulatory potential. The interplay among different ncRNAs could also yield insights into their degradation processes, which is crucial for their development as biomarkers or therapeutic targets. Despite these advances, our understanding of the precise mechanisms governing ncRNA functions remains limited, highlighting a significant gap in current knowledge of how these molecules contribute to gene regulation (Srijyothi et al. 2018). Consequently, future research is recommended to delve deeper into the stability and degradation of ncRNAs, to unravel their multifaceted roles in gene expression and regulation.

1.3.5 Conclusion

Noncoding regions of plant genomes are increasingly recognized as integral to the complexity and functionality of genetic systems. These regions, encompassing introns, intergenic sequences, regulatory elements, and various noncoding RNAs (ncRNAs), serve as critical regulators of gene expression, genomic stability, and evolutionary innovation. Recent advances underscore their role in modulating transcriptional dynamics and responding to environmental and biological stresses, particularly through the activity of ncRNAs like long noncoding RNAs (lncRNAs) and small RNAs. Comparative analyses across plant species reveal the conservation of certain noncoding elements while highlighting their diversity as a source of functional innovation (Yadav et al. 2023). Tools such as the PMG map have further refined our understanding of plant mitochondrial genomes by visualizing noncoding features and their complex genomic architectures (Zhang et al. 2024a, b). Additionally, research on microproteins encoded by smORFs within lncRNAs suggests their critical role in developmental processes and stress adaptation (Fesenko et al. 2021). Despite significant strides, many facets of noncoding DNA remain elusive, necessitating interdisciplinary efforts to uncover their full potential for applications in crop improvement and disease resistance (Sahraei et al. 2025). By integrating computational, experimental, and evolutionary perspectives, this chapter positions noncoding regions as a cornerstone for future genomic innovations.

References

Akcesme B, Islam N, Lekic D, Cutuk R, Basovic N (2025) Analysis of Alzheimer's disease associated deleterious non-synonymous single nucleotide polymorphisms and their impacts on protein structure and function by performing in-Silico methods. Neurogenetics 26(1):1–15

Ali N, Singh S, Garg R (2025) Unlocking crops' genetic potential: advances in genome and Epigenome editing of regulatory regions. Curr Opin Plant Biol 83:102669

Alonso D, Mondragón A (2021) Mechanisms of catalytic RNA molecules. Biochem Soc Trans 49(4):1529–1535

Ariel F, Romero-Barrios N, Jégu T, Benhamed M, Crespi M (2015) Battles and hijacks: noncoding transcription in plants. Trends Plant Sci 20(6):362–371

Bao W, Zhang W, Qiuying Yang Y, Zhang BH, Minghong G, Xue Y, Cheng Z (2006) Diversity of Centromeric repeats in two closely related wild Rice species, Oryza Officinalis and Oryza Rhizomatis. Mol Gen Genomics 275:421–430

Bartonicek N, Clark MB, Quek XC, Torpy JR, Pritchard AL, Maag JLV, Gloss BS, Crawford J, Taft RJ, Hayward NK et al (2017) Intergenic disease-associated regions are abundant in novel transcripts. Genome Biol 18:1–16

Bhatti GK, Khullar N, Sidhu IS, Navik US, Reddy AP, Hemachandra Reddy P, Bhatti JS (2021) Emerging role of non-coding RNA in health and disease. Metab Brain Dis 36:1119–1134

Cao W, Limei Yang M, Zhuang HL, Wang Y, Zhang Y, Ji J (2024) Plant non-coding RNAs: the new frontier for the regulation of plant development and adaptation to stress. Plant Physiol Biochem:108435

Capitanchik C, Wilkins OG, Wagner N, Gagneur J, Ule J (2024) From computational models of the splicing code to regulatory mechanisms and therapeutic implications. Nat Rev Genet:1–20

Chand Jha U, Nayyar H, Mantri N, Siddique KHM (2021) Non-coding RNAs in legumes: their emerging roles in regulating biotic/abiotic stress responses and plant growth and development. Cells 10(7):1674

Chatterjee D, Zhang Z, Lin P-Y, Wang P-H, Sidhu GK, Yennawar NH, Hsieh J-WA, Chen P-Y, Song R, Meyers BC et al (2025) Maize unstable factor for Orange1 encodes a nuclear protein that affects redox accumulation during kernel development. Plant Cell 37(1):koae301

Chen L-L, Narry Kim V (2024) Small and long non-coding RNAs: past, present, and future. Cell 187(23):6451–6485

Chen W-H, de Meaux J, Lercher MJ (2010) Co-expression of Neighbouring genes in Arabidopsis: separating chromatin effects from direct interactions. BMC Genomics 11:1–8

Chen Y-H, Jun L, Yang X, Huang L-C, Zhang C-Q, Liu Q-Q, Li Q-F (2023) Gene editing of non-coding regulatory DNA and its application in crop improvement. J Exp Bot 74(19):6158–6175

Chou P-C, Huang Y-C, Sebastian Y (2024) Mechanisms of epigenetic inheritance in post-traumatic stress disorder. Life 14(1):98

Cordin O, Hahn D, Beggs JD (2012) Structure, function and regulation of Spliceosomal RNA helicases. Curr Opin Cell Biol 24(3):431–438

Corsaro D, Venditti D (2020) Putative group I introns in the eukaryote nuclear internal transcribed spacers. Curr Genet 66(2):373–384

Csank C, Taylor FM, Martindale DW (1990) Nuclear pre-MRNA introns: analysis and comparison of intron sequences from Tetrahymena Thermophila and other eukaryotes. Nucleic Acids Res 18(17):5133–5141

Das S, Rojas MPZ, Tran EJ (2024) Novel insights on the positive correlation between sense and antisense pairs on gene expression. RNA 15(4):e1864

de Boer CG, Taipale J (2024) Hold out the genome: a roadmap to solving the cis-regulatory code. Nature 625(7993):41–50

di Fagagna FA (2014) A direct role for small non-coding RNAs in DNA damage response. Trends Cell Biol 24(3):171–178

Dombrovska O, Qiu Y-L (2004) Distribution of introns in the mitochondrial gene Nad1 in land plants: phylogenetic and molecular evolutionary implications. Mol Phylogenet Evol 32(1):246–263

Domínguez-Rosas E, Hernandez-Oñate MA et al (2023) Plant long non-coding RNAs: identification and analysis to unveil their physiological functions. Front Plant Sci 14:1275399

Fahad M, Tariq L, Muhammad S, Liang W (2024) Underground communication: long noncoding RNA signaling in the plant rhizosphere. Plant Commun 5:100927

Fesenko I, Shabalina SA, Mamaeva A, Knyazev A, Glushkevich A, Lyapina I, Ziganshin R, Kovalchuk S, Kharlampieva D, Lazarev V et al (2021) A vast Pool of lineage-specific microproteins encoded by long non-coding RNAs in plants. Nucleic Acids Res 49(18):10328–10346

Finnegan L (1971) Virtual conference, 17th September 2021. Ir J Med Sci 191(Suppl 3):S65–S79

Galli M, Feng F, Gallavotti A (2020) Mapping regulatory determinants in plants. Front Genet 11:591194

Garcia LE, Virginia Sanchez-Puerta M (2024) Mitochondrial splicing efficiency is lower in Holoparasites than in free-living plants. Plant Cell Physiol 65(12):2018–2029

Gil N, Ulitsky I (2020) Regulation of gene expression by cis-acting long non-coding RNAs. Nat Rev Genet 21(2):102–117

Göös H, Kinnunen M, Salokas K, Tan Z, Liu X, Yadav L, Zhang Q, Wei G-H, Varjosalo M (2022) Human transcription factor protein interaction networks. Nat Commun 13(1):766

Gullotta G, Korte A, Marquardt S (2023) Functional variation in the non-coding genome: molecular implications for food security. J Exp Bot 74(7):2338–2351
He F, Zhang X, Jinyong H, Turck F, Dong X, Goebel U, Borevitz J, de Meaux J (2012) Genome-wide analysis of cis-regulatory divergence between species in the Arabidopsis genus. Mol Biol Evol 29(11):3385–3395
Hu Y, Yan S, Yan H, Jingping S, Cui Z, Li J, Wang S, Sun Y, Li W, Gao S (2025) PacBio full-length transcriptome analysis reveals the role of TRNA-like structures in RNA processing. Cell Signal 125:111515
Ibarra-Laclette E, Lyons E, Hernández-Guzmán G, Pérez-Torres CA, Carretero-Paulet L, Chang T-H, Lan T, Welch AJ, Juárez MJA, Simpson J et al (2013) Architecture and evolution of a minute plant genome. Nature 498(7452):94–98
Ignatieva EV, Matrosova EA (2021) Disease-associated genetic variants in the regulatory regions of human genes: mechanisms of action on transcription and genomic resources for dissecting these mechanisms. Vavilov J Genet Breed 25(1):18
Jaenisch R, Bird A (2003) Epigenetic regulation of gene expression: how the genome integrates intrinsic and environmental signals. Nat Genet 33(3):245–254
Kolovos P, Knoch TA, Grosveld FG, Cook PR, Papantonis A (2012) Enhancers and silencers: An integrated and simple model for their function. Epigenetics Chromatin 5:1–8
Kugel JF, Goodrich JA (2012) Non-coding RNAs: key regulators of mammalian transcription. Trends Biochem Sci 37(4):144–151
Kukushkina IV, Makhnovskii PA, Nefedova LN, Milyaeva PA, Kuzmin IV, Lavrenov AR, Kim AI (2020) Analysis of transcriptome of drosophila melanogaster strains with disrupted control of gypsy retrotransposon transposition. Russ J Genet 56:562–571
Lakhotia SC (2018) Central Dogma, selfish DNA and noncoding Rnas: a historical perspective
Li Y (2023) Non – coding RNA performs its biological function by interacting with macromolecules. Int J Mol Sci 24(22):16246
Liu P, Zhang X, Zhang F, Miaoze X, Ye Z, Wang K, Liu S, Han X, Cheng Y, Zhong K et al (2021) A virus-derived SiRNA activates plant immunity by interfering with ROS scavenging. Mol Plant 14(7):1088–1103
López-Jiménez E, Andrés-León E (2021) The implications of NcRNAs in the development of human diseases. Non-Coding RNA 7(1):17
Lu Y, Chan Y-T, Tan H-Y, Li S, Wang N, Feng Y (2020) Epigenetic regulation in human cancer: the potential role of epi-drug in cancer therapy. Mol Cancer 19:1–16
Martin W, Koonin EV (2006) Introns and the origin of nucleus – cytosol compartmentalization. Nature 440(7080):41–45
Matsui M, Corey DR (2017) Non-coding RNAs as drug targets. Nat Rev Drug Discov 16(3):167–179
Mattick JS, Amaral PP, Carninci P, Carpenter S, Chang HY, Chen L-L, Chen R, Dean C, Dinger ME, Fitzgerald KA et al (2023) Long non-coding RNAs: definitions, functions, challenges and recommendations. Nat Rev Mol Cell Biol 24(6):430–447
Megarioti AH, Kouvelis VN (2020) The coevolution of fungal mitochondrial introns and their homing endonucleases (GIY-YIG and LAGLIDADG). Genome Biol Evol 12(8):1337–1354
Moafian Z, Maghrouni A, Soltani A, Hashemy SI (2021) Cross-talk between non-coding RNAs and PI3K/AKT/MTOR pathway in colorectal cancer. Mol Biol Rep 48(5):4797–4811
Moosavi A, Ardekani AM (2016) Role of epigenetics in biology and human diseases. Iran Biomed J 20(5):246
Nagarajan R, Gill KS (2018) Evolution of rubisco Activase gene in plants. Plant Mol Biol 96:69–87
Narberhaus F (1999) Negative regulation of bacterial heat shock genes. Mol Microbiol 31(1):1–8
Nelson CE, Hersh BM, Carroll SB (2004) The regulatory content of intergenic DNA shapes genome architecture. Genome Biol 5:1–15
Nemeth K, Bayraktar R, Ferracin M, Calin GA (2024) Non-coding RNAs in disease: from mechanisms to therapeutics. Nat Rev Genet 25(3):211–232
Nielsen H, Johansen SD (2009) Group I introns: moving in new directions. RNA Biol 6(4):375–383

Pagiatakis C, Musolino E, Gornati R, Bernardini G, Papait R (2021) Epigenetics of aging and disease: a brief overview. Aging Clin Exp Res 33:737–745
Pang B, van Weerd JH, Hamoen FL, Snyder MP (2023) Identification of non-coding silencer elements and their regulation of gene expression. Nat Rev Mol Cell Biol 24(6):383–395
Paul A-L, Ferl RJ (1998) Permeabilized Arabidopsis protoplasts provide new insight into the chromatin structure of plant alcohol dehydrogenase genes. Dev Genet 22(1):7–16
Peela SCM (2025) Basics of genetics and genomics. In: Deep learning in genetics and genomics. Elsevier, pp 1–33
Poliseno L, Lanza M, Pandolfi PP (2024) Coding, or non-coding, that is the question. Cell Res 34(9):609–629
Ricker RD, Kaji A (1992) Studies on antisense inhibition of translation in vitro anomalies and reevaluation. FEBS Lett 309(3):363–370
Rogers SO (2019) Integrated evolution of ribosomal RNAs, introns, and intron nurseries. Genetica 147:103–119
Sahraei S, Mahdinezhad N, Emamjomeh A, Kavousi K, Solouki M, Delledonne M (2025) Differentiation of long non-coding RNA expression profiles in three fruiting stages of grape. Gene 934:149029
SanMiguel P, Vitte C (2009) The LTR-retrotransposons of maize. In: Handbook of maize: genetics and genomics. Springer, pp 307–327
Santosh B, Varshney A, Yadava PK (2015) Non-coding RNAs: biological functions and applications. Cell Biochem Funct 33(1):14–22
Schrumpfova PP, Vychodilova I et al (2014) Telomere repeat binding proteins are functional components of Arabidopsis telomeres and interact with telomerase. Plant J 77(5):770
Segert JA, Gisselbrecht SS, Bulyk ML (2021) Transcriptional silencers: driving gene expression with the brakes on. Trends Genet 37(6):514–527
Shukla C, Datta B (2024) G-Quadruplexes in long non-coding RNAs and their interactions with proteins. Int J Biol Macromol 278:134946
Siddique HR, Ansari MA (2024) Role of epigenetics in drug-resistance: a personalized approach to cancer therapy. In: Personalized epigenetics. Elsevier, pp 511–530
Slack FJ, Chinnaiyan AM (2019) The role of non-coding RNAs in oncology. Cell 179(5):1033–1055
Srijyothi L, Ponne S, Prathama T, Ashok C, Baluchamy S (2018) Roles of non-coding RNAs in transcriptional regulation. In: Transcriptional and post-transcriptional regulation, vol 55
Staiger D, Zecca L, Wieczorek DA, Kirk KA, Eckstein L (2003) The circadian clock regulated RNA-binding protein AtGRP7 autoregulates its expression by influencing alternative splicing of its own pre-MRNA. Plant J 33(2):361–371
Statello L, Guo C-J, Chen L-L, Huarte M (2021) Gene regulation by long non-coding RNAs and its biological functions. Nat Rev Mol Cell Biol 22(2):96–118
Szymański M, Barciszewski J (2009) Noncoding RNAs in biology and disease. ELS
Tanabe N, Noshi M, Mori D, Nozawa K, Tamoi M, Shigeoka S (2019) The basic helix-loop-helix transcription factor, BHLH11 functions in the iron-uptake system in *Arabidopsis thaliana*. J Plant Res 132:93–105
Thiemann A, Junjie F, Seifert F, Grant-Downton RT, Schrag TA, Pospisil H, Frisch M, Melchinger AE, Scholten S (2014) Genome-wide meta-analysis of maize Heterosis reveals the potential role of additive gene expression at Pericentromeric loci. BMC Plant Biol 14:1–14
Tian L, Liu H, Ren L, Lixia K, Liuji W, Li M, Wang S, Zhou J, Song X, Zhang J et al (2018) MicroRNA 399 as a potential integrator of photo-response, phosphate homeostasis, and sucrose signaling under long day condition. BMC Plant Biol 18:1–13
Wahl MC, Will CL, Lührmann R (2009) The spliceosome: design principles of a dynamic RNP machine. Cell 136(4):701–718
Watson JD, Crick FHC (1953) The structure of DNA. In: Cold Spring Harbor symposia on quantitative biology, vol 18, pp 123–131
Weake VM, Workman JL (2010) Inducible gene expression: diverse regulatory mechanisms. Nat Rev Genet 11(6):426–437

Wu Y-L, Lin Z-J, Li C-C, Lin X, Shan S-K, Guo B, Zheng M-H, Li F, Yuan L-Q, Li Z-h (2023) Epigenetic regulation in metabolic diseases: mechanisms and advances in clinical study. Signal Transduct Target Ther 8(1):98

Yadav VK, Jalmi SK, Tiwari S, Kerkar S (2023) Deciphering shared attributes of plant long non-coding RNAs through a comparative computational approach. Sci Rep 13(1):15101

Yadav A, Mathan J, Dubey AK, Singh A (2024) The emerging role of non-coding RNAs (NcRNAs) in plant growth, development, and stress response signaling. Non-Coding RNA 10(1):13

Yang Y, Singer SD, Liu Z (2011) Petunia AGAMOUS enhancer-derived chimeric promoters specify a carpel-, stamen-, and petal-specific expression pattern sufficient for engineering male and female sterility in tobacco. Plant Mol Biol Report 29:162–170

Yogi S, Sangeeth A, Chimata AV, Padma R, Singh A (2025) Genetic and molecular basis of neurodegenerative diseases. In: Genome editing for neurodegenerative diseases. Elsevier, pp 19–45

Yoon J, Cho L-H, Kim SL, Choi H, Koh H-J, An G (2014) The BEL 1-type Homeobox gene SH 5 induces seed shattering by enhancing abscission-zone development and inhibiting lignin biosynthesis. Plant J 79(5):717–728

Yoshihisa T (2014) Handling TRNA introns, archaeal way and eukaryotic way. Front Genet 5:213

Yuan L, Han Y, Zhao J, Zhang Y, Sun Y (2023) Recognition and cleavage mechanism of intron-containing pre-TRNA by human TSEN endonuclease complex. Nat Commun 14(1):6071

Zaretskaya MV, Lebedeva ON, Fedorenko OM (2023) Peculiarities of the transcriptional activity of long non-coding RNAs (COOLAIR, COLDAIR, and COLDWRAP) during the Vernalization of the plant *Arabidopsis thaliana* of northern natural populations. Russ J Genet 59(8):817–823

Zhang Y-C, Chen Y-Q (2017) Epigenetic regulation by noncoding RNAs in plant development. Plant Epigenetics:183–198

Zhang P, Wenyi W, Chen Q, Chen M (2019) Non-coding RNAs and their integrated networks. J Integr Bioinform 16(3):20190027

Zhang X, Chen H, Ni Y, Bin W, Li J, Burzyński A, Liu C (2024a) Plant mitochondrial genome map (PMGmap): a software tool for the comprehensive visualization of coding, noncoding and genome features of plant mitochondrial genomes. Mol Ecol Resour 24:e13952

Zhang Y, Dong Q, Wang Z, Liu Q, Haopeng Y, Sun W, Cheema J, You Q, Ding L, Cao X et al (2024b) A fine-scale Arabidopsis chromatin landscape reveals chromatin conformation-associated transcriptional dynamics. Nat Commun 15(1):3253

Zhao Z, Zang S, Zou W, Pan Y-B, Yao W, You C, Que Y (2022) Long non-coding RNAs: new players in plants. Int J Mol Sci 23(16):9301

Zheng C, Ye M, Sang M, Rongling W (2019) A regulatory network for MiR156-SPL module in *Arabidopsis thaliana*. Int J Mol Sci 20(24):6166

Chapter 2
Structural and Functional Diversity of Noncoding RNA

Vikash Kumar Yadav

Abstract Noncoding RNAs (ncRNAs) have emerged as critical regulators of gene expression, genome organization, and cellular homeostasis, challenging the earlier view of noncoding regions as "junk DNA." This chapter provides a systematic overview of the classification, biogenesis, structural features, and functional roles of ncRNAs, focusing on plants. ncRNAs are mainly categorized into small (<200 nt) and long (>200 nt) classes, including microRNAs, small interfering RNAs, PIWI-interacting RNAs, small nucleolar RNAs, long noncoding RNAs, circular RNAs, and others. Particular emphasis focuses on the roles of ncRNAs in transcriptional and posttranscriptional regulation, epigenetic modification, genome stability, nuclear architecture, development, and stress responses. Structural motifs, evolutionary conservation, and posttranscriptional modifications are highlighted as key determinants of ncRNA function. Through specific examples, this chapter underscores the regulatory versatility and evolutionary significance of ncRNAs in shaping the complexity of eukaryotic genomes.

Keywords Epigenetic modification · Evolutionary conservation · Genome stability · Noncoding RNA · Transcriptional regulation

2.1 Introduction

The idea that noncoding genome regions were "junk DNA" took hold when it became apparent that large portions of eukaryotic genomes did not encode proteins. This perception was reinforced by the observation that the number of protein-coding genes did not correlate with organismal complexity, leading to the assumption that much of the genome was functionally unimportant. However, advances in genomics and transcriptomics have significantly overturned this view. Pioneering efforts such

V. K. Yadav (✉)
School of Biological Sciences and Biotechnology, Goa University, Taleigao, Goa, India
e-mail: vikash@unigoa.ac.in

S. Tiwari et al. (eds.), *Decoding Non-Coding RNA*, Concepts and Strategies in Plant Sciences, https://doi.org/10.1007/978-981-96-9859-2_2

as the ENCODE (Encyclopedia of DNA Elements) project revealed that a substantial portion of the genome is transcribed into noncoding RNAs (ncRNAs), many exhibiting tissue-specific and developmentally regulated expression patterns, suggesting functional relevance (Feingold et al. 2004). In plants, genome-wide studies have similarly shown extensive transcription of noncoding regions, often linked to development, stress responses, and epigenetic regulation (Kaikkonen and Adelman 2018; Mercer et al. 2009; Yadav et al. 2022). This paradigm shift from viewing noncoding DNA as useless remnants to recognizing it as a rich source of essential regulatory elements represents a fundamental change in our understanding of genome function.

Noncoding RNAs (ncRNAs) are RNA molecules transcribed from the genome that do not code for proteins but play crucial roles in gene regulation and genome stability (Rinn and Chang 2012). Once dismissed as mere transcriptional noise, ncRNAs are now recognized for their involvement in various cellular processes, including transcription, RNA processing, chromatin organization, and epigenetic modification (Kaikkonen and Adelman 2018; Mercer et al. 2009; Rinn and Chang 2012). Based on the size, they are broadly categorized into small ncRNAs (typically <200 nucleotides) and long noncoding RNAs (lncRNAs, >200 nucleotides), each with distinct biogenesis pathways and mechanisms of action (Hombach and Kretz 2016; Mercer et al. 2009). These molecules function at various regulatory levels, transcriptional, posttranscriptional, and epigenetic, often through interactions with other RNAs, DNA, or proteins. Their roles span all domains of life, influencing development, stress responses, genome evolution, and disease. In plants and animals, ncRNAs have emerged as key components of complex regulatory networks, challenging the long-standing, protein-centric view of gene regulation (Hombach and Kretz 2016; Sun et al. 2018). Their regulatory versatility and functional diversity underscore the significance of ncRNAs in expanding the genome's functional capacity far beyond its protein-coding regions.

2.2 Classification and Structural Features of Noncoding RNAs

Based on their function and expression patterns, ncRNAs are classified into two general categories:

(i) *Housekeeping ncRNAs* are constitutively expressed across most cell types and are essential for core cellular functions (Zhang et al. 2019). This group includes ribosomal RNAs (rRNAs), transfer RNAs (tRNAs), small nuclear RNAs (snRNAs), and small nucleolar RNAs (snoRNAs). These RNAs are highly conserved across species and structurally stable, underscoring their indispensable role in cellular homeostasis (Mattick et al. 2023).

(ii) *Regulatory ncRNAs* are expressed in a tissue-, developmental stage-, or stimulus-specific manner and primarily modulate gene expression at transcriptional, posttranscriptional, or epigenetic levels (Bhogireddy et al. 2021). This category includes microRNAs (miRNAs), small interfering RNAs (siRNAs), PIWI-interacting RNAs (piRNAs), long noncoding RNAs (lncRNAs), and circular RNAs (circRNAs). They regulate developmental transitions, stress responses, chromatin remodeling, and defense against transposable elements.

However, in addition to these functional classes, ncRNAs are primarily categorized based on size, as discussed in the section below.

(i) *Small noncoding RNAs* (sncRNAs): <200 nucleotides
(ii) *Long noncoding RNAs* (lncRNAs): >200 nucleotides

2.2.1 Small Noncoding RNAs (sncRNAs)

Small noncoding RNAs (sncRNAs), typically ranging from 18 to 35 nucleotides, play critical roles in gene regulation at both the transcriptional and posttranscriptional levels. Key types include miRNAs, siRNAs, piRNAs, snRNAs, and snoRNAs (Table 2.1). These RNAs function through sequence-specific recognition of RNA or chromatin targets and exhibit tightly regulated expression across developmental and environmental contexts.

2.2.1.1 MicroRNAs (miRNAs)

MicroRNAs (miRNAs) are a conserved class of endogenous small RNAs, generally 20–24 long nucleotides, involved in posttranscriptional gene silencing in eukaryotes. They are transcribed as primary transcripts and undergo stepwise processing by Drosha and Dicer-like enzymes (or their equivalents) to generate mature miRNAs (Table 2.1) (Singh et al. 2018). These are incorporated into Argonaute (AGO) proteins, forming the RNA-induced silencing complex (RISC), which guides sequence-specific targeting of messenger RNAs (mRNAs), leading to mRNA cleavage or translational repression.

In plants, miRNAs often exhibit near-perfect complementarity to their targets, resulting in direct cleavage of mRNAs. In contrast, animal miRNAs typically bind with partial complementarity, leading to translational inhibition (Table 2.1). Functionally, miRNAs regulate diverse biological processes such as developmental timing, organ formation, hormone signaling, nutrient homeostasis, and stress responses (Kim et al. 2024). For example, miR156 and miR172 regulate developmental phase transitions, while miR398 and miR399 mediate responses to oxidative stress and phosphate starvation, respectively (Waheed and Zeng 2020). miRNAs are highly dynamic, exhibiting tissue-specific and developmental-stage-specific

Table 2.1 This shows the different classes of sncRNAs, their biogenesis, and their primary functions

Class	Size (nt)	Biogenesis pathway	Function	Key references
MicroRNAs	~20–24	Processed from pri-miRNAs by Drosha and dicer	Posttranscriptional regulation via mRNA degradation or translational repression	Kim et al. (2024)
Small interfering RNAs	~20–24	Derived from long double-stranded RNAs by dicer	mRNA cleavage, antiviral defense, transposon silencing	Xie and Yu (2015), Vaucheret and Voinnet (2024)
PIWI-interacting RNAs	~24–31	Dicer-independent; primary processing and amplification	Transposon silencing in germline cells	Wang et al. (2022), Iwasaki et al. (2025)
Small nucleolar RNAs	~60–300	Processed from introns or independent transcription	RNA modification (2'-O-methylation, pseudouridylation) of rRNA, snRNA	Webster and Ghalei (2023)
tRNA-derived small RNAs	~18–40	Cleavage of mature or precursor tRNAs by angiogenin and others	Stress response, gene regulation, retrotransposon control	Liu et al. (2021)
rRNA-derived small RNAs	~20–30	Derived from the cleavage of ribosomal RNAs	Emerging roles in gene regulation and stress responses	Kalakoti et al. (2024)
Small nuclear RNAs	~100–300	Transcribed by RNA polymerase II/III; snRNP assembly	Pre-mRNA splicing (spliceosome components)	Cheng et al. (2024), Webster and Ghalei (2023)

expression patterns. Due to their evolutionary conservation and specificity, miRNAs are promising candidates for applications in crop improvement, molecular diagnostics, and therapeutic interventions.

2.2.1.2 Small Interfering RNAs (siRNAs)

Small interfering RNAs (siRNAs) are ~21–24 nucleotides long double-stranded RNAs that mediate gene silencing via RNA interference (RNAi). They arise from long double-stranded RNA precursors derived from transposable elements, antisense transcripts, or viral RNA, processed by Dicer-like (DCL) enzymes in conjunction with RNA-dependent RNA polymerases (Table 2.1) (Singh et al. 2018). Mature small interfering RNAs (siRNAs) are loaded into Argonaute (AGO) proteins, forming RNA-induced silencing complexes (RISCs) that mediate gene silencing at either the transcriptional or posttranscriptional level (Singh et al. 2018).

In plants, siRNAs are central to the RNA-directed DNA methylation (RdDM) pathway, which induces cytosine methylation and histone modifications at target loci. This mechanism is crucial for silencing transposable elements and maintaining genome integrity, especially during reproduction and under stress. Among the most abundant are 24-nt heterochromatic siRNAs, which ensure epigenomic stability (Hombach and Kretz 2016). Plants also produce transacting siRNAs (tasiRNAs) and phased siRNAs (phasiRNAs) derived from the miRNA-cleaved precursors (Fei et al. 2013), which regulate specific gene families associated with the development and immune responses. Functionally, siRNAs act as molecular guardians of genome stability, mediating fine-tuned regulation of gene expression in response to developmental cues and environmental stimuli (Vaucheret and Voinnet 2024). Their sequence specificity and silencing capacity make them valuable tools in reverse genetics, functional genomics, and crop engineering, particularly in designing RNAi-based resistance to pathogens and targeted epigenome editing (Vaucheret and Voinnet 2024).

2.2.1.3 PIWI-Interacting RNAs (piRNAs)

PIWI-interacting RNAs (piRNAs) are a specialized class of small noncoding RNAs, typically 24–31 nucleotides long, that associate with PIWI proteins, a subfamily within the Argonaute protein family (Table 2.1). Unlike miRNAs and siRNAs, piRNAs are produced independently of Dicer from long single-stranded precursor transcripts originating from discrete genomic regions known as piRNA clusters (Wang et al. 2022). Primarily characterized in the germline of animals, piRNAs serve a crucial function in repressing transposable elements, thus preserving genome stability during gametogenesis (Iwasaki et al. 2025).

The piRNA pathway operates through two main mechanisms: (i) the primary processing pathway that generates mature piRNAs directly from precursor transcripts (Iwasaki et al. 2025), and (ii) the ping-pong amplification cycle which enhances silencing through reciprocal cleavage of complementary transposon RNAs, creating a feed-forward loop. Once loaded onto PIWI proteins, piRNAs guide them to their target RNAs, leading to transcript degradation or chromatin-level transcriptional silencing by recruiting epigenetic modifiers. In addition to transposon suppression, piRNAs have emerging roles in epigenetic regulation, germ cell development, and fertility (Iwasaki et al. 2025; Wang et al. 2022). Although canonical piRNAs are absent in plants, piRNA-like molecules and PIWI-domain proteins have been identified in some unicellular eukaryotes and basal metazoans, suggesting an ancient evolutionary origin (Iwasaki et al. 2025). The sequence diversity and rapid turnover of piRNA clusters reflect their coevolution with mobile genetic elements.

2.2.1.4 Small Nucleolar RNAs (snoRNAs)

Small nucleolar RNAs (snoRNAs) are a conserved group of noncoding RNAs, typically 60–300 nucleotides long, that primarily function in guiding chemical modifications of ribosomal RNAs (rRNAs) and small nuclear RNAs (Webster and Ghalei 2023). snRNAs are broadly categorized in two prominent families: C/D box snoRNAs, which mediate 2′-O-methylation, and H/ACA box snoRNAs, which catalyze pseudouridylation (Table 2.1). A related subclass, small Cajal body-specific RNAs (scaRNAs), localizes to Cajal bodies and modifies snRNAs involved in pre-mRNA splicing (Cheng et al. 2024).

Most snoRNAs are encoded within the introns of protein-coding or noncoding host genes and released following intron splicing and processing. Beyond their canonical roles in RNA modification, recent studies suggest their broader regulatory functions in gene expression modulation, cellular stress responses, and possible contributions to pathologies such as cancer and neurodegenerative disorders (Cheng et al. 2024; Webster and Ghalei 2023). Despite decades of study, the full scope of snoRNA functions in cell biology and disease remains incompletely understood, making them an ongoing subject of active research.

2.2.1.5 tRNA-Derived Small RNAs (tsRNAs)

tRNA-derived small RNAs (tsRNAs) represent a diverse group of regulatory noncoding RNAs generated from precise cleavage of mature or precursor tRNAs. tsRNAs, ranging from 18 to 40 nucleotides in length, are generally classified into tRNA halves (tiRNAs) and tRNA-derived fragments (tRFs) based on their size and cleavage origin (Table 2.1). Their biogenesis can be triggered by stress-responsive nucleases such as angiogenin or under physiological conditions via enzymes like Dicer (Liu et al. 2021). While initially considered degradation by-products, tsRNAs are now recognized as regulatory functional molecules regulating gene expression, modulating translation, repressing retrotransposons, and mediating stress responses. Furthermore, tsRNAs have been implicated in critical processes such as stem cell differentiation, immune system modulation, and disease pathogenesis, including cancer and metabolic disorders (Muthukumar et al. 2024).

2.2.1.6 rRNA-Derived Small RNAs (rsRNAs)

rRNA-derived small RNAs (rsRNAs) are a newly recognized class of regulatory noncoding RNAs, typically 20–30 nucleotides in length, generated through precise ribosomal RNA (rRNA) cleavage (Table 2.1). These molecules are derived from various regions of rRNA subunits, including 18S, 5.8S, and 28S, and their production often occurs in a context-specific and regulated manner. These were considered mere degradation by-products and are now understood to have functional roles in gene regulation, cellular stress responses, and potentially in ribosome biogenesis (Lambert et al. 2019).

Although the detailed mechanisms governing their biogenesis remain elusive, existing evidence suggests their involvement in specific ribonucleases and stress-responsive cleavage pathways (Singh et al. 2018). Functionally, rsRNAs have been shown to associate with Argonaute proteins, influencing mRNA stability and modulating translation, similar to microRNA-mediated regulation. Importantly, altered expression of rsRNAs has been linked to various pathological conditions, including cancer and neurodegenerative diseases, suggesting that these small RNAs contribute to regulatory networks far beyond their origin in the ribosome (Kalakoti et al. 2024; Lambert et al. 2019).

2.2.1.7 Small Nuclear RNAs (snRNAs)

Small nuclear RNAs (snRNAs) are a conserved group of noncoding RNAs, generally 100–300 nucleotides long, essential for pre-mRNA splicing in eukaryotes. The major spliceosomal snRNAs U1, U2, U4, U5, and U6 form the core RNA components of the spliceosome, a dynamic ribonucleoprotein complex responsible for intron recognition and excision (Table 2.1) (Terns and Terns 2018; Webster and Ghalei 2023). Each snRNA combines with specific proteins to form small nuclear ribonucleoproteins (snRNPs), which guide the sequential steps of spliceosome assembly, catalytic splicing reactions, and disassembly.

Beyond their central role in pre-mRNA splicing, certain snRNAs, such as U7, participate in specialized RNA processing events, including the 3′-end maturation of histone pre-mRNAs. snRNAs are transcribed by either RNA polymerase II (e.g., U1, U2, U4, U5) or RNA polymerase III (e.g., U6), and the transcripts undergo various posttranscriptional modifications, such as 2′-O-methylation and pseudouridylation. These modifications are governed by small Cajal body-specific RNAs (Terns and Terns 2018; Webster and Ghalei 2023). Defective snRNA processing has been associated with various human disorders, including spinal muscular atrophy and certain malignancies, highlighting their indispensable role in gene expression regulation.

2.2.2 Long Noncoding RNAs (lncRNAs)

Long noncoding RNAs (lncRNAs) are a heterogeneous class of transcripts longer than 200 nucleotides that do not encode functional proteins but exert regulatory effects at multiple levels on gene expression (Table 2.2). Transcribed primarily by RNA polymerase II, lncRNAs often undergo 5′ capping, splicing, and polyadenylation, similar to mRNAs. They can originate from diverse genomic contexts, including intergenic regions, introns, antisense strands, and enhancers, and are characterized by low sequence conservation but high spatiotemporal specificity

Table 2.2 Table summarizing the major classes of long noncoding RNAs

Class	Genomic origin	Function	Key references
Long intergenic noncoding RNAs	Intergenic regions	Transcriptional regulation, chromatin remodeling	Ransohoff et al. (2017)
Intronic lncRNAs	Derived from introns of protein-coding genes	Regulation of host gene expression, splicing	Ransohoff et al. (2017), Zhang et al. (2013)
Antisense lncRNAs	Opposite strand of protein-coding genes	Transcriptional interference, chromatin modification	Werner et al. (2024), Villegas and Zaphiropoulos (2015)
Enhancer RNAs	Active enhancer regions	Enhance transcription of target genes	Chen et al. (2023), Ye et al. (2020)
Circular RNAs	Back-splicing of exons or introns	miRNA sponging, transcriptional regulation	Misir et al. (2022), Liu and Chen (2022)
Pseudogene-derived lncRNAs	Transcription of pseudogene sequences	Compete with endogenous mRNAs and regulate translation	Milligan and Lipovich (2015), Schoeftner et al. (2021)

(Kaikkonen and Adelman 2018; Rinn and Chang 2012). lncRNAs influence transcriptional regulation, posttranscriptional processing, epigenetic remodeling, and chromatin organization, contributing to cellular identity, development, and disease progression (Kaikkonen and Adelman 2018; Rinn and Chang 2012).

2.2.2.1 Long Intergenic Noncoding RNAs (lincRNAs)

Long intergenic noncoding RNAs (lincRNAs) form a major subclass of lncRNAs, transcribed from genomic intervals that do not overlap with protein-coding genes (Table 2.2). Though lacking significant open reading frames, these transcripts are structurally similar to mRNAs and undergo typical RNA polymerase II processing (Ransohoff et al. 2017). Functionally, lincRNAs act as molecular scaffolds, decoys, or guides to regulate gene expression through diverse mechanisms (Mercer et al. 2009; Palos et al. 2022). A key role involves the recruitment of chromatin-modifying complexes, such as Polycomb Repressive Complex 2 (PRC2), to specific genomic loci, enabling epigenetic silencing (Ransohoff et al. 2017; Rinn and Chang 2012). Others modulate RNA-binding proteins or serve as competing endogenous RNAs (ceRNAs) sequestering microRNA activity. Despite the identification of thousands of lincRNAs, only a fraction has been functionally characterized, and their regulatory mechanisms often display strong cell type and context specificity. Aberrant lincRNA expression has been implicated in development, immune responses, oncogenesis, and neurological disorders (Plewka and Raczynska 2022).

2.2.2.2 Intronic lncRNAs

Intronic lncRNAs are transcribed entirely from the intronic sequences, independent of their exonic regions. Intronic lncRNAs can be transcribed in a sense or antisense direction and typically possess their own promoters, distinguishing them from intron-retained fragments (Table 2.2). Like other lncRNAs, they are capped, spliced, and polyadenylated (Ransohoff et al. 2017; Zhang et al. 2013). Intronic lncRNAs participate in diverse regulatory processes, including modulation of host gene expression, mRNA splicing, transcript stability, and chromatin accessibility. Some act locally (in cis) by influencing nearby genes, while others function at distant sites (in trans), often via interactions with chromatin-modifying enzymes or RNA-binding proteins (Mercer et al. 2009; Rinn and Chang 2012). Although many remain uncharacterized, emerging evidence implicates intronic lncRNAs in developmental regulation, cellular stress responses, and diseases such as cancer and neurodegeneration.

2.2.2.3 Antisense lncRNAs

Antisense long noncoding RNAs (antisense lncRNAs) are transcribed from the DNA strand opposite to protein-coding or noncoding genes and overlap partially or entirely with them (Xu 2018). Typically, longer than 200 nucleotides, antisense lncRNAs can regulate gene expression at the transcriptional, posttranscriptional, and epigenetic levels (Table 2.2) (Villegas and Zaphiropoulos 2015). They can influence the stability, splicing, or translation of their sense mRNA counterparts through RNA-RNA duplex formation or by recruiting chromatin-modifying complexes to specific loci (Villegas and Zaphiropoulos 2015; Werner et al. 2024; Xu 2018). Some antisense lncRNAs act in cis to regulate neighboring genes, while others function in trans-modulating distant genes or global pathways (Villegas and Zaphiropoulos 2015). Dysregulation of antisense lncRNAs has been associated with developmental abnormalities, neurological disorders, and cancer, highlighting their significant regulatory potential (Li et al. 2019). However, their functional roles exhibit high cell type and context specificity remains to be well characterized.

2.2.2.4 Enhancer RNAs (eRNAs)

Enhancer RNAs (eRNAs) are a class of noncoding RNAs transcribed from active enhancer regions of the genome, typically ranging from 0.5 to 5 kb in length. They are generally produced bidirectionally by RNA polymerase II, are often nonpolyadenylated, and exhibit short half-lives (Table 2.2) (Ye et al. 2020). eRNAs facilitate enhancer-promoter communication by stabilizing chromatin loops, recruiting transcriptional coactivators, or modulating the activity of RNA-binding proteins and chromatin remodelers (Ding et al. 2018). Initially considered transcriptional noise, recent functional studies have shown that many eRNAs are required to transcribe nearby genes efficiently. eRNA production strongly correlates with enhancer activation markers such as H3K27ac and the binding of transcription factors like p300 and Mediator complex components (Chen et al. 2023). Emerging evidence links eRNAs to cell identity, development, and disease, particularly in cancer and inflammatory responses (Ding et al. 2018). However, their regulatory function depends on the RNA molecule itself or the act of transcription, which remains an active area of investigation.

2.2.2.5 Circular RNAs (circRNAs)

Circular RNAs (circRNAs) are unique noncoding RNAs with a covalently closed loop structure without 5′ caps or 3′ poly(A) tails. They are preferentially generated through a back-splicing mechanism where a downstream splice donor joins an upstream splice acceptor (Table 2.2 Liu and Chen 2022). circRNAs are highly stable, often more resistant to exonucleases than linear RNAs, and display tissue-specific and developmentally regulated expression. Functionally, they act as microRNA sponges, interact with RNA-binding proteins, regulate transcription, and, in some cases, can even be translated into functional peptides. circRNAs are enriched in neural tissues and are implicated in synaptic plasticity, cell proliferation, and immune regulation processes (Liu and Chen 2022; Misir et al. 2022). Aberrant circRNA expression is associated with various diseases, including cancer, cardiovascular disorders, and neurological conditions. Despite growing interest, many aspects of circRNA biogenesis, turnover, and function remain to be elucidated, making them an active area of research (Misir et al. 2022).

2.2.2.6 Pseudogene-Derived lncRNAs

Pseudogene-derived lncRNAs are transcribed from pseudogene loci genomic sequences resembling protein-coding genes but rendered nonfunctional due to mutations (Table 2.2). Although often noncoding, these lncRNAs can regulate gene expression through diverse mechanisms, such as acting as competing endogenous RNAs that sequester microRNAs, thereby modulating the stability and translation of target mRNAs (Milligan and Lipovich 2015; Schoeftner et al. 2021). Some pseudogene-derived lncRNAs also influence chromatin remodeling and transcriptional regulation, or they even form sense–antisense pairs with their ancestral protein-coding genes to regulate them in cis or trans (Milligan and Lipovich 2015). The expression of pseudogene-derived lncRNAs is often tissue-specific and dynamically regulated during development and disease, suggesting they have evolved functional roles beyond mere genomic relics. However, the biological relevance of these lncRNAs remains uncharacterized and represents an emerging field of study.

2.3 Structural Features Associated with ncRNAs

Noncoding RNAs (ncRNAs) are characterized not only by their lack of protein-coding potential but also by a rich array of structural features integral to their stability, localization, and molecular functions (Hombach and Kretz 2016; Kaikkonen and Adelman 2018; Yadav et al. 2023). These attributes underpin their diverse roles in gene regulation, chromatin organization, and RNA metabolism.

The stem-loop or hairpin structure is a common structural motif in many small ncRNAs, miRNAs, siRNAs, and snoRNAs (Singh et al. 2018). In miRNAs, the precursor hairpin is essential for recognition and processing by Dicer, ultimately generating the mature miRNA (Singh et al. 2018). Similarly, snoRNAs adopt conserved secondary structures that guide site-specific RNA modifications by recruiting ribonucleoprotein complexes to rRNAs and snRNAs (Webster and Ghalei 2023).

Long ncRNAs often contain repetitive elements such as short interspersed nuclear elements (SINEs) and long interspersed nuclear elements (LINEs). These repeat sequences contribute to RNA stability, nuclear retention, and chromatin targeting (Chattopadhyay et al. 2021). Tandem repeats are characteristic of specific ncRNA subclasses, including those transcribed from centromeric and telomeric regions, where they support nuclear architecture and genome integrity (Carrieri et al. 2012; Chattopadhyay et al. 2021).

Secondary structural elements such as bulges, internal loops, pseudoknots, and multibranch junctions are widespread among ncRNAs and facilitate specific interactions with RNA-binding proteins, other RNAs, and chromatin (Chillón and Marcia 2020; Kung et al. 2013). Some ncRNAs adopt complex tertiary conformations, allowing them to perform particular molecular recognition tasks and catalytic activity similar to that of proteins. Notable examples include ribozymes and riboswitches, where the structure is directly coupled to catalytic or regulatory activity (Chillón and Marcia 2020; Kung et al. 2013). Circular RNAs (circRNAs), which adopt covalently closed-loop structures without free 5′- or 3′-ends, exhibit exceptional resistance to exonucleolytic degradation and are often more stable than their linear counterparts (Liu and Chen 2022).

Posttranscriptional RNA modifications such as N6-methyladenosine (m^6A), pseudouridine (Ψ), and 2′-O-methylation provide an additional regulatory layer to ncRNA function (Nachtergaele and He 2017; Yue et al. 2015). These modifications are dynamically controlled by dedicated writer, reader, and eraser proteins, forming the basis of the epi-transcriptomic code (Nachtergaele and He 2017). These chemical modifications influence RNA folding, stability, molecular interactions, and subcellular localization.

Additionally, specific sequence motifs are also critical for ncRNA function. For instance, the C/D box and H/ACA box motifs in snoRNAs determine their association with specific protein complexes and direct them to nucleolar sites of rRNA modification (Massenet et al. 2017). Similarly, Sm-binding sites in snRNAs are essential for snRNP assembly and nuclear import. Many ncRNAs also harbor localization signals that direct them to distinct subcellular compartments. For example, snoRNAs to the nucleolus, snRNAs to nuclear speckles, and specific lncRNAs to chromatin-enriched foci (Kufel and Grzechnik 2019). Together, these structural features confer upon ncRNAs the molecular precision and adaptability required for their diverse and often context-specific functions in gene regulation and nuclear organization.

2.4 Conservation and Evolution of Structural Motifs in ncRNAs

The evolutionary conservation of noncoding RNAs (ncRNAs) is often more evident in their structural motifs than in their primary sequences. This structural conservation reflects the functional importance of specific RNA folds that mediate interactions with proteins, other RNAs, and chromatin (Rivas 2021; Walter Costa 2023). Across diverse taxa, conserved secondary and tertiary elements such as stem loops, hairpins, and pseudoknots play essential roles in RNA maturation, stability, and gene regulation (Rivas 2021; Walter Costa 2023).

Canonical examples include the conserved stem-loop cloverleaf structure of tRNAs, the C/D and H/ACA box motifs in snoRNAs, and the Sm-binding site in snRNAs, all of which are critical for RNA processing and their functions (Massenet et al. 2017). The seed region in miRNAs, essential for target recognition, also shows strong conservation across species. These motifs enable ncRNAs to perform precise biochemical functions, such as guiding site-specific RNA modifications or regulating mRNA translation and stability (Johnsson et al. 2014; Leypold and Speicher 2021).

In contrast, long noncoding RNAs (lncRNAs) display rapid sequence evolution and often exhibit lineage- or species-specific expression patterns. However, despite limited primary sequence conservation, many lncRNAs retain conserved structural domains, particularly those mediating protein interactions. This suggests that evolutionary selection acts more strongly on functional RNA structure and sequence homology (Johnsson et al. 2014; Leypold and Speicher 2021). A well-known example is the triple-helix structure at the 3′-end of the mammalian lncRNA MALAT1, which has functional analogs in plants, stabilizing nonpolyadenylated transcripts (Arun et al. 2020).

The evolution of novel ncRNA functions often arises via exon shuffling, domain acquisition, or duplication, enabling the preservation of essential core structures while incorporating new interaction capabilities (Johnsson et al. 2014; Mercer et al. 2009). Moreover, ncRNAs and their associated proteins often coevolve in many cases, resulting in optimized binding interfaces and coordinated regulatory functions. Comparative genomics has identified a set of deeply conserved ncRNAs, such as rRNAs and U6 snRNA, and select miRNAs that trace back to the last eukaryotic common ancestor (Belair et al. 2018).

Further, lineage-specific genomic innovations also shape the structural diversity of ncRNAs. In plants, transposable element-derived ncRNAs, including siRNAs and lncRNAs, contribute significantly to regulatory complexity and often play a role in stress responses and display clade-specific expression patterns (Hadjiargyrou and Delihas 2013; Wang et al. 2017). Additionally, structures such as R-loops and G-quadruplexes have emerged as evolutionarily conserved yet dynamic motifs involved in chromatin remodeling, transcriptional regulation, and genome stability (Hombach and Kretz 2016; Mattick et al. 2023). Thus, ncRNAs' evolutionary landscape is marked by structural conservation under functional constraint and sequence-level plasticity. This duality enables ncRNAs to diversify, specialize, and integrate into regulatory networks while retaining the molecular architectures necessary for their biological roles.

2.5 Functional Roles of Noncoding RNAs

Noncoding RNAs (ncRNAs) are important regulators of gene expression and genome function across eukaryotes. They act through diverse mechanisms at multiple regulatory levels, including transcriptional, posttranscriptional regulations, and epigenetic modifications, and maintain genome integrity (Kaikkonen and Adelman 2018; Mercer et al. 2009). In plants, ncRNAs are essential in developmental processes, stress responses, and adaptation to environmental stimuli (Abdulraheem et al. 2024; Sun et al. 2018).

2.5.1 Transcriptional Regulation

Several ncRNAs regulate gene expression at the transcriptional level, mainly by interacting with chromatin-modifying complexes and transcription factors. Long noncoding RNAs (lncRNAs), for example, can act as molecular scaffolds recruiting chromatin-modifying proteins to specific genomic loci, thereby influencing histone modifications and DNA methylation patterns (Cao et al. 2024; Chekanova 2021). A well-known example is the COLDAIR lncRNA in *Arabidopsis thaliana*, which mediates vernalization-induced repression of FLOWERING LOCUS C (FLC) by recruiting PRC2 (Kim et al. 2017). Similarly, small interfering RNAs (siRNAs) produced via RNA-directed DNA methylation (RdDM) pathways guide de novo DNA methylation to silence transposons and repetitive elements (Xie and Yu 2015).

2.5.2 Posttranscriptional Regulation

Small ncRNAs, such as miRNAs and siRNAs, guide sequence-specific gene silencing by targeting mRNAs for cleavage or translational repression. In plants, miRNAs typically bind to their mRNA targets with near-perfect complementarity, enabling AGO-mediated endonucleolytic cleavage (Fei et al. 2013; Nachtergaele and He 2017). For instance, miR156 targets SPL transcription factors, influencing phase transitions and leaf morphology (Wang et al. 2009).

2.5.3 Epigenetic Regulation

ncRNAs contribute substantially to establishing and maintaining epigenetic marks. siRNAs derived from transposable elements or repetitive sequences participate in RNA-directed DNA methylation (RdDM), a hallmark of plant-specific epigenetic silencing (Abdulraheem et al. 2024; Xie and Yu 2015). In Arabidopsis, 24-nt siRNAs guide DRM2 to target loci through scaffold RNAs produced by Pol V, establishing a feedback loop for heterochromatin formation (Wierzbicki et al. 2009). lncRNAs also modulate chromatin dynamics, acting as molecular scaffolds or decoys for epigenetic regulators (Mercer et al. 2009).

2.5.4 Gene Silencing

Noncoding RNAs mediate gene silencing through two principal mechanisms: transcriptional gene silencing (TGS) and posttranscriptional gene silencing (PTGS). Small interfering RNAs (siRNAs) and microRNAs (miRNAs) primarily mediate PTGS by directing the degradation or translational repression of target mRNAs (Sijen et al. 2001). TGS involves siRNA-directed DNA methylation and histone modifications to repress transcription initiation (Xie and Yu 2015). These mechanisms are essential for silencing transposable elements, regulating gene dosage, and defending against viral pathogens.

2.5.5 Genome Stability

ncRNAs maintain genome integrity by silencing transposable elements and maintaining heterochromatin structure. siRNAs derived from TEs direct DNA methylation and histone modifications to prevent transposon mobilization (Hadjiargyrou and Delihas 2013; Khanduja et al. 2016). Additionally, emerging reports suggest the role of ncRNAs in DNA damage sensing and repair. In mammals, damage-induced lncRNAs promote the recruitment of repair proteins, while plant-specific examples remain limited, similar roles are emerging (Hadjiargyrou and Delihas 2013; Michelini et al. 2017).

2.5.6 *Nuclear Organization*

Recent studies suggest that ncRNAs contribute to nuclear architecture by allowing the formation and maintenance of nuclear bodies such as Cajal bodies, nucleoli, and speckles. lncRNAs and small nuclear RNAs (snRNAs) act as scaffolds or organizational guides for assembling these subnuclear compartments (Palihati and Saitoh 2024; Quinodoz and Guttman 2022). In Arabidopsis, lncRNAs have been implicated in regulating chromatin loop formation and interchromosomal interactions, although the underlying mechanisms are still not elucidated (Zhao et al. 2018).

2.5.7 *Roles in Plant Development and Stress Responses*

In plants, ncRNAs are crucial for developmental processes like embryogenesis, organogenesis, flowering, and hormone signaling (Bhogireddy et al. 2021; Panchal et al. 2023; Yadav et al. 2022, 2023). For example, miR160 and miR167 target auxin response factors (ARFs), modulating auxin-mediated developmental pathways (Jodder 2020). Likewise, lncRNAs such as LDMAR (in rice) are essential for reproductive development under long-day conditions (Caixia et al. 2020). Under abiotic and biotic stress conditions, ncRNAs mediate rapid transcriptomic reprogramming (Sun et al. 2018). miR398, for example, modulates oxidative stress response by targeting CSD1 and CSD2, thus enhancing plant stress tolerance (Leng et al. 2017).

2.6 Conclusions

Over the past two decades, noncoding RNAs have transformed our understanding of gene regulation, genome architecture, and cellular function. These diverse RNA species, ranging from well-characterized miRNAs and siRNAs to emerging classes like tsRNAs and rsRNAs, play central roles in shaping molecular processes across the kingdom. The structural versatility of ncRNAs underpins a remarkable functional diversity, enabling precise control over transcription, chromatin dynamics, RNA stability, and translation. As research continues to uncover new noncoding RNA species and mechanisms, it is becoming increasingly clear that these molecules are key to decoding the complexity of life itself. A deeper understanding of their functions holds great promise for advancing basic biology and opening new frontiers in diagnostics, therapeutics, and sustainable agriculture.

Acknowledgments The author acknowledges the Goa University for support.

Funding This work was supported by a grant from the Department of Biotechnology (DBT), India (Project Code, BT/PR**52132**/BSA/**33/23/2024**) and Seed money grant from Goa University (GU/D-RDRM/Seed Money/SBSB/VKY/**7074/2025/374**).

References

Abdulraheem MI, Xiong Y, Moshood AY, Cadenas-Pliego G, Zhang H, Hu J (2024) Mechanisms of plant epigenetic regulation in response to plant stress: recent discoveries and implications. Plants 13(2):163. https://doi.org/10.3390/PLANTS13020163

Arun G, Aggarwal D, Spector DL (2020) MALAT1 long non-coding RNA: functional implications. Non-Coding RNA 6(2):22. https://doi.org/10.3390/NCRNA6020022

Belair C, Sim S, Wolin SL (2018) Non-coding RNA surveillance: the ends justify the means. Chem Rev 118(8):4422–4447

Bhogireddy S, Mangrauthia SK, Kumar R, Pandey AK, Singh S, Jain A, Budak H, Varshney RK, Kudapa H (2021) Regulatory non-coding RNAs: a new frontier in regulation of plant biology. Funct Integr Genomics 21(3):313–330. https://doi.org/10.1007/S10142-021-00787-8

Caixia G, Xiuwen Z, Hubo L, Mlekwa UA, Yu G, Jie X (2020) Roles of lncRNAs in Rice: advances and challenges. Rice Sci 27(5):384–395. https://doi.org/10.1016/J.RSCI.2020.03.003

Cao W, Yang L, Zhuang M, Lv H, Wang Y, Zhang Y, Ji J (2024) Plant non-coding RNAs: the new frontier for the regulation of plant development and adaptation to stress. Plant Physiol Biochem 208:108435. https://doi.org/10.1016/J.PLAPHY.2024.108435

Carrieri C, Cimatti L, Biagioli M, Beugnet A, Zucchelli S, Fedele S, Pesce E, Ferrer I, Collavin L, Santoro C, Forrest ARR, Carninci P, Biffo S, Stupka E, Gustincich S (2012) Long non-coding antisense RNA controls Uchl1 translation through an embedded SINEB2 repeat. Nature 491(7424):454–457

Chattopadhyay P, Srinivasa Vasudevan J, Pandey R (2021) Non-coding RNAs: modulators and modulatable players during infection-induced stress response. Brief Funct Genomics 20(1):28–41. https://doi.org/10.1093/BFGP/ELAA026

Chekanova JA (2021) Plant long non-coding RNAs in the regulation of transcription. Essays Biochem 65(4):751–760. https://doi.org/10.1042/EBC20200090

Chen Q, Zeng Y, Kang J, Hu M, Li N, Sun K, Zhao Y (2023) Enhancer RNAs in transcriptional regulation: recent insights. Front Cell Dev Biol 11:1205540. https://doi.org/10.3389/FCELL.2023.1205540/BIBTEX

Cheng Y, Wang S, Zhang H, Lee JS, Ni C, Guo J, Chen E, Wang S, Acharya A, Chang TC, Buszczak M, Zhu H, Mendell JT (2024) A non-canonical role for a small nucleolar RNA in ribosome biogenesis and senescence. Cell 187(17):4770–4789. https://doi.org/10.1016/J.CELL.2024.06.019

Chillón I, Marcia M (2020) The molecular structure of long non-coding RNAs: emerging patterns and functional implications. Crit Rev Biochem Mol Biol 55(6):662–690

Ding M, Liu Y, Liao X, Zhan H, Liu Y, Huang W (2018) Enhancer RNAs (eRNAs): new insights into gene transcription and disease treatment. J Cancer 9(13):2334–2340. https://doi.org/10.7150/JCA.25829

Fei Q, Xia R, Meyers BC (2013) Phased, secondary, small interfering RNAs in post-transcriptional regulatory networks. Plant Cell 25(7):2400–2415. https://doi.org/10.1105/TPC.113.114652

Feingold EA, Good PJ, Guyer MS, Kamholz S, Liefer L, Wetterstrand K, Collins FS, Gingeras TR, Kampa D, Sekinger EA, Cheng J, Hirsch H, Ghosh S, Zhu Z, Patel S, Piccolboni A, Yang A, Tammana H, Bekiranov S et al (2004) The ENCODE (ENCyclopedia of DNA elements) project. Science 306(5696):636–640

Hadjiargyrou M, Delihas N (2013) The intertwining of transposable elements and non-coding RNAs. Int J Mol Sci 14(7):13307–13328. https://doi.org/10.3390/IJMS140713307
Hombach S, Kretz M (2016) Non-coding RNAs: classification, biology and functioning. Adv Exp Med Biol 937:3–17. https://doi.org/10.1007/978-3-319-42059-2_1
Iwasaki YW, Siomi MC, Siomi H (2025) PIWI-interacting RNA: its biogenesis and functions. Annu Rev Biochem 18:15. https://doi.org/10.1146/annurev-biochem-060614-034258
Jodder J (2020) miRNA-mediated regulation of auxin signaling pathway during plant development and stress responses. J Biosci 45(1). https://doi.org/10.1007/s12038-020-00062-1
Johnsson P, Lipovich L, Grandér D, Morris KV (2014) Evolutionary conservation of long non-coding RNAs; sequence, structure, function. Biochim Biophys Acta Gen Subj 1840(3):1063–1071. https://doi.org/10.1016/J.BBAGEN.2013.10.035
Kaikkonen MU, Adelman K (2018) Emerging roles of non-coding RNA transcription. Trends Biochem Sci 43(9):654–667
Kalakoti G, Vivek A, Kamboj A, Singh A, Chakraborty S, Kumar S (2024) Comprehensive profiling of rRNA-derived small RNAs in *Arabidopsis thaliana* using rsRNAfinder pipeline. MethodsX 12:102494. https://doi.org/10.1016/J.MEX.2023.102494
Khanduja JS, Calvo IA, Joh RI, Hill IT, Motamedi M (2016) Nuclear non-coding RNAs and genome stability. Mol Cell 63(1):7–20. https://doi.org/10.1016/J.MOLCEL.2016.06.011
Kim DH, Xi Y, Sung S (2017) Modular function of long non-coding RNA, COLDAIR, in the vernalization response. PLoS Genet 13(7):e1006939. https://doi.org/10.1371/JOURNAL.PGEN.1006939
Kim H, Lee YY, Kim VN (2024) The biogenesis and regulation of animal microRNAs. Nat Rev Mol Cell Biol 26(4):276–296
Kufel J, Grzechnik P (2019) Small nucleolar RNAs tell a different tale. Trends Genet 35(2):104–117. https://doi.org/10.1016/J.TIG.2018.11.005
Kung JTY, Colognori D, Lee JT (2013) Long non-coding RNAs: past, present, and future. Genetics 193(3):651–669. https://doi.org/10.1534/GENETICS.112.146704
Lambert M, Benmoussa A, Provost P (2019) Small non-coding RNAs derived from eukaryotic ribosomal RNA. Non-Coding RNA 5(1):16. https://doi.org/10.3390/NCRNA5010016
Leng XP, Wang P, Zhu X, Li X, Zheng T, Shangguan L, Fang J (2017) Ectopic expression of CSD1 and CSD2 targeting genes of miR398 in grapevine is associated with oxidative stress tolerance. Funct Integr Genomics 17(6):697–710. https://doi.org/10.1007/S10142-017-0565-9/TABLES/1
Leypold NA, Speicher MR (2021) Evolutionary conservation in non-coding genomic regions. Trends Genet 37(10):903–918. https://doi.org/10.1016/J.TIG.2021.06.007
Li L, Zhuang Y, Zhao X, Li X (2019) Long non-coding RNA in neuronal development and neurological disorders. Front Genet 10(JAN):436598. https://doi.org/10.3389/FGENE.2018.00744/XML/NLM
Liu CX, Chen LL (2022) Circular RNAs: characterization, cellular roles, and applications. Cell 185(12):2016–2034. https://doi.org/10.1016/J.CELL.2022.04.021
Liu B, Cao J, Wang X, Guo C, Liu Y, Wang T (2021) Deciphering the tRNA-derived small RNAs: origin, development, and future. Cell Death Dis 13(1):1–13. https://doi.org/10.1038/s41419-021-04472-3
Massenet S, Bertrand E, Verheggen C (2017) Assembly and trafficking of box C/D and H/ACA snoRNPs. RNA Biol 14(6):680–692
Mattick JS, Amaral PP, Carninci P, Carpenter S, Chang HY, Chen LL, Chen R, Dean C, Dinger ME, Fitzgerald KA, Gingeras TR, Guttman M, Hirose T, Huarte M, Johnson R, Kanduri C, Kapranov P, Lawrence JB, Lee JT et al (2023) Long non-coding RNAs: definitions, functions, challenges and recommendations. Nat Rev Mol Cell Biol 24(6):430–447. https://doi.org/10.1038/s41580-022-00566-8
Mercer TR, Dinger ME, Mattick JS (2009) Long non-coding RNAs: insights into functions. Nat Rev Genet 10(3):155–159

Michelini F, Pitchiaya S, Vitelli V, Sharma S, Gioia U, Pessina F, Cabrini M, Wang Y, Capozzo I, Iannelli F, Matti V, Francia S, Shivashankar GV, Walter NG, Di Fagagna FDA (2017) Damage-induced lncRNAs control the DNA damage response through interaction with DDRNAs at individual double-strand breaks. Nat Cell Biol 19(12):1400–1411. https://doi.org/10.1038/ncb3643

Milligan MJ, Lipovich L (2015) Pseudogene-derived lncRNAs: emerging regulators of gene expression. Front Genet 6(FEB):117941. https://doi.org/10.3389/FGENE.2014.00476/BIBTEX

Misir S, Wu N, Yang BB (2022) Specific expression and functions of circular RNAs. Cell Death Differ 29(3):481–491. https://doi.org/10.1038/s41418-022-00948-7

Muthukumar S, Li CT, Liu RJ, Bellodi C (2024) Roles and regulation of tRNA-derived small RNAs in animals. Nat Rev Mol Cell Biol 25(5):359–378. https://doi.org/10.1038/s41580-023-00690-z

Nachtergaele S, He C (2017) The emerging biology of RNA post-transcriptional modifications. RNA Biol 14(2):156–163

Palihati M, Saitoh N (2024) RNA in chromatin organization and nuclear architecture. Curr Opin Genet Dev 86:102176. https://doi.org/10.1016/J.GDE.2024.102176

Palos K, Nelson Dittrich AC, Yu L, Brock JR, Railey CE, Wu HYL, Sokolowska E, Skirycz A, Hsu PY, Gregory BD, Lyons E, Beilstein MA, Nelson ADL (2022) Identification and functional annotation of long intergenic non-coding RNAs in Brassicaceae. Plant Cell 34(9):3233–3260. https://doi.org/10.1093/PLCELL/KOAC166

Panchal A, Maurya J, Seni S, Singh RK, Prasad M (2023) An insight into the roles of regulatory ncRNAs in plants: an abiotic stress and developmental perspective. Plant Physiol Biochem 201:107823. https://doi.org/10.1016/J.PLAPHY.2023.107823

Plewka P, Raczynska KD (2022) Long intergenic non-coding RNAs affect biological pathways underlying autoimmune and neurodegenerative disorders. Mol Neurobiol 59(9):5785–5808. https://doi.org/10.1007/S12035-022-02941-0

Quinodoz SA, Guttman M (2022) Essential roles for RNA in shaping nuclear organization. Cold Spring Harb Perspect Biol 14(5):a039719. https://doi.org/10.1101/CSHPERSPECT.A039719

Ransohoff JD, Wei Y, Khavari PA (2017) The functions and unique features of long intergenic non-coding RNA. Nat Rev Mol Cell Biol 19(3):143–157. https://doi.org/10.1038/nrm.2017.104

Rinn JL, Chang HY (2012) Genome regulation by long non-coding RNAs. Annu Rev Biochem 81:145–166

Rivas E (2021) Evolutionary conservation of RNA sequence and structure. RNA 12(5):e1649

Schoeftner S, Scarola M, Benetti R (2021) Chromatin regulation at parental gene promoters by pseudogene sense lncRNAs. Methods Mol Biol 2324:203–217. https://doi.org/10.1007/978-1-0716-1503-4_13

Sijen T, Vijn I, Rebocho A, Van Blokland R, Roelofs D, Mol JNM, Kooter JM (2001) Transcriptional and post-transcriptional gene silencing are mechanistically related. Curr Biol 11(6):436–440. https://doi.org/10.1016/S0960-9822(01)00116-6

Singh A, Gautam V, Singh S, Sarkar Das S, Verma S, Mishra V, Mukherjee S, Sarkar AK (2018) Plant small RNAs: advancement in the understanding of biogenesis and role in plant development. Planta 248(3):545–558. https://doi.org/10.1007/S00425-018-2927-5

Sun X, Zheng H, Sui N (2018) Regulation mechanism of long non-coding RNA in plant response to stress. Biochem Biophys Res Commun 503(2):402–407. https://doi.org/10.1016/J.BBRC.2018.07.072

Terns MP, Terns RM (2018) Small nucleolar RNAs: versatile trans-acting molecules of ancient evolutionary origin. Gene Expr 10(1–2):17

Vaucheret H, Voinnet O (2024) The plant siRNA landscape. Plant Cell 36(2):246–275. https://doi.org/10.1093/PLCELL/KOAD253

Villegas VE, Zaphiropoulos PG (2015) Neighboring gene regulation by antisense long non-coding RNAs. Int J Mol Sci 16(2):3251–3266. https://doi.org/10.3390/IJMS16023251

Waheed S, Zeng L (2020) The critical role of miRNAs in regulation of flowering time and flower development. Genes 11(3):319. https://doi.org/10.3390/GENES11030319

Walter Costa MB (2023) Evolutionary conservation of RNA secondary structure. Methods Mol Biol 2586:121–146. https://doi.org/10.1007/978-1-0716-2768-6_8/PROTOCOL

Wang JW, Czech B, Weigel D (2009) miR156-regulated SPL transcription factors define an endogenous flowering pathway in *Arabidopsis thaliana*. Cell 138(4):738–749. https://doi.org/10.1016/J.CELL.2009.06.014

Wang D, Qu Z, Yang L, Zhang Q, Liu ZH, Do T, Adelson DL, Wang ZY, Searle I, Zhu JK (2017) Transposable elements (TEs) contribute to stress-related long intergenic non-coding RNAs in plants. Plant J 90(1):133–146

Wang X, Ramat A, Simonelig M, Liu MF (2022) Emerging roles and functional mechanisms of PIWI-interacting RNAs. Nat Rev Mol Cell Biol 24(2):123–141. https://doi.org/10.1038/s41580-022-00528-0

Webster SF, Ghalei H (2023) Maturation of small nucleolar RNAs: from production to function. RNA Biol 20(1):715–736. https://doi.org/10.1080/15476286.2023.2254540

Werner A, Kanhere A, Wahlestedt C, Mattick JS (2024) Natural antisense transcripts as versatile regulators of gene expression. Nat Rev Genet 25(10):730–744. https://doi.org/10.1038/s41576-024-00723-z

Wierzbicki AT, Ream TS, Haag JR, Pikaard CS (2009) RNA polymerase v transcription guides ARGONAUTE4 to chromatin. Nat Genet 41(5):630–634

Xie M, Yu B (2015) siRNA-directed DNA methylation in plants. Curr Genomics 16(1):23–31. https://doi.org/10.2174/1389202915666141128002211

Xu J (2018) Antisense RNA: the new favorite in genetic research. J Zhejiang Univ Sci B 19(10):739–749. https://doi.org/10.1631/JZUS.B1700594

Yadav VK, Sawant SV, Yadav A, Jalmi SK, Kerkar S (2022) Genome-wide analysis of long non-coding RNAs under diel light exhibits role in floral development and the circadian clock in *Arabidopsis thaliana*. Int J Biol Macromol 223:1693–1704. https://doi.org/10.1016/J.IJBIOMAC.2022.09.295

Yadav VK, Jalmi SK, Tiwari S, Kerkar S (2023) Deciphering shared attributes of plant long non-coding RNAs through a comparative computational approach. Sci Rep 13(1):1–13

Ye R, Cao C, Xue Y (2020) Enhancer RNA: biogenesis, function, and regulation. Essays Biochem 64(6):883–894. https://doi.org/10.1042/EBC20200014

Yue Y, Liu J, He C (2015) RNA N6-methyladenosine methylation in post-transcriptional gene expression regulation. Genes Dev 29(13):1343–1355. https://doi.org/10.1101/GAD.262766.115

Zhang Y, Zhang XO, Chen T, Xiang JF, Yin QF, Xing YH, Zhu S, Yang L, Chen LL (2013) Circular Intronic long non-coding RNAs. Mol Cell 51(6):792–806

Zhang P, Wu W, Chen Q, Chen M (2019) Non-coding RNAs and their integrated networks. J Integr Bioinform 16(3)

Zhao X, Li J, Lian B, Gu H, Li Y, Qi Y (2018) Global identification of Arabidopsis lncRNAs reveals the regulation of MAF4 by a natural antisense RNA. Nat Commun 9(1):1–12. https://doi.org/10.1038/s41467-018-07500-7

Chapter 3
Promoter Regulatory Regions of Abiotic Stress Responses

P. P. Amritha, A. T. Meharban, T. G. Ajith, and Jasmine M. Shah

Abstract Promoters are the connecting link between perceiving of the environmental changes/stress and the triggered-gene expression. Plants are highly dependent on these switch-like sequences because they greatly facilitate rapid adjustability and dynamic environmental adaption to the non-locomotive plants. Promoters recognize external signals via stress-responsive transcription factors (TFs), noncoding RNAs (ncRNAs), and epigenetic chemical changes of DNA/histones. These factors have binding sites on/near the promoters. All these factors work in coordination to drive the promoter to facilitate dynamic regulation of various genes, with the intention for providing optimum survival conditions. Also, the promoters of gene encoding these regulatory factors (TFs, and ncRNAs) are also in turn regulated by the same mechanism, thereby indicating the occurrence of complex networks of pathways. Abiotic stresses trigger the expression of many of these regulatory factors. Various abiotic stress-responsive TF families, and reports on abiotic-stress-induced ncRNAs and other epigenetic modifications in plants are discussed in this chapter. Transgenerational epigenetic changes due to abiotic stresses are also described. Recent high throughput information on genome, transcriptome, and epigenome sequencing in multiple plants demands efficient bioinformatics analysis, aided by artificial intelligence, to dig out maximum information. Deeper understanding of the complex regulatory webs is required to find reliable solutions to mitigate abiotic stress-induced damages in plants. This would aid in developing desirable synthetic promoters, overexpression strategies, and genome/epigenome editing in plants.

Keywords Abiotic stress · Epigenetic · Noncoding RNA (Nc RNA) · Promoters · Transcription factors

---

P. P. Amritha · A. T. Meharban · T. G. Ajith · J. M. Shah (✉)
Department of Plant Science, Central University of Kerala, Kasaragod, Kerala, India
e-mail: jasmine@cukerala.ac.in

S. Tiwari et al. (eds.), *Decoding Non-Coding RNA*, Concepts and Strategies in Plant Sciences, https://doi.org/10.1007/978-981-96-9859-2_3

3.1 Introduction

Promoters are the sensory connections between an external signal and the gene response. RNA polymerases, along with certain transcription factors (TFs), bind to the promoters to initiate transcription. Apart from the general TFs, promoters are also the locations for the binding of special TFs, which participate in regulating transcription. Additionally, promoters also harbor binding sites for epigenetic regulatory factors (ERFs) such as DNA/histone modifying enzymes (and Bergman 2009) and ncRNAs, which also control transcription (Holoch and Moazed 2015). The special TFs and ERFs in turn are responsive to factors such as age, tissue and environmental stresses (Feil and Fraga 2012; Issa 2003). Recent reports also convey that the proximal ends of certain plant promoters code for about >200 bp-long noncoding (lnc) RNAs (Domínguez-Rosas et al. 2023). Therefore, a promoter is like a flute that tunes gene expression via multiple gene-regulatory factors (Fig. 3.1). Plants being sessile organisms, the promoter function is of more significance because it is the gateway to survival and adaptation via regulated gene expression. It is interesting and informative to observe how plants perceive stress, and under the absence of locomotory-escape mechanism result in an orchestrated gene expression via the promoter. Such studies aid in the development of biotic/abiotic stress resilient crops. This chapter elaborates the mechanisms of promoter regulatory features that respond to abiotic stresses.

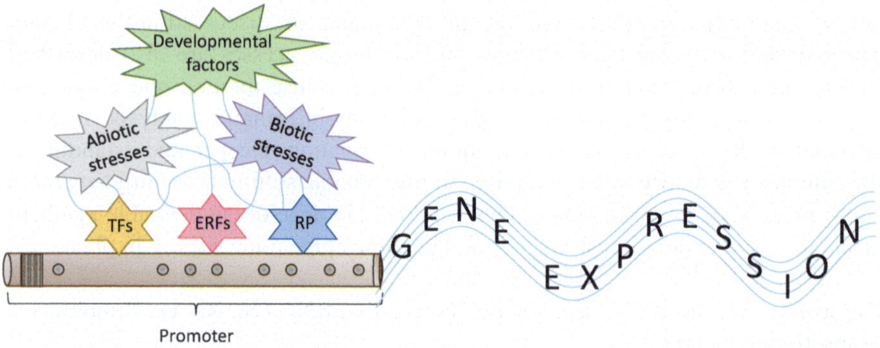

Fig. 3.1 An image of a flute representing the promoter that decides the tune of gene expression. TFs, ERFs, and RP are transcription factors, epigenetic regulatory factors, and RNA polymerase, respectively, which bind to the promoter for tuning gene regulation in response to various environmental (biotic/abiotic) parameters/stresses, and developmental factors including age, time, tissue, and others

3.2 Abiotic Stress Responsive Transcription Factors (ASRTFs)

These are special TFs that regulate gene expression in response to abiotic stresses. Special TFs are TFs, which are not mandatory for routine transcription, but are required for tuning (activating or repressing) gene expression under special circumstances (Martin et al. 2010). These have binding sites on the promoter/genic regions, RNA polymerases, or other TFs. The plant ASRTFs comprise of more than 60 members (Yoon et al. 2020) that belong to the families such as Myeloblastosis viral oncogene homolog (MYB), MYC Proto-Oncogene (MYC), NAM, ATAF1/2, and CUC2 (NAC), Basic helix–loop–helix (bHLH), apetala 2/ethylene response factor (AP2/ERF), basic leucine zipper (bZIP), and WRKY (Hoang et al. 2017; Yoon et al. 2020; Mizoi et al. 2012). These respond to abiotic stresses such as drought, temperature, radiations, salt, mutagens, nutrient deficiency, toxicity, waterlogging, and others. Many previous reviews have elaborately tabulated various ASRTFs, their source, and abiotic stress-responsive (ASR) functions (Hoang et al. 2017; Lata et al. 2011; Liu et al. 2024; Yoon et al. 2020). Some of these ASRTF families such as WRKY, MYB, DREB, bHLH, ERF, and NAC, overlap for other stresses such as biotic stresses also (Liu et al. 2024).

Myeloblastosis viral oncogene homolog (MYB): It is an extensive plant TF family, first identified in the plant *Zea mays* (Paz-Ares et al. 1987). The structure of MYB protein is comprised of three regions such as the DNA structure binding domain (DBD), the transcriptional activation domain, and the negative regulatory region. Based on the variation in the conserved DBD, MYB family is classified into four classes such as 1R-, R2R3-, 3R- and 4R-MYB protein. MYB plays a multifunctional role against drought stress through inducing different pathways including biosynthesis of flavonoids, wax, and cutin and physiological changes such as stomatal opening and their formation. Also, MYB TFs regulate responses toward abiotic stresses such as salt, cold, and heat stress tolerance. Similar to drought, MYB contributes to salt stress through cutin, wax, and suberin biosynthesis regulation, which enhances antioxidative capacity. Cold stress-induced MYB response in turn regulates the C-repeat-binding factors (CBFs) that are important transcription factors in the cold- and heat-induced calcium signaling regulation. Since MYB family is the largest among others it showed diverse functions such as ABA signaling, biotic stress responses, plant development, controlling cell shape, petal morphogenesis, trichome development, proliferation and differentiation of cells, and nutrient deficiency.

NAM, ATAF1/2, and CUC2 (NAC): These TFs are a large TF super family of about 117 *NAC* genes, identified from *Arabidopsis*. Many plants express more than a hundred numbers of NAC TFs (Shao et al. 2015). These TFs are included in diverse functional groups depending on their response to biotic/abiotic stresses, plant development, senescence, and secondary wall formation. A typical NAC TF is composed of the N-terminal NAC domain (150 amino acids) and the C-terminal transcription regulatory (TR) region (Tran et al. 2010). Expression of NAC genes is affected by various factors such as the binding of stress-responsive elements on their promoter, miRNA-mediated posttranscriptional regulation, and post-translational regulations (ubiquitination and phosphorylation) (Marques et al. 2017). NAC TFs are mainly involved in drought, and salinity tolerance. Some NAC factors express a downstream effect of plant hormones. For instance, expression of *ZmNAC55* was induced by abscisic acid in maize plant (Mao et al. 2016). Some NAC TFs function independently (Shahnejat-Bushehri et al. 2012). In *Arabidopsis*, overexpression of genes encoding ANAC019, ANAC055, and ANAC072 lead to the upregulation of several stress-inducible genes including gene encoding glyoxylase I (Tran et al. 2004). Functions of many NAC genes was characterized by the over-expression technique, which often lead to some obvious negative effect on the transgenic plant phenotype such as dwarfism, flowering delay, reduced yield, and increased susceptibility to fungal/bacterial infections (Shao et al. 2015).

Basic helix–loop–helix (bHLH): These TFs are the next largest family in *Arabidopsis*. This includes 225 and 211 members in *Arabidopsis* and rice, respectively (Castilhos et al. 2014). A typical bHLH TF is comprised of the HLH region and the basic domain, both of these being of highly conserved. These domains recognize the E and G boxes of target gene (Castilhos et al. 2014). These TFs mainly respond to drought stress by getting actively involved in the development of stomata, formation of root hairs, and regulation of hormone metabolism (Pillitteri and Torii 2007; Zhao et al. 2008; Abe et al. 2003). Overexpression of *bHLH1122* in *Arabidopsis* make imparted resistance to drought, cold, and osmotic stresses (Liu et al. 2014). MYC2, an important bHLH TF, regulates the expression of jasmonic acid- (JA) responsive genes. In sweet potato, salt stress exposure led to upregulated expression of genes related to JA biosynthesis and signaling pathway gene expression (Zhang et al. 2017). Also, MYC2 activation tolerates salt stress via proline biosynthesis regulation (Verma et al. 2020). The bHLH TFs respond to variety stress conditions. Functionally annotated TFs were mainly classified into 32 subfamilies among which half of them respond to abiotic stresses. These subfamilies comprise TFs that respond to abiotic stresses such as drought, salt, cold, heat, and iron deficiency (Qian et al. 2021).

APETALA 2 (AP2)/ethylene responsive factor (ERF): This is also a large TF family that plays multiple roles including developmental stages, environmental stress responses, and primary/secondary metabolism in plants (Feng et al. 2020). AP2 is a core DNA binding domain, which is highly conserved. Based on the number of AP2 domain, there are five subfamilies— AP2, ERF, and dehydration-responsive element binding (DREB), related to ABI3/VP (RAV), and Soloist (Debbarma et al. 2019; Licausi et al. 2013). AP2/ERF TFs respond to various abiotic stresses. Drought tolerance in *Arabidopsis* was imparted by the overexpression of *AtERF019* via regulation of physiological changes such as delayed flowering, plant growth and senescence in *Arabidopsis* (Scarpeci et al. 2017). AtERF105 is required for cold tolerance and AtERF15 for resistance to pathogens (Bolt et al. 2017; Zhang et al. 2015). AP2/ERF induces diverse stress responses in crops such as rice, tomato, and wheat also. Overexpression of *OsAP37* and *OsEREBP* in rice enhanced tolerance to drought (Oh et al. 2009; Jisha et al. 2015). In wheat, an

ERF-TaERF3 function resulted in increased salinity and drought tolerance (Rong et al. 2014). Increased expression of TERF2/LeERF2 in tomato enhanced tolerance to freezing (Zhijin and Rongfeng 2010).

Basic leucine zipper (bZIP): It is also a large TF group, involved in regulating various biological processes, and response to abiotic/biotic stresses. There are 75 bZIP TFs reported in *Arabidopsis*, and these are classified into ten subfamilies, A, B, C, D, E, F, G, H, I, and S. These TFs have the conserved basic region, and the leucine zipper domain. The basic domain binds to specific DNA *cis*-elements (Guo et al. 2024). bZIP TFs interact with stress-responsive genes. These actively regulate gene responses to cold (Sun et al. 2011), and drought stress (Yang et al. 2019). In *Arabidopsis*, *AtbZIP63* regulates the proline dehydrogenase activity and makes plant endurant to osmotic stress (Veerabagu et al. 2024), which is an after effect of drought/salt exposure. Similar to this different bZIP TFs showed diverse abiotic stress responses that are summarized in the table.

WRKY: These TFs are unique for plants and the name indicates the amino acid sequence (WRKYGQK) that is highly conserved. WRKY TFs comprise the DNA binding domain at the N-terminus and a zinc-finger structure at the C-terminal (Ma and Hu 2024). WRKYs play important roles under biotic/abiotic stresses. WRKYs regulate ABA signaling that is required for drought stress response. For instance, GsWRKY20 and TaWRKY146 impart drought stress tolerance through regulating ABA-induced stomatal opening. IgWRKY32, IgWRKY50, GhWRKY33, GhWRKY21, and OsWRKY5 are examples of other WRKYs involved in ABA regulation (Zhang et al. 2022a, b; Wang et al. 2019, 2021; Lim et al. 2022). WRKYs are also salt/cold stress-responsive. *Trans* expression of GarWRKY17a, GarWRKY104, and PeWRKY83 imparted tolerance to salt in *Arabidopsis* (Fan et al. 2015; Wu et al. 2017). Increased expression of CsWRKY46 in cucumber and CdWRKY2 in bermudagrass resulted in increased cold stress tolerance (Zhang et al. 2016; Huang et al. 2022).

3.3 Epigenetic Regulatory Factors (ERFs)

Promoter regulation due to ERFs is due to chemical modification of the DNA/histones and due to ncRNAs (Fig. 3.2). These changes can either increase or decrease the expression of genes.

DNA modification: It is a critical epigenetic modification in plants. Eukaryotic DNA can carry methylated nucleotides such as 5-methylcytosine and 6-methyladinine. In plants, methylation of DNA is established by the RNA- or DNA-directed methylation pathways, and the DNA methylation/demethylation is regulated dynamically (He et al. 2011). DNA methylation at fifth position of the carbon (5-methylcytosine) is the most abundant DNA modification in plants. It is a reversible modification but has the potential to be stable for a few numbers of generations. Though all cells of an organism share the same DNA sequence, methylation serves a versatile mechanism for regulating variable gene expression (Chen

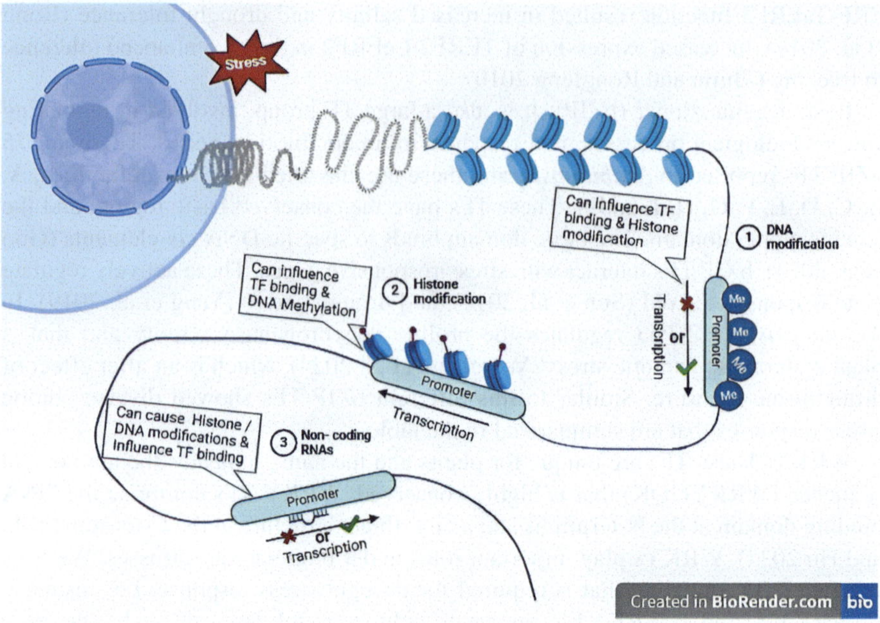

Fig. 3.2 Types of promoter regulation due to various epigenetic regulatory factors

et al. 2010). DNA methylation is commonly linked to gene silencing, X-chromosome inactivation, imprinting, and regulation of transposon activity that serve as important epigenetic markers (Law and Jacobsen 2010; Jones 2012). Expression of the class-C floral homeotic gene *ect-pMADS3* was regulated by RNA-directed DNA methylation (RdDM) at CG sites in *Petunia*, indicating that DNA methylation can have diverse roles in different genomic regions (Shibuya et al. 2009). Methylcytosines in promoter sequences mostly repressed transcription, unlike those in the coding regions. In plants, DNA methylation regulates many tissues-specific promoters. Methylation in the promoter sequence prevents transcription activator binding while enhancing the binding of transcription repressors thereby leading to transcription inactivation or down regulation (Chen et al. 2010; Zhang et al. 2016; Lucibelli et al. 2022). Promoter DNA methylation can also repress transcription indirectly by promoting the repressive H3K9me2 histone methylation, and/or inhibiting histone acetylation, which is a transcription-permissive histone modification (Zhu et al. 2016). Certain promoter methylations attract methylcytosine-binding proteins such as KRYPTONITE (KYP), histone H3 lysine 9 methyltransferases (H3K9), and VARIANT IN METHYLATION 1 (VIM1), which also cause remodeling of the chromatin (Woo et al. 2008). There are examples where methylation sometimes promotes transcription, as seen with the *A. thaliana ROS1* (Lei et al. 2015). Promoter DNA methylation mostly occurs at CG or CG-rich regions, which is dependent on MET 1 and DRM 2 enzymes. DNA methylation of CG sites, maintained by MET 1, likely requires additional RdDM aided by DRM2, at specific loci (Berdasco et al.

2008). The interaction of DNA methyltransferases (DNMTs) with site-specific TFs also regulates gene expression. *RARβ2* gene was silenced due to DNMT3a recruitment to the promoter by the TF PML–RAR (Hervouet et al. 2009; Di croce et al. 2002). In rice, while promoter methylation led to transcription repression, gene-body methylation enhanced it (Li et al. 2012). Furthermore, methylation at the transcription termination regions was found to suppress gene expression more strongly than promoter methylation. DNA methylation is also involved in the silencing of transposable elements (Ramakrishnan et al. 2021). Altered methylation of the transposable element located in the promoter region of *VTE3(1)* regulated its expression (Quadrana et al. 2014).

Abiotic stresses induce DNA methylation, which cause epigenetic variations in plants (Yang et al. 2020). Stress-induced promoter hypermethylation was seen in the progeny of stressed *A. thaliana* plants, and this correlated with a decrease in the corresponding gene expression, indicating that methylation can stress-induce hypermethylation and can repress transcription (Bilichak et al. 2012). Hypomethylation of the promoter DNA of *Arabidopsis FWA* gene that is a suppressor of flowering; is required for the normal *FWA* expression (Kinoshita et al. 2007). In fact, this gene was identified from the late-flowering epigenetic mutant plants. Expression of the COLD RESPONSIVE (COR) pathway genes is activated by the promoter-binding of TFs such as the C REPEAT BINDING FACTOR (CBF), as well as promoter DNA demethylation (Qian et al. 2024). Loci-specific demethylation could be correlated to the accumulation of various abiotic/biotic stress-induced transcripts in tobacco plants (Wada et al. 2003).

Histone modification: These regulate gene expression, thereby influencing growth and development, across eukaryotes. Histone posttranslational modifications (PTMs) are invovled in chromatin organization and gene expression regulation. Chemical groups are added/removed on the histones by histone-modifying enzymes such as histone methyltransferases (HMTs) and histone demethylases. Expression of these enzymes varies throughout different developmental stages, thereby influencing the chromosomal organization (Zhang 2008; Liu et al. 2010). Histone modifications such as the addition/removal of acetyl, methyl, ubiquitin, and phosphate groups occur at specific amino acid residues of the N-terminal histone tails (Shin et al. 2022). Histone H3 methylation at four sites (K4, K9, K27, and K36), and acetylation at four sites (K9, K14, K18, and K23) was identified in *Arabidopsis* (Zhang et al. 2007). Histone acetylation loosens the histones-DNA association, thereby activating transcription. On the other hand, histone methylation can either activate or repress gene expression based on the effector proteins and their complexes that they recruit. For instance, while H3K9 methylation can activate as well as repress gene expression, its dimethylation is repressive in nature, and its acetylation is associated with increased expression (Feng and Jacobsen 2011). H3K9me2 and H3K27me2 dimethylation together with DNA hypermethylation are linked with transcriptional repression at promoters; H3K4me2 dimethylation and H3K9ac acetylation along with DNA hypomethylation are associated with transcription activation (Iizuka and Smith 2003). While high levels of acetylation at H3K9 were associated with actively transcribed genes and rarely with transposons, its methylation was predominantly seen at transposable elements and, to a lesser extent at genes that were poorly transcribed, or genes not associated with transposable elements (Zhou et al. 2010). Also, H3K9ac was more frequent around the transcription and translation start sites, and H3K9me2 was distributed across coding regions (Zhou et al. 2010).

Biotic/abiotic stresses decreased the levels of H3K4me3 and H3K36me3 at certain regulatory loci in tomato (Zhang et al. 2024). Altered stress tolerance was observed due to the loss of function of two histone methyltransferases, SET Domain Group 33 (SDG33) and SDG34 (Bvindi et al. 2022). When the salt stress-responsive AtMYB44 repressors are released, nucleosomes are removed from the promoters of *HAI1*, *PP2C gene like ABI1*, and *ABI2* genes, thereby activating their expression (Nguyen et al. 2019). Under these conditions, there is a significant increase in H3ac and H3K4me3 around the transcription start sites. Activation of the *FLOWERING LOCUS C* (*FLC*) gene is associated with H3K4me3 tri-methylation and H3K9ac acetylation in its promoter, suggesting their combined regulation (Pien et al. 2008; Wang et al. 2006). Drought stress alters histone modification status and nucleosome occupancy in the coding regions as well. H3K4me3 and H3K9ac enrichment in the coding region, and associated gene activation was induced by drought stress in the *Arabidopsis* genes *RD20*, *RAP2.4*, *RD29A*, and *RD29B* (Kim et al. 2008). However, H3K23ac and H3K27ac enrichment was observed in *RD29B*, *RD20,* and *RAP2.4*, but not on *RD29A*. Same study reported that nucleosome occupancy reduced under drought stress in *RD20* and *RAP2.4*, but not in *RD29A* and *RD29B*. Salt stress resulted in enrichment and decrease of the repressive and permissive chromatin marks, H3K9me2 and H3K9ac, respectively, on the promoter and coding regions of the *Arabidopsis ROS1* gene (Bilichak et al. 2012).

Noncoding RNAs: These include all RNAs other than the mRNAs. ASR NcRNAs that interact with promoter elements include the short RNAs (sRNAs) and long RNAs (lncRNAs). While sRNAs comprise of short-interfering RNA (siRNA) and microRNA (miRNA), and are 20–24 nucleotide long, the lncRNAs are about >200 nucleotide long (Ma et al. 2022; Waititu et al. 2020). The miRNAs are generated from precursor RNAs transcribed by *mir* genes or even lncRNAs; siRNAs are either derived from the cleavage for miRNA-induced double stranded RNAs, or are generated due to external sources such as viruses. LncRNAs are transcribed from multiple sources such as promoters and exons of coding genes, intergenic regions, introns, and pseudogenes, some being antisense to coding regions (Domínguez-Rosas et al. 2023; Statello et al. 2021). While the miRNAs are transcribed by RNA polymerase II (Voinnet 2009), and siRNA by an RNA-dependent RNA polymerase, the lncRNAs are transcribed by RNA polymerase II, and II-related IV and V (Ariel et al. 2014; Ariel et al. 2015; Matzke et al. 2009). These ncRNAs influence gene expression via transcriptional and posttranscriptional regulation. Promoter regulation is via the former mechanism, where they cause methylation/demethylation of cytosine in the DNA or histone (Matzke et al. 2009; Waititu et al. 2020). Such methylations though mostly decrease gene expression, occasional increased expression is also seen (Waititu et al. 2020). The methylation/demethylation in plants can be symmetric (at CG and CHG motifs) as well as asymmetric (at CHH motif). While the symmetric changes are maintained by DNA methyltransferases, the asymmetric ones are due to RNA-directed DNA methylation (RdDM) guided by a 24-nucleotide siRNA known as heterochromatic small interfering RNA (hetsiRNA) (Creasey et al. 2014). The transposon promoters are also regulated by the methylation/demethylation via ncRNA (Creasey et al. 2014). The chromatin-level regulation due to ncRNAs is sensitive to biotic/abiotic/developmental factors and promotes phenotypic diversity and physiological plasticity (Matzke et al. 2009).

MiRNAs and TFs have an intricate relation involving transcriptional as well as posttranscriptional regulation (Zhang et al. 2022a, b). While the transcription of coding genes can be regulated by TFs and miRNAs, the expression of genes encoding these two regulatory factors are also under the control of other TFs and miRNAs (Fig. 3.3). This regulation could be transcriptional and/or posttranscriptional (Samad et al. 2017). Many instances of TFs regulating promoters of various abiotic stress-responsive genes are in turn regulated by miRNA-based posttranscriptional regulation of the TFs. For example, in apple (*Malus domestica*), salt-resistance was imparted by the TF MdWRKY100, whose promoter was positively regulated by another TF MdSPL13, which was posttranscriptionally negatively regulated by the miRNA *mdm-MIR156a*; while *MdSPL13* overexpression enhanced salt tolerance, *MIR156a* overexpression weakened it (Ma et al. 2021).

LncRNA bind to multiple factors such as DNA, RNA, or proteins within the nucleus, cytoplasm, and other organelles and thus have broad-ranged functions. Plant lncRNAs are associated with the ASR gene promoters in multiple intriguing ways. Some of these can be transcribed from promoters of other coding genes. For example, the promoter of cold-responsive *Arabbidpsis FLC* gene transcribes the *COLDWRAP* lncRNA, which is required for repressing *FLC* during vernalisation (Kim and Sung 2017). Certain lncRNAs form DNA-RNA loops (R-loops) and change chromosomal confirmation, thereby influencing gene expression. For

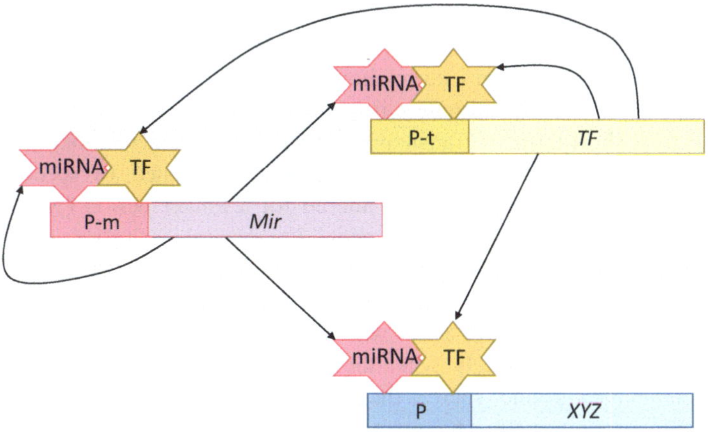

Fig. 3.3 Transcription factors (TFs) and miRNAs regulating promoters of *Mir*, TF-encoding, and other coding genes (denoted as *XYZ*). P-m, P-t, and P indicate promoters of *Mir*, TF-encoding, and other coding genes. The miRNA and TFs shown are representations in general

example, the *Arabidopsis APOLO* (*AUXIN-REGULATED PROMOTERLOOP*) lncRNA forms R-loop with the neighboring promoter of *PINOID (PID)* gene that encodes a root gravitropism regulating kinase (Ariel et al. 2014). The binding of lncRNA to promoters can either enhance or reduce gene expression. For example, salt tolerance was imparted in *Betula platyphylla* by *LncY1* associated with the promoters of two TF-encoding genes, *BpCDF3* and *BpMYB96*, inhibiting and inducing their expressions, respectively (Jia et al. 2023). Certain lncRNAs interact with TFs and facilitate their binding to other promoters. For example, the lncRNA *VAS* (*TaVRN1 alternative splicing*) associated with TaRF2b facilitating its docking, along with TaRF2a, to the *TaVRN1* promoter thereby increasing its expression (Xu et al. 2021). *TaVRN1* is involved in the vernalisation of hexaploid wheat.

With the advent and developments in whole transcriptome sequencing over the years, numerous ASR ncRNAs have been discovered from various plants such as *Arabidopsis*, sugar beet, *B. platyphylla*, rice, maize, *Brachypodium*, wheat, soybean, *Brassica*, and tomato (Jha et al. 2020; Ma et al. 2022; Nejat and Mantri 2017). Many of these can impart abiotic stress resilience to crops by overexpression/silencing (Basso et al. 2019). Nevertheless, the regulatory functions and targets of numerous ncRNAs remain undeciphered.

3.4 Transgenerational Inheritance in Abiotic Stress-Responsive Gene Promoters

Plants exposed to abiotic/biotic stresses have the capability to pass information about this experience to their progeny generations, thereby imparting an improved tolerance to stress. The heritable epigenetic marks include changes in DNA methylation status and chromatin structure (Boyko and Kovalchuk 2011). Important processes such as replication, transcription, repair, and recombination are dependent on the appropriate dynamicity of chromatin modification. The information transmission by epigenetic factors can lead to short-/long-term inheritance. ASR signaling metabolites and TFs that accumulate in plant cells upon an encounter to stresses contribute to a short-term or transient memory. Long term memory most likely involves epigenetic modifications in the chromatin related to specific stress-responsive genes whose expression is altered for a longer period (Nguyen et al. 2022); hence called as "stress memory genes." For example, salt stress induced heritable reduced expression of the *SU(VAR)3–9* homologs, *SUVH5*, *SUVH6*, and *SUVH8*, due to promoter hypermethylation in *Arabidopsis* (Bilichak et al. 2012). Abnormal expression of *FWA* gene was seen in DNA methylation mutants of *Arabidopsis* plants. Interestingly, when the promoter region of *FWA* was demethylated in a mutant background (such as the *ddm1* mutant), this demethylated state was inherited transgenerationally (Kinoshita et al. 2007). Even upon restoration of *DDM1*, this depletion of DNA methylation was passed on for five generations. Inverted repeats homologous to certain promoter regions could reinstate the methylation status in wild-type segregants. The silenced state of FWA was stably heritable even if the inverted repeat cassette was segregated out, thereby underscoring the combined involvement of sRNAs and DNA methylation in sustaining the heritability (Kinoshita et al. 2007).

3.5 Computational Predication of Promoter Regulatory Regions

One of the principal mechanisms for coordinating transcription spatially and temporally is through the recognition of *cis*-regulatory elements. These are short DNA sequence motifs located across promoters, mostly flanking the transcriptional start site (TSS), and few beyond also. They function by recruiting transcription factors that, in turn, assemble the transcriptional initiation machinery. The ability to predict these binding sites de novo has become a cornerstone of functional genomics research. Understanding gene-specific *cis*-regulatory control can elucidate global transcriptional networks and enhance our comprehension of the "blueprint of life" necessary for a properly functioning biological system (Haudry et al. 2013; Wittkopp and Kalay 2012). Various online tools and prediction programs delve into analyzing the co-expressed gene promoters in order to understand prediction strategies for *cis*-regulatory elements, and their implications in transcriptional regulation. *Cis*-element prediction methods are broadly categorized into two, intrinsic strategies and an extrinsic approach involving comparative genomics.

The intrinsic methods include "Word Counting" and "Gibbs Sampling." In the former approach, promoters of coexpressed genes are segmented into "k-mer words" such as six-nucleotide chunks, and the frequency of each k-mer is calculated. These frequencies are compared to a background distribution generated from randomly selected promoter sequences. Statistical tests identify significantly overrepresented k-mers as potential *cis*-elements. Tools like Promomer implement this method and provide deterministic results, ensuring consistent outputs across runs (Austin et al. 2016). In Gibbs sampling, tools like MotifSampler, identifies k-mers probabilistically. Initially, random start positions for k-mers in each promoter sequence are selected, and a frequency matrix for nucleotides at each position is computed. The process iteratively refines k-mer alignments to converge on a consensus sequence. Unlike word counting, this method accounts for wobble nucleotides, introducing variability in predicted *cis*-elements (Bailey et al. 2009). In extrinsic methods such as the Phylogenetic Footprinting, comparative genomics identifies conserved regions in orthologous promoters across species. This requires an optimal evolutionary distance between orthologs to preserve meaningful sequence conservation. For example, 90,000 noncoding *Arabidopsis* sequences were identified to be conserved across Brassicaceae (Haudry et al. 2013).

While certain tools are for functional characterization, certain others are for predicting regulatory regions. The former type includes "ePlant" and "Cistome" that are designed to identify and explore known protein-DNA interactions and transcription factor binding sites. These tools rely on curated databases and computational algorithms to provide insights into regulatory mechanisms (Austin et al. 2016; Waese et al. 2017). "MEME" and "Promomer" are highly effective in predicting *cis*-elements within promoters of coexpressed genes. These tools leverage statistical and probabilistic methods to uncover motifs that regulate gene expression (Bailey et al. 2009). Several computational tools facilitate the identification and analysis of promoter regulatory regions. "Promomer" identifies statistically overrepresented k-mers in coexpressed gene promoters. The tool exhaustively compares k-mer frequencies between coexpressed and randomly sampled promoter clusters, providing robust statistical insights (Austin et al. 2016). "TF2Network" predicts common transcriptional regulators for coexpressed genes using position weight matrices and hypergeometric testing. It incorporates experimentally determined interactions of proteins with other proteins and/or DNA (Kulkarni et al. 2018). "JASPAR" provides position-specific scoring matrices for transcription factors, enabling the mapping of motifs to gene promoters. This tool aids in testing hypotheses about motif enrichment in specific promoter sets (Fornes et al. 2020). "MEME" applies an expectation-maximization algorithm to predict *cis*-regulatory elements. Its versatility supports species-agnostic analyses of promoter sequences (Bailey et al. 2009). MEME is a versatile tool that can be applied to identify *cis*-regulatory elements across various plant species, making it a powerful resource for comparative promoter analysis. "Cistome" analyzes motif overrepresentation, visualizes motif locations within promoters, and generates sequence logos for enriched motifs (Austin et al. 2016). "ePlant" hosts an interaction viewer that maps protein-DNA interactions, facilitating regulatory network exploration (Waese et al. 2017). Integration of computational tools and genomic data has revolutionized our ability to predict and understand *cis*-regulatory elements in promoter regions. By applying these tools to model organisms like *A. thaliana*, researchers can uncover conserved regulatory motifs and their roles in transcriptional networks. Future advancements in sequencing and computational power will likely expand our capacity to decode complex regulatory systems.

3.6 Applications of Decoding Promoter Regions Responsive to Abiotic Stresses

Availability of complete genome/transcriptome sequences from various plants has improved our understanding of various regulatory factors involved in abiotic/biotic stresses. As ASRTFs can regulate multiple genes in a pathway, they make an ideal candidate for imparting abiotic stress resilience in crops. There are many reports in transgenic plants such as *Arabidopsis*, tobacco, rice, potato, wheat, peanut, and others where overexpression of various ASRTFs provide the plants with abiotic stress tolerance, and these are previously tabulated elaborately in other reviews (Lata et al.

2011; Hoang et al. 2017). Many target genes for miRNAs responsive to abiotic stresses including drought, salt, heavy metals, high temperature, and cold have been previously reported (Ma and Hu 2023). While more previous studies are on post-transcriptional regulation due to miRNAs, those on transcriptional regulation are needed (Ma and Hu 2023). Genome editing of such sequences can aid in altering the miRNA-based gene regulation, toward improvising abiotic stress tolerance. Novel synthetic promoters can be developed based on the *cis* regulatory elements desirable for imparting better abiotic stress resilience (Ali and Kim 2019).

3.7 Conclusions

Promoters are one of the niche sites of gene regulation as they harbor a feast of *cis*-regulatory elements. The ASR regulatory elements facilitate gene regulation due to various ASRTFs, ncRNAs, and DNA/histone epigenetic modifications. The promoters of ASRTF-encoding and ncRNA genes are also under the influence of ASRTFs, ncRNAs, and other epigenetic factors such as DNA/histone modifications. Recent developments in genome/transcriptome/epigenome sequencing though has unraveled the functions of many regulatory connections in plants; so far only the tip of the iceberg has become visible. Future deployment of artificial intelligence will aid in unraveling more elements of the orchestrated regulation, and their role under various abiotic stresses. This information has aided in developing strategies to mitigate abiotic stress-induced damage by deploying synthetic promoters, overexpression strategies and genome editing. Considering the on-going drastic global climate change, deeper research in this area is a requisite to enhance climate resilience in crop plants.

References

Abe H, Urao T, Ito T et al (2003) *Arabidopsis* AtMYC2 (bHLH) and AtMYB2 (MYB) function as transcriptional activators in abscisic acid signaling. Plant Cell 15:63–78. https://doi.org/10.1105/tpc.006130

Ali S, Kim WC (2019) A fruitful decade using synthetic promoters in the improvement of transgenic plants. Front Plant Sci 10:1433. https://doi.org/10.3389/fpls.2019.01433

Ariel F, Jegu T, Latrasse D et al (2014) Noncoding transcription by alternative RNA polymerases dynamically regulates an auxin-driven chromatin loop. Mol Cell 55:383–396. https://doi.org/10.1016/j.molcel.2014.06.011

Ariel F, Romero-Barrios N, Jegu T et al (2015) Battles and hijacks: noncoding transcription in plants. Trends Plant Sci 20:362–371. https://doi.org/10.1016/j.tplants.2015.03.003

Austin RS, Shu H, Waese J et al (2016) New BAR tools for mining expression data exploring *cis*-elements in *Arabidopsis thaliana*. Plant J 88:490–504. https://doi.org/10.1111/tpj.13261

Bailey TL, Boden M, Buske FA et al (2009) MEME SUITE: tools for motif discovery and searching. Nucleic Acids Res 37:202–208. https://doi.org/10.1093/nar/gkp335

Basso MF, Ferreira PCG, Kobayashi AK et al (2019) MicroRNAs and new biotechnological tools for its modulation and improving stress tolerance in plants. Plant Biotechnol J 17:1482–1500. https://doi.org/10.1111/pbi.13116

Berdasco M, Alcazar R, Garcia-Ortiz MV et al (2008) Promoter DNA hypermethylation and gene repression in undifferentiated *Arabidopsis* cells. PLoS One 3:3306. 1010.1371/journal.pone.0003306

Bilichak A, Ilnystkyy Y, Hollunder J et al (2012) The progeny of *Arabidopsis thaliana* plants exposed to salt exhibit changes in DNA methylation, histone modifications and gene expression. PLoS One 7:30515. https://doi.org/10.1371/journal.pone.0030515

Bolt S, Zuther E, Zintl S et al (2017) *ERF105* is a transcription factor gene of *Arabidopsis thaliana* required for freezing tolerance and cold acclimation. Plant Cell Environ 40:108–120. https://doi.org/10.1111/pce.12838

Boyko A, Kovalchuk I (2011) Genome instability and epigenetic modification-heritable responses to environmental stress. Curr Opin Plant Biol 14:260–266. https://doi.org/10.1016/j.pbi.2011.03.003

Bvindi C, Lee S, Tang L et al (2022) Improved pathogen and stress tolerance in tomato mutants of SET domain histone 3 lysine methyltransferases. New Phytol 235:1957–1976. https://doi.org/10.1111/nph.18277

Castilhos G, Lazzarotto F, Spagnolo-Fonini L et al (2014) Possible roles of basic helix-loop-helix transcription factors in adaptation to drought. Plant Sci 223:1–7. https://doi.org/10.1016/j.plantsci.2014.02.010

Cedar H, Bergman Y (2009) Linking DNA methylation and histone modification: patterns and paradigms. Nat Rev Genet 10:295–304. https://doi.org/10.1038/nrg2540

Chen M, Lv S, Meng Y (2010) Epigenetic performers in plants. Develop Growth Differ 52:555–566. https://doi.org/10.1111/j.1440-169X.2010.01192.x

Creasey KM, Zhai J, Borges F et al (2014) miRNAs trigger widespread epigenetically activated siRNAs from transposons in *Arabidopsis*. Nature 508:411–415. https://doi.org/10.1038/nature13069

Debbarma J, Sarki YN, Saikia B, Boruah HPD, Singha DL, Chikkaputtaiah C (2019) Ethylene response factor (ERF) family proteins in abiotic stresses and CRISPR-Cas9 genome editing of ERFs for multiple abiotic stress tolerance in crop plants: a review. Mol Biotechnol 61(2):153–172. https://doi.org/10.1007/s12033-018-0144-x

Domínguez-Rosas E, Hernández-Oñate MÁ, Fernandez-Valverde S-L et al (2023) Plant long non-coding RNAs: identification and analysis to unveil their physiological functions. Front Plant Sci 14:1275399 https://doi.org/10.3389/fpls.2023.1275399

Di Croce L, Raker VA, Corsaro M et al (2002) Methyltransferase recruitment and DNA hypermethylation of target promoters by an oncogenic transcription factor. Science 295:1079–1082. https://doi.org/10.1126/science.1065173

Fan XQ, Guo Q, Xu P et al (2015) Transcriptome-wide identification of salt-responsive members of the *WRKY* gene family in *Gossypium aridum*. PLoS One 10:126–148. https://doi.org/10.1371/journal.pone.0126148

Feil R, Fraga M (2012) Epigenetics and the environment: emerging patterns and implications. Nat Rev Genet 13:97–109. https://doi.org/10.1038/nrg3142

Feng S, Jacobsen SE (2011) Epigenetic modifications in plants: an evolutionary perspective. Curr Opin Plant Biol 14:179–186. https://doi.org/10.1016/j.pbi.2010.12.002

Feng K, Hou XL, Xing GM et al (2020) Advances in AP2/ERF super-family transcription factors in plant. Crit Rev Biotechnol 40:750–776. https://doi.org/10.1080/07388551.2020.1768509

Fornes O, Castro-Mondragon JA, Khan A et al (2020) JASPAR 2020: update of the open-access database of transcription factor binding profiles. Nucleic Acids Res 48:87–92. https://doi.org/10.1093/nar/gkz1001

Guo Z, Dzinyela R, Yang L, Hwarari D (2024) bZIP transcription factors: structure, modification, abiotic stress responses and application in plant improvement. Plants (Basel) 13:20–58. https://doi.org/10.3390/plants13152058

Haudry A, Platts AE, Vello E et al (2013) An atlas of over 90,000 conserved noncoding sequences provides insight into crucifer regulatory regions. Nat Genet 45:891–898. https://doi.org/10.1038/ng.2684

He XJ, Chen T, Zhu JK (2011) Regulation and function of DNA methylation in plants and animals. Cell Res 21:442–465. https://doi.org/10.1038/cr.2011.23

Hervouet E, Vallette FM, Cartron PF (2009) Dnmt3/transcription factor interactions as crucial players in targeted DNA methylation. Epigenetics 4:1–13. https://doi.org/10.4161/epi.4.7.9883

Hoang XLT, Nhi DNH, Thu NBA et al (2017) Transcription factors and their roles in signal transduction in plants under abiotic stresses. Curr Genomics 18:483–497. https://doi.org/10.2174/1389202918666170227150057

Holoch D, Moazed D (2015) RNA-mediated epigenetic regulation of gene expression. Nat Rev Genet 16:71–84. https://doi.org/10.1038/nrg3863

Huang XB, Cao LW, Fan JB et al (2022) CdWRKY2-mediated sucrose biosynthesis and CBF-signalling pathways coordinately contribute to cold tolerance in bermudagrass. Plant Biotechnol J 20:660–675. https://doi.org/10.1111/pbi.13745

Iizuka M, Smith MM (2003) Functional consequences of histone modifications. Curr Opin Genet Dev 13:154–160. https://doi.org/10.1016/s0959-437x(03)00020-0

Issa JP (2003) Age-related epigenetic changes and the immune system. Clin Immunol 109:103–108. https://doi.org/10.1016/s1521-6616(03)00203-1

Jha UC, Nayyar H, Jha R et al (2020) Long non-coding RNAs: emerging players regulating plant abiotic stress response and adaptation. BMC Plant Biol 20:466. https://doi.org/10.1186/s12870-020-02595-x

Jia Y, Zhao H, Niu Y et al (2023) Identification of birch lncRNAs and mRNAs responding to salt stress and characterization of functions of lncRNA. Horti Res 10:277. https://doi.org/10.1093/hr/uhac277

Jisha V, Dampanaboina L, Vadassery J et al (2015) Overexpression of an AP2/ERF type transcription factor OsEREBP1 confers biotic and abiotic stress tolerance in rice. PLoS One 10:127–831. https://doi.org/10.1371/journal.pone.0127831

Jones PA (2012) Functions of DNA methylation: islands, start sites, gene bodies and beyond. Nat Rev Genet 13:484–492. https://doi.org/10.1038/nrg3230

Kim DH, Sung S (2017) Vernalization-triggered intragenic chromatin loop formation by long non-coding RNAs. Dev Cell 40:302–312. https://doi.org/10.1016/j.devcel.2016.12.021

Kim J-M, To TK, Ishida J et al (2008) Alterations of lysine modifications on the histone H3 N-tail under drought stress conditions in *Arabidopsis thaliana*. Plant Cell Physiol 49:1580–1588. https://doi.org/10.1093/pcp/pcn133

Kinoshita Y, Saze H, Kinoshita T et al (2007) Control of FWA gene silencing in *Arabidopsis thaliana* by SINE-related direct repeats. Plant J 49:38–45. https://doi.org/10.1111/j.1365-313X.2006.02936.x

Kulkarni SR, Vaneechoutte D, Van de Velde J et al (2018) TF2Network: predicting transcription factor regulators and gene regulatory networks in Arabidopsis using publicly available binding site information. Nucleic Acids Res 46: e31. https://doi.org/10.1093/nar/gkx1279

Lata C, Yadav A, Pras M (2011) Role of Plant Transcription Factors in Abiotic Stress Tolerance. In; Shanker A (ed) Abiotic Stress Response in Plants - Physiological, Biochemical and Genetic Perspectives. InTech. London. https://doi.org/10.5772/23172

Law JA, Jacobsen SE (2010) Establishing, maintaining and modifying DNA methylation patterns in plants and animals. Nat Rev Genet 11:204–220. https://doi.org/10.1038/nrg2719

Lei M, Zhang H, Julian R et al (2015) Regulatory link between DNA methylation and active demethylation in *Arabidopsis*. Proc Natl Acad Sci USA 112:3553–3557. https://doi.org/10.1073/pnas.1502279112

Li X, Zhu J, Hu F et al (2012) Single-base resolution maps of cultivated and wild rice methylomes and regulatory roles of DNA methylation in plant gene expression. BMC Genomics 13:300. https://doi.org/10.1186/1471-2164-13-300

Licausi F, Ohme-Takagi M, Perata P (2013) APETALA 2/ethylene responsive factor (AP 2/ERF) transcription factors: mediators of stress responses and developmental programs. New Phytol 199:639–649. https://doi.org/10.1111/nph.12291

Lim C, Kang K, Shim Y et al (2022) Inactivating transcription factor OsWRKY5 enhances drought tolerance through abscisic acid signaling pathways. Plant Physiol 188:1900–1916. https://doi.org/10.1093/plphys/kiab492

Liu C, Lu F, Cui X et al (2010) Histone methylation in higher plants. Annu Rev Plant Biol 61:395–420. https://doi.org/10.1146/annurev.arplant.043008.091939

Liu WW, Tai HH, Li SS et al (2014) bHLH122 is important for drought and osmotic stress resistance in *Arabidopsis* and in the repression of ABA catabolism. New Phytol 201:1192–1204. https://doi.org/10.1111/nph.12607

Liu F, Xi M, Liu T et al (2024) The central role of transcription factors in bridging biotic and abiotic stress responses for plants' resilience. New Crops 1:100005. https://doi.org/10.1016/j.ncrops.2023.11.003

Lucibelli F, Valoroso MC, Aceto S (2022) Plant DNA methylation: an epigenetic mark in development, environmental interactions, and evolution. Int J Mol Sci 23:8299. https://doi.org/10.3390/ijms23158299

Ma Z, Hu L (2023) MicroRNA: a dynamic player from signalling to abiotic tolerance in plants. Int J Mol Sci 24:11364. https://10.3390/ijms241411364

Ma Z, Hu L (2024) WRKY transcription factor responses and tolerance to abiotic stresses in plants. Int J Mol Sci 25:6845. https://doi.org/10.3390/ijms25136845

Ma Y, Xue H, Zhang F et al (2021) The *miR156/SPL* module regulates apple salt stress tolerance by activating *MdWRKY100* expression. Plant Biotechnol J 19:311–323. https://doi.org/10.1111/pbi.13464

Ma X, Zhao F, Zhou B (2022) The characters of non-coding RNAs and their biological roles in plant development and abiotic stress response. Int J Mol Sci 23:4124. https://doi.org/10.3390/ijms23084124

Mao H, Yu L, Han R et al (2016) ZmNAC55, a maize stress-responsive NAC transcription factor, confers drought resistance in transgenic *Arabidopsis*. Plant Physiol Biochem 105:55–66. https://doi.org/10.1016/j.plaphy.2016.04.018

Marques DN, Dos Reis SP, De Souza CR (2017) Plant NAC transcription factors responsive to abiotic stresses. Plant Gene 11:170–179. https://doi.org/10.1016/j.plgene.2017.06.003

Martin C, Ellis N, Rook F (2010) Do transcription factors play special roles in adaptive variation? Plant Physiol 154:506–511. https://doi.org/10.1104/pp.110.161331

Matzke M, Kanno T, Daxinger L et al (2009) RNA-mediated chromatin-based silencing in plants. Curr Opin Cell Biol 21:367–376. https://doi.org/10.1016/j.ceb.2009.01.025

Mizoi J, Shinozaki K, Yamaguchi-Shinozaki K (2012) AP2/ERF family transcription factors in plant abiotic stress responses. Biochim Biophys Acta 1819:86–96. https://doi.org/10.1016/j.bbagrm

Nejat N, Mantri N (2017) Emerging roles of long non-coding RNAs in plant response to biotic and abiotic stresses. Crit Rev Biotechnol 38:93–105. https://doi.org/10.1080/07388551.2017.1312270

Nguyen NH, Jung C, Cheong JJ (2019) Chromatin remodeling for the transcription of *type 2C protein phosphatase* genes in response to salt stress. Plant Physiol Biochem 141:325–331. https://doi.org/10.1016/j.plaphy.2019.06.012

Nguyen NH, Vu NT, Cheong JJ (2022) Transcriptional stress memory and transgenerational inheritance of drought tolerance in plants. Int J Mol Sci 23:12918. https://doi.org/10.3390/ijms232112918

Oh SJ, Kim YS, Kwon CW et al (2009) Overexpression of the transcription factor AP37 in rice improves grain yield under drought conditions. Plant Physiol 150:1368–1379. https://doi.org/10.1104/pp.109.137554

Pien S, Fleury D, Mylne JS et al (2008) *ARABIDOPSIS TRITHORAX1* dynamically regulates *FLOWERING LOCUS C* activation *via* histone 3 lysine 4 trimethylation. Plant Cell 20:580–588. https://doi.org/10.1105/tpc.108.058172

Paz-Ares J, Ghosal D, Wienand U et al (1987) The regulatory c1 locus of Zea mays encodes a protein with homology to myb proto-oncogene products and with structural similarities to transcriptional activators. EMBO J 6: 3553-8. https://doi.org/10.1002/j.1460-2075.1987.tb02684

Pillitteri LJ, Torii KU (2007) Breaking the silence: three bHLH proteins direct cell-fate decisions during stomatal development. BioEssays 29:861–870. https://doi.org/10.1002/bies.20625

Qian Y, Zhang T, Yu Y et al (2021) Regulatory mechanisms of bHLH transcription factors in plant adaptive responses to various abiotic stresses. Front Plant Sci 12:677611. https://doi.org/10.3389/fpls.2021.677611

Qian Z, He L, Li F (2024) Understanding cold stress response mechanisms in plants: an overview. Front Plant Sci 15:1443317. https://doi.org/10.3389/fpls.2024.1443317

Quadrana L, Almeida J, Asís R et al (2014) Natural occurring epialleles determine vitamin E accumulation in tomato fruits. Nat Commun 5:4027. https://doi.org/10.1038/ncomms5027

Ramakrishnan M, Satish L, Kalendar R et al (2021) The dynamism of transposon methylation for plant development and stress adaptation. Int J Mol Sci 22:11387. https://doi.org/10.3390/ijms222111387

Rong W, Qi L, Wang A, Ye X, Du L, Liang H, Xin Z, Zhang Z (2014) The ERF transcription factor TaERF3 promotes tolerance to salt and drought stresses in wheat. Plant Biotechnol J 12(4):468–479. https://doi.org/10.1111/pbi.12153

Samad AFA, Sajad M, Nazaruddin N et al (2017) MicroRNA and transcription factor: key players in plant regulatory network. Front Plant Sci 8:565. https://doi.org/10.3389/fpls.2017.00565

Scarpeci TE, Frea VS, Zanor MI et al (2017) Overexpression of *AtERF019* delays plant growth and senescence, and improves drought tolerance in *Arabidopsis*. J Exp Bot 68:673–685. https://doi.org/10.1093/jxb/erw429

Shao H, Wang H, Tang X (2015) NAC transcription factors in plant multiple abiotic stress responses: progress and prospects. Front Plant Sci 6:902. https://doi.org/10.3389/fpls.2015.00902

Shahnejat-Bushehri S, Mueller-Roeber B, Balazadeh S (2012) Arabidopsis NAC transcription factor JUNGBRUNNEN1 affects thermomemory-associated genes and enhances heat stress tolerance in primed and unprimed conditions. Plant Signal Behav 7: 1518–1521. https://doi.org/10.4161/psb.22092

Shibuya K, Fukushima S, Takatsuji H (2009) RNA-directed DNA methylation induces transcriptional activation in plants. Proc Natl Acad Sci USA 106:1660–1665. https://doi.org/10.1073/pnas.0809294106

Shin H, Choi WL, Lim JY et al (2022) Epigenome editing: targeted manipulation of epigenetic modifications in plants. Genes Genomics 44:307–315. https://doi.org/10.1007/s13258-021-01199-5

Statello L, Guo CJ, Chen LL et al (2021) Gene regulation by long non-coding RNAs and its biological functions. Nat Rev Mol Cell Biol 22: 96-118. https://doi.org/10.1038/s41580-020-00315-9

Sun X-L, Li Y, Cai H et al (2011) *Arabidopsis* bZIP1 transcription factor binding to ABRE *cis*-element regulates abscisic acid signal transduction. Acta Agrono Sinica 37:612–619. https://doi.org/10.1016/s1875-2780(11)60016-3

Tran LSP, Nakashima K, Sakuma Y et al (2004) Isolation and functionalanalysis of *Arabidopsis* stress-inducible NAC transcription factors that bind to a drought-responsive *cis*-element in the early responsive to dehydration stress1 promoter. Plant Cell 16:2481–2498. https://doi.org/10.1105/tpc.104.022699

Tran LSP, Nishiyama R, Yamaguchi-Shinozaki K et al (2010) Potential utilization of NAC transcription factors to enhance abiotic stress tolerance in plants by biotechnological approach. GM Crops 1:32–39. https://doi.org/10.4161/gmcr.1.1.10569

Veerabagu M, Kirchler T, Elgass K et al (2024) The interaction of the *Arabidopsis* response regulator ARR18 with bZIP63 mediates the regulation of *PROLINE DEHYDROGENASE* expression. Mol Plant 7:1560–1577. https://doi.org/10.1093/mp/ssu074

Verma D, Jalmi SK, Bhagat PK et al (2020) A bHLH transcription factor, MYC2, imparts salt intolerance by regulating proline biosynthesis in *Arabidopsis*. FEBS J 287:2560–2576. https://doi.org/10.1111/febs.15157

Voinnet O (2009) Origin, biogenesis, and activity of plant microRNAs. Cell 136:669–687. https://doi.org/10.1016/j.cell.2009.01.046
Wada Y, Ohya H, Yamaguchi Y et al (2003) Preferential *de novo* methylation of cytosine residues in non-CpG sequences by a domains rearranged DNA methyltransferase from tobacco plants. J Biol Chem 278:42386–42393. https://doi.org/10.1074/jbc.M303892200
Waese J, Fan J, Pasha A et al (2017) ePlant: visualizing and exploring multiple levels of data for hypothesis generation in plant biology. Plant Cell 29:1806–1821. https://doi.org/10.1105/tpc.17.00073
Waititu JK, Zhang C, Liu J, Wang H (2020) Plant non-coding RNAs: origin, biogenesis, mode of action and their roles in abiotic stress. Int J Mol Sci 21:8401. https://doi.org/10.3390/ijms21218401
Wang J, Tian L, Lee HS, Chen ZJ (2006) Nonadditive regulation of *FRI* and *FLC* loci mediates flowering-time variation in *Arabidopsis* allopolyploids. Genetics 173:965–974. https://doi.org/10.1534/genetics.106.056580
Wang NN, Xu SW, Sun YL et al (2019) The cotton WRKY transcription factor (GhWRKY33) reduces transgenic *Arabidopsis* resistance to drought stress. Sci Rep 9:724. https://doi.org/10.1038/s41598-018-37035-2
Wang JY, Wang LJ, Yan Y et al (2021) *GhWRKY21* regulates ABA-mediated drought tolerance by fine-tuning the expression of *GhHAB* in cotton. Plant Cell Rep 40:2135–2150. https://doi.org/10.1007/s00299-020-02590-4
Wittkopp P, Kalay G (2012) *Cis*-regulatory elements: molecular mechanisms and evolutionary processes underlying divergence. Nat Rev Genet 13:59–69. https://doi.org/10.1038/nrg3095
Woo HR, Dittmer TA, Richards EJ (2008) Three SRA-domain methylcytosine-binding proteins cooperate to maintain global CpG methylation and epigenetic silencing in *Arabidopsis*. PLoS Genet 4:100–156. https://doi.org/10.1371/journal.pgen.1000156
Wu M, Liu HL, Han GM et al (2017) A moso bamboo *WRKY* gene *PeWRKY83* confers salinity tolerance in transgenic *Arabidopsis* plants. Sci Rep 7:11721. https://doi.org/10.1038/s41598-017-10795-z
Xu S, Dong Q, Deng M et al (2021) The vernalization-induced long non-coding RNA VAS functions with the transcription factor TaRF2b to promote *TaVRN1* expression for flowering in hexaploid wheat. Mol Plant 14:1525–1538. https://doi.org/10.1016/j.molp.2021.05.026
Yang S, Xu K, Chen S et al (2019) A stress-responsive bZIP transcription factor *OsbZIP62* improves drought and oxidative tolerance in rice. BMC Plant Biol 19:260. https://doi.org/10.1186/s12870-019-1872-1
Yang X, Sanchez R, Kundariya H et al (2020) Segregation of an *MSH1* RNAi transgene produces heritable non-genetic memory in association with methylome reprogramming. Nat Commun 11:2214. https://doi.org/10.1038/s41467-020-16036-8
Yoon Y, Seo DH, Shin H et al (2020) The role of stress-responsive transcription factors in modulating abiotic stress tolerance in plants. Agronomy 10:788. https://doi.org/10.3390/agronomy10060788
Zhang X (2008) The epigenetic landscape of plants. Science 320:489–492. https://doi.org/10.1126/science.1153996
Zhang K, Sridhar VV, Zhu J et al (2007) Distinctive core histone post-translational modification patterns in *Arabidopsis thaliana*. PLoS One 2:1210. https://doi.org/10.1371/journal.pone.0001210
Zhang H, Huang L, Dai Y, Liu S, Hong Y, Tian L, Huang L, Cao Z, Li D, Song F (2015) *Arabidopsis* AtERF15 positively regulates immunity against *pseudomonas syringae* pv. Tomato DC3000 and *Botrytis cinerea*. Front Plant Sci 6:686. https://doi.org/10.3389/fpls.2015.00686
Zhang Y, Yu HJ, Yang XY et al (2016) CsWRKY46, a WRKY transcription factor from cucumber, confers cold resistance in transgenic-plant by regulating a set of cold-stress responsive genes in an ABA-dependent manner. Plant Physiol Biochem 108:478–487. https://doi.org/10.1016/j.plaphy.2016.08.013

Zhang H, Zhang Q, Zhai H et al (2017) Transcript profile analysis reveals important roles of jasmonic acid signalling pathway in the response of sweet potato to salt stress. Sci Rep 7:1–12. https://doi.org/10.1038/srep40819

Zhang J, Huang D, Zhao X et al (2022a) Drought-responsive WRKY transcription factor genes *IgWRKY50* and *IgWRKY32* from *Iris germanica* enhance drought resistance in transgenic *Arabidopsis*. Front Plant Sci 13:983600. https://doi.org/10.3389/fpls.2022.983600

Zhang F, Yang J, Zhang N et al (2022b) Roles of microRNAs in abiotic stress response and characteristics regulation of plant. Front Plant Sci 13:919243. https://doi.org/10.3389/fpls.2022.919243

Zhang P, He Y, Huang S (2024) Unlocking epigenetic breeding potential in tomato and potato. aBIOTECH 5:507–518. https://doi.org/10.1007/s42994-024-00184-2

Zhao M, Morohashi K, Hatlestad G et al (2008) The TTG1-bHLH-MYB complex controls trichome cell fate and patterning through direct targeting of regulatory loci. Development 135:1991–1999. https://doi.org/10.1242/dev.016873

Zhijin Z, Rongfeng H (2010) Enhanced tolerance to freezing in tobacco and tomato overexpressing transcription factor TERF2/LeERF2 is modulated by ethylene biosynthesis. Plant Mol Biol 73:241–249. https://doi.org/10.1007/s11103-010-9609-4

Zhou J, Wang X, He K et al (2010) Genome-wide profiling of histone H3 lysine 9 acetylation and dimethylation in *Arabidopsis* reveals correlation between multiple histone marks and gene expression. Plant Mol Biol 72:585–595. https://doi.org/10.1007/s11103-009-9594-7

Zhu H, Wang GH, Qian J (2016) Transcription factors as readers and effectors of DNA methylation. Nat Rev Genet 17:551–565. https://doi.org/10.1038/nrg.2016.83

Chapter 4
Long Non-coding RNAs (lncRNAs) in Abiotic Stress Adaptation

Prabhatam Mishra, Priyamvada Mishra, Sanchita Gupta, Mehar H. Asif, and Sribash Roy

Abstract Plants being sessile, need to develop strategies to respond to variable environmental conditions. They have evolved various complex mechanisms to identify and activate signaling pathways at various stages of their development. Recent studies suggest epigenetic changes including DNA methylation, miRNA, and lncRNA play important roles in plant adaptation in varying environmental conditions. . The long non-coding RNAs are more than in 200 nt in length and have provided a new insight into plant adaption. lncRNAs regulate the plants' responses at epigenomic and chromatin reorganization levels, transcriptional and post-transcriptional levels. Various lncRNA are functionally characterized having roles in plant growth and developments like regulation of flowering, development of lateral roots, and different abiotic stresses. Recent advances in next-generation sequencing (NGS) techniques have enabled the plant researchers to discover various lncRNAs and their roles in plant biotic and abiotic stresses. lncRNAs are known to be involved in various abiotic stresses, like excess heat, drought, cold, salinity, and CO_2 stress. We have briefly described the role of lncRNAs in plant abiotic stress responses and their future aspects in agriculture.

Keywords Epigenomics · Long non-coding RNA · Abiotic stress · Stress · Posttranscriptional regulation

P. Mishra · P. Mishra · S. Gupta · M. H. Asif
CSIR-National Botanical Research Institute, Lucknow, UP, India

S. Roy (✉)
Department of Plant Sciences, School of Life Sciences, University of Hyderabad, Hyderabad, Telangana, India
e-mail: sribashroy@uohyd.ac.in

S. Tiwari et al. (eds.), *Decoding Non-Coding RNA*, Concepts and Strategies in Plant Sciences, https://doi.org/10.1007/978-981-96-9859-2_4

4.1 Introduction

About 90% of the plant genome is transcribed into RNA, out of which only 1–2% is translated into proteins (mRNA). Ribosomal RNA constitutes about 80–90% of total RNA. Remaining RNAs are generally non-coding RNAs. The amount of non-coding RNAs (ncRNA) varies depending on specific type but lower than ribosomal and messenger RNA. These RNAs do not code for any functional proteins, and known as transcriptional noise or junk RNAs (Ford Doolittle 2013; Ariel et al. 2015). The non-coding RNAs those are more than 200 nucleotides in length are called long non-coding RNAs (lncRNAs). These lncRNAs predominantly show regulatory control over gene expression by modulating various mechanisms at epigenetic, transcriptional, posttranscriptional (Dykes and Emanueli 2017) translational, and post-translational levels (Bond et al. 2011). They may play roles in various essential biological processes, including chromatin remodeling, transcriptional activation (Heo and Sung 2011), transcriptional interference, RNA processing, and mRNA translation. Long non-coding RNAs (lncRNAs) play significant roles in plant growth and development in responses to biotic as well as abiotic stress in plants. The long non-coding RNAs (lncRNAs) are characterized by a poly A tail and transcribed by RNA polymerase II (Pol II). Besides PolII, other plant specific RNA polymerases like RNA polymerase IV and V are also known to have roles in lncRNA production (Chen et al. 2018a, b; Chekanova 2015). On the basis of their interactions with protein coding genes, the lncRNAs may be divided into the long intergenic lncRNA (lincRNAs; derived from region in between the two coding genes in sense orientation), intronic lncRNA (produced from the intronic region of the gene in any orientation with respect to coding genes), NATs lncRNA (originate in the antisense orientation with respect to the sense strand of coding genes (Liu et al. 2012). LncRNA can serve as scaffolds for enzymes that are involved in histone modification, DNA methylation, and chromatin remodeling resulting in up- or down-regulation of gene expression. They may affect mRNA decay, work as sponges for miRNA binding sites, produce siRNA and ultimately regulate gene expression at posttranscriptional levels.

4.2 Mechanism of Action of lncRNAs

Long non-coding RNAs (lncRNAs) perform diverse and complex biological functions such as interactions with micro and macro molecules to regulate the functions of gene on the protein modification levels and chromatin remodeling (Ma et al. 2013). Transcription, secondary structure formation, and splicing pattern of lncRNA are regulated by RNA modification, histone modification, and DNA methylation. The known mechanisms of actions have already been reviewed extensively (Sanchita et al. 2020). The lncRNAs are classified into four types based on their regulatory mechanisms:

1. **Signaling lncRNAs:** These lncRNA acts as signals or signaling molecules bind with the transcription factors to regulate the functions of protein coding genes (Swiezewski et al. 2009).
2. **Decoylnc-RNAs:** These lncRNAs act as decoys to indirectly regulate the expression of protein coding genes. These lncRNAs are recruited as RNA binding proteins during transcription of protein coding gene-like transcription factors (Franco-Zorrilla et al. 2007).
3. **Guide lncRNAs:** These lncRNAs act as guides of RNA binding proteins (ribonucleoproteins) to regulate the gene expression by chromatin modifying enzymes. These type of lncRNAs act in *cis* and *trans* modes to regulate the gene expression (Wood et al. 2006; Swiezewski et al. 2009; Heo and Sung 2011).
4. **Scaffold lncRNA:** These types of lncRNA act as scaffolds and comprise of binding domains for the multiple ribonucleoproteins to form ribonucleoprotein complexes and hence regulation the gene expression (Swiezewski et al. 2009).

4.3 Biosynthesis of lncRNA in Plants

The nuclear RNA polymerase enzyme in plants synthesizes lncRNAs that undergo similar posttranscriptional changes. It has been discovered that all five of the RNA polymerases found in plants—Pol I, II, III, and the two Pol II-related, plant-specific RNA polymerases, Pol IV and Pol V—transcribe different lncRNA products (Wierzbicki et al. 2021; Bohmdorfer et al. 2016). These products are involved in RNA-directed DNA methylation and the control of transposable elements in plants. To guarantee proper structure, location, and function, lncRNAs go through RNA capping, splicing, polyadenylation, and nuclear export after transcription (Marquardt et al. 2014). Thus, like protein-coding genes, lncRNAs are made up of exons, introns, and 5′ and 3′ untranslated regions (UTRs). Nonetheless, the quantity and effectiveness of polyadenylation in lncRNAs are generally lower compared to mRNAs (Cao et al. 2018). Nevertheless, lncRNAs can produce several splice variants from a single gene and exhibit a notable level of alternative splicing (AS). The alternative splicing of lncRNAs has been investigated and studied through ultra-deep RNA-seq analysis in response to cold treatment, revealing a dynamic expression of lncRNAs (Calixto et al. 2019). Despite lncRNAs having a smaller number of exons (Fesenko et al. 2021) and being expressed at reduced levels compared to mRNAs (Song et al. 2021), studies have shown that they are also expressed in specific patterns related to different tissues and cell types in various plant species, including rice (Zhang et al. 2014a, b) and grapevine (Bhatia et al. 2019). In grapevine, various lncRNAs that are differentially expressed in leaves, inflorescences, and berries were identified, showing regulation of lncRNAs that is specific to spatial, temporal, and developmental stages (Bhatia et al. 2019). The spatial-temporal expression of lncRNAs was also identified in maize, with over 90% being specifically expressed in particular tissues or at specific developmental stages (Kim et al. 2017). The expression profiling of cell-specific lncRNAs in the *Arabidopsis* root revealed that merely 25% of intergenic lncRNAs (lincRNAs) were present in over half of the cell types (Li et al. 2016). The localization of lncRNAs within cells may offer significant insights into their functions. The positioning of lncRNAs within cells may give significant insights into their roles. The majority of lncRNAs were discovered to be mainly situated in the nucleus, which relates to their roles in chromatin organization and the regulation of gene transcription (Carlevaro-Fita and Johnson 2019; Bridges et al. 2021). RNA fluorescence in situ hybridization (FISH) and lncRNA promoter-driven reporter methods can be used to identify the subcellular localization of newly discovered lncRNAs. So far, specific lncRNAs have also been identified in the cytoplasm of *Drosophila* and mammals like humans and mice (Carlevaro-Fita and Johnson 2019, Bridges et al. 2021). Nonetheless, research providing strong evidence on plants remains scarce. The distribution of lncRNAs among the genomes of various plant species revealed varied patterns. An uneven distribution of lncRNAs across chromosomes was observed in grapevine (Bhatia et al. 2019); conversely, they are uniformly distributed in the maize genome, exhibiting a marginally reduced density on chromosome 1 (Kim et al. 2017). Depending on their orientation (sense or antisense) and their locations concerning other genes (intergenic, intronic, and coding regions), the lncRNAs can be categorized as (1) lincRNAs (located in intergenic regions), (2) intronic ncRNAs (incRNAs), (3) natural antisense transcripts (NATs, present in many protein-coding genes), and (4) sense lncRNAs (Fig. 4.1). LncRNAs may also be categorized based on their biochemical pathways, sequence or structural conservation, roles, and genomic positions (St et al. 2015). The biosynthesis and mechanism of action of lncRNAs is shown in Fig. 4.1.

4.4 Functional Roles of lncRNAs

4.4.1 Abiotic Stresses

Being sessile, plants are very sensitive and need to adapt under abiotic stresses impeding plant growth and development. Evolutionary processes have equipped plants with mechanisms to withstand various natural stresses like drought, light, heat, temperature, cold, salinity, heavy metals, nutrient deficiency, hypoxia, etc. (Fig. 4.2). These stressors significantly compromise the fundamental physiological functions of crop plants, resulting in diminished productivity and, ultimately, plant mortality. In response, plants have developed intricate mechanisms, including stress perception, epigenetic modifications, and the regulation of transcription and translation, to cope with these challenges. The extensive literature has elucidated the diverse regulatory functions of long non-coding RNAs (lncRNAs) in plants. Recent progress in high-throughput transcriptome sequencing and expression profiling using next-generation sequencing (NGS) techniques has given rise to the discovery of various non-coding RNAs, particularly long non-coding RNAs (lncRNAs). They are believed to perform pivotal functions in plant immunity and in responding to environmental changes.

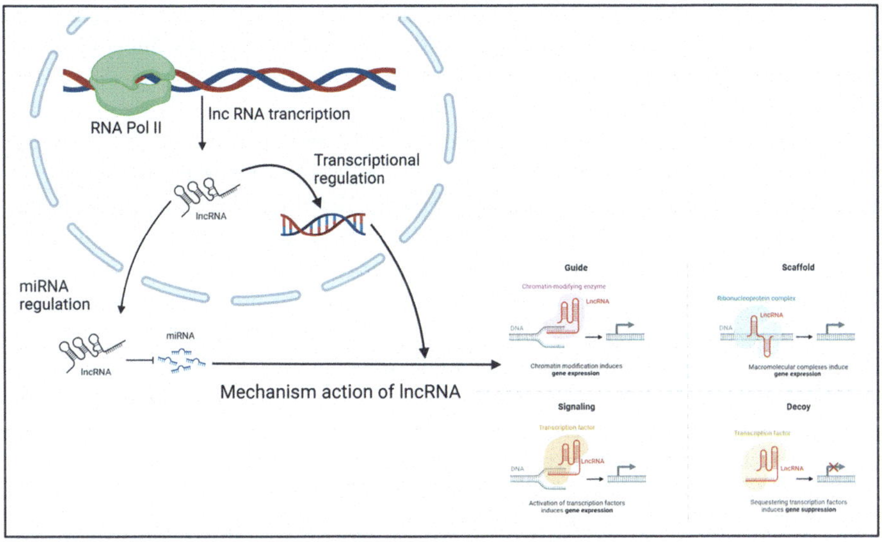

Fig. 4.1 The schematic representation of biosynthesis and mechanism of action of lncRNAs

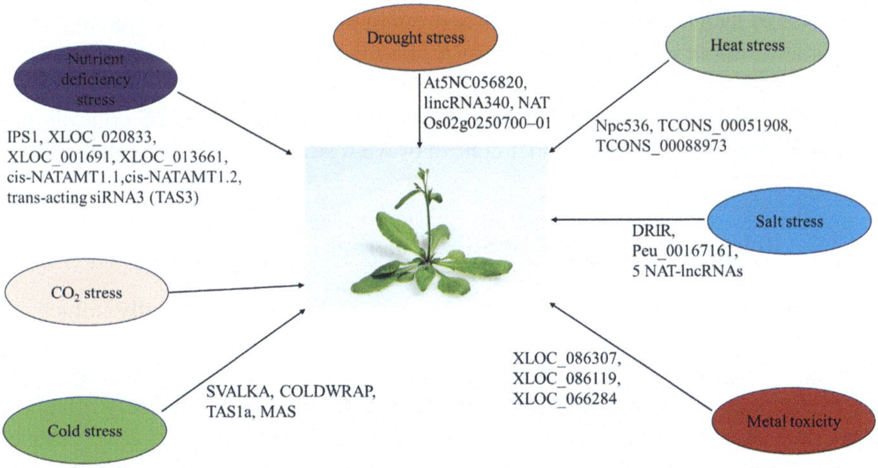

Fig. 4.2 Various abiotic stresses and representative lncRNAs in different stresses in plants

4.4.1.1 Role of LncRNAs in Drought Stress Responses

Recent findings have shown the involvement of long noncoding RNAs (lncRNAs) in response to drought stress in plants. This is demonstrated through the utilization of co-expression networks involving lncRNAs, miRNAs, protein-coding genes, and transcription factors (Zhang et al. 2018; Ding et al. 2019; Pang et al. 2019). Significant examples of lncRNAs responsive to drought stress have been documented across diverse plant species, including six in *Arabidopsis* (Amor et al.

2009), 504 in *Populus* spp. (Shuai et al. 2014),98 in *Oryza sativa* (Chung et al. 2016), 664 in *Zea mays* (Zhang et al. 2014a, b), 19 in foxtail millet (Qi et al. 2013),185 in Cassava (Li et al. 2017), and 1597 in switchgrass (Zhang et al. 2018). The impact of lncRNAs on the drought stress response involves intricate mechanisms such as endogenous target mimicry (eTM), modulation through antisense transcription, chromatin modulation, or direct regulation of the transcription of various drought-responsive genes. A recently known DRIRlncRNA overexpressing line in *Arabidopsis thaliana* resulted in enhanced tolerance to both drought and salt stress in the transgenic plants. DRIRlncRNA also regulates genes associated with ABA signaling, water transport, and various stress-relief processes. Recently Niu et al. 2021 revealed a soybean abiotic-stress-related lncRNA, lncRNA77580 that plays a regulatory role in the expression of nearby genes. The overexpression line of lncRNA77580 showed the enhanced drought tolerance in soybean plants. The natural antisense transcripts (NATs) are a class of lncRNAs reported in *Solanum tuberosum* called *StFLORE* involved in drought tolerance playing a role in regulating water loss by influencing stomatal growth and diurnal opening (Ramirez-Parra et al. 2005). In *Arabidopsis thaliana*, a lncRNA named *DOG1* (1GOD) plays roles in salt and drought stress, seed germination *DELAY OF GERMINATION* 1 suppress the expression of 1GOD under controlled growth conditions. The stress phytohormone (ABA) and drought stress inhibit the 1GOD transcription and this results in high level of *DOG1* expression results in drought tolerance (Yatusevich et al. 2017). One other drought stress tolerant lncRNA identified in *Arabidopsis,* namely *CDF5 LONG NONCODING RNA* (*FLORE),* that promotes flowering (Henriques et al. 2017). In potato, StCDF1 with the help of NAT lncRNA *StFLORE* controls the stomatal opening and closing under control of ABA (Ramírez Gonzales et al. 2021).

4.4.1.2 Role of LncRNA in Heat Stress Tolerance

Heat stress represents a prominent abiotic stress factor which markedly influences plant growth, physiology, metabolic processes, development, and yield performance (Zhang et al. 2019). The ongoing increase in global temperatures is causing detrimental impacts on plant phenology and adaptation processes, adversely affecting crop yields and posing challenges to global food security (Piao et al. 2019). Plants employ a range of mechanisms, encompassing adaptive, biochemical, and molecular responses, to counteract the effects of heat stress (Zhang et al. 2019; Janni et al. 2020). The rapidly increasing research in field of lncRNAs evidence the significant role of lncRNAs as a molecular regulatory layer provides valuable insights into the regulatory mechanisms underlying the heat stress response in crop plants. Xin et al. (2011) have identified 54 heats responsive lncRNAs in wheat among them there washtree heat-responsive differentially expressed lncRNAs (DElncRNAs) from different genomic locations like lncRNA TCONS_00004594, lncRNAs TCONS_00048391 and TCONS_00010856 playing a crucial role in heat stress tolerance (Wang et al. 2019). In *Arabidopsis* a heat responsive lincRNA called *MARS* (MARneral *SILENCING*), present inside the cluster of marneral biosynthetic genes and regulate epigenetic changes in response to ABA (Roulé et al. 2022). Few other heat responsive NATs were also reported from *Arabidopsis* and *Brassica,* namely *MIR398* genes, *NAT398b* and *NAT398c,* respectively, and shows effect on their respective miRNAs (Li et al. 2020).

4.4.1.3 LncRNA Involved in Cold Stress Tolerance

The stress induced by low temperatures is a significant abiotic factor that poses challenges to the growth and yield of plants. In *Arabidopsis thaliana*, the well-known regulatory locus FLOWERING LOCUS C (FLC) is a widely recognized that epigenetically controls the timing of flowering (Whittaker and Dean 2017). FLOWERING LOCUS C (FLC) serves as a suppressor of flowering under cold conditions (Bastow et al. 2004). The active involvement of long non-coding RNAs (lncRNAs) in suppressing the expression of the FLC locus through Polycomb-mediated epigenetic regulation during vernalization under cold stress is a well-established mechanism for regulating cold acclimation in *Arabidopsis thaliana* (Swiezewski et al. 2009; Heo and Sung 2011; Csorba et al. 2014). During the initial stages of cold stress, the suppression of the FLC locus is regulated by cold induced long antisense intragenic RNAs (COOLAIR) that is an alternatively spliced non-coding RNA transcribed in the antisense orientation of the FLC gene. Similarly, cold-assisted intronic noncoding RNA (COLDAIR) that is transcribed from intron1 of the FLC gene, plays a role in recruiting the Polycomb Repressive Complex 2 (PRC2). This recruitment aids in chromatin modification by increasing the presence of the repressive histone mark H3K27me3 at the FLC locus, leading to the suppression of FLC locus expression (Heo and Sung 2011). In *Arabidopsis*, another cold responsive lncRNA SVALKA and cryptic antisense CBF1 (asCBF1) long non-coding RNAs (lncRNAs) induced by cold stress (Kindgren et al. 2018) and modulate cold acclimation by inhibiting the transcription of the CBF1 gene through RNA polymerase II (RNAPII) collision, a process originating from the lncRNAs SVALKA and CBF1. Zhao et al. (2018) have shown that involvement of long non-coding RNAs (lncRNAs) in regulating the cold stress responses in *Arabidopsis*, through strand-specific RNA-sequencing (ssRNA-seq) identified 4050 natural antisense transcript (NAT) lncRNAs and 2460 long intergenic non-coding RNAs (lincRNAs) as lncRNAs responsive to cold conditions. MAS activated the transcription of the associated cold-responsive MADS AFFECTING FLOWERING 4 (MAF4), an FLC family member, by engaging the WDR5a complex. Similarly, in *Brassica rapa*, three paralogs of flowering locus C (FLC) genes, that is, BrFLC1, BrFLC2. and BrFLC3, functioning as floral repressors during vernalization, have been documented (Shea et al. 2019a, b, Itabashi et al. 2018). The participation of natural antisense transcripts (NATs) at the FLC2 locus of *Brassica rapa* under cold stress has been observed by Li et al. 2016. Transcriptome analysis using RNA-seq of leaves from both control and cold-treated *Brassica rapa* have identified a total of 2088 long non-coding RNAs (lncRNAs) (Shea et al. 2019a, b). Out of the total, there were 1444 long intergenic noncoding RNAs (lincRNAs), 551 natural antisense transcripts (NATs), and 93 intronic noncoding RNAs (incRNAs). Among 2088 long non-coding RNAs (lncRNAs), 549exhibited significant changes in their expression levels in response to a four-week cold treatment. A comprehensive investigation of cold-responsive long non-coding RNAs (lncRNAs) in grapevine, conducted through RNA-seq analysis, identified 284 newly upregulated lncRNAs, 182 newly downregulated lncRNAs, and 242 differentially expressed lncRNAs (DElncRNAs) targeting 326 protein-coding genes. To uncover genome-wide long non-coding RNAs (lncRNAs) responsive to cold and drought in cassava, a total of 318 lncRNAs were identified (Li et al. 2017).

4.4.1.4 Role of lncRNA in Salt Stress Tolerance

Our understanding of the molecular mechanisms controlling plants' response to salinity stress has improved as evidence of the regulatory roles of long non-coding RNAs (lncRNAs) that allow plants to tolerate salt stress has emerged (Amor et al. 2009; Qin et al. 2017). It has been determined that a number of lncRNAs in *Medicago truncatula* act during osmotic and salinity stress. TCONS_00046739 that controls cytochrome P450 in roots; TCONS_00097188 that controls photosynthesis by up-regulating the Medtr6g006990 gene; TCONS_00047650 that controls the upregulation of the Medtr3g069280 gene that codes for phosphatidylinositol-specific phospholipase C; TCONS_00116877 that helps to control the upregulation of the Medtr7g094600 gene that codes for glutathione peroxidase in roots; and TCONS_00020253 that controls the expression of the Na+/H+ exchanger gene Medtr1g081900 in roots (Wang et al. 2015). Through deep transcriptome analysis of RNA libraries generated from leaf and root tissues, 1710 lncRNAs responsive to combined salinity and boron stress were examined in the Lluteño landrace of maize (Huanca-Mamani et al. 2018). Therefore, by affecting oxidative stress through the control of genes encoding antioxidant enzymes, the discovered lncRNAs offered tolerance to salinity stress. They also controlled additional regulatory genes linked to salt stress as well as a variety of Na+/H+ exchanger genes. The *Arabidopsis thaliana* lncRNA DRIR controls the production of P5CS1, RD29A, RD29B, AtrbohB, FUT4, ANNAT7, and NAC3, which provide drirD mutant and DRIR-overexpressing lines their resistance to salt.

4.4.1.5 Role of lncRNAs in Heavy Metal Toxicity in Plants

Heavy metals are recognized as environmental contaminants owing to their toxic properties, persistence in the environment, and tendency to accumulate in living organisms. The environmentally significant and most hazardous heavy metals and metalloids encompass chromium (Cr), nickel (Ni), copper (Cu), zinc (Zn), cadmium (Cd), lead (Pb), mercury (Hg), and arsenic (As). Long non-coding RNAs (lncRNAs) might play a pivotal role in the consumption of heavy metals into the plant system (Ali et al. 2019). Long non-coding RNAs (lncRNAs) could potentially play a crucial role in regulating the uptake of heavy metals into the plant system. RNA-seq profiling revealed 301 cadmium-responsive long non-coding RNAs (lncRNAs) in *Brassica napus*, among which 67 acted as endogenous target mimics (eTMs) for 36 cadmium-responsive microRNAs (miRNAs) (Feng et al. 2016). Four lncRNAs were recognized as precursors for miR824, miR167d, miR156d, and miR156e in response to Cd stress. Functional validation through qRT-PCR analysis of TCONS_00091906, TCONS_00033487, and TCONS_00097191 lncRNAs under cadmium stress confirmed their significant role as target mimics for EL628609, TC182597, and TC203372 mRNAs that are involved in cadmium uptake and detoxification (Zhou et al. 2012). Similarly, Chen et al. 2018a, b conducted a comprehensive genome-wide survey of long non-coding RNAs (lncRNAs) through RNA deep transcriptome sequencing, revealing both up- and down-regulation of lncRNAs in response to cadmium (Cd) (Chen et al. 2018a, b). Additionally, the functional analysis of differentially expressed lncRNAs shed light on their regulatory role in target genes associated with cysteine and methionine metabolism under Cd stress, given the increasing evidence of arsenic (As) toxicity in rice (Tang et al. 2019).

4.4.1.6 Role of lncRNAs in CO_2 Stress Tolerance

There are very few studies showing effect of CO_2 on lncRNA. Huang et al. (2019) showed that there were 54 long intronic non-coding RNAs (lincRNAs) that were upregulated, while 59 lincRNAs were downregulated in *Phaeodactylum tricornutum* under elevated CO_2 conditions. The majority of lincRNA are 600–700nts in length. Using deep sequencing in marine *Nannochloropsis oceanica.* Wei et al. (2022) found that 51 lncRNAs were differentially expressed between low and high CO_2 treatment out of which 33 were upregulated and 18 were downregulated lncRNAs. The predicted target of lncRNA and lncRNA-mRNA co-location analysis suggested that these lncRNA are mainly involved in gene expression regulation, protein expression and epigenetic regulation. They also suggested that these lncRNA might be involved in alternative splicing event under CO_2 fluctuation (Wei et al. 2022).

4.4.1.7 Role of lncRNAs in Nutrient Deficiency Stress

Nutrient deficiency imposes a great stress on plants. In *Oryza sativa* Pi starvation induced lncRNA, *induced by phosphate starvation1 (IPS1)* and its homolog named *At4* was also identified in *Arabidopsis*. Two recently discovered lncRNAs in maize in response to phosphate starvation named *Pi-DEFICIENCY-INDUCED LNCRNA 1 (PILNCR1)* and *Pi-DEFICIENCY-INDUCED LNCRNA 2 (PILNCR2)* (Qingguo et al. 2018). In poplars an antisense lncRNA was identified in N-deficiency (Chen et al. 2020). TPSI1 (tomato phosphate starvation-induced 1)/Mt4 gene family was identified in tomato under Pi starvation conditions (Li et al. 2016). In trifolium orange leaves under boron deficiency, 721 downregulated and 729 upregulated lncRNAs were identified.

4.5 Conclusion and Future Prospects

Long non-coding RNAs (lncRNAs) have emerged as key regulatory molecules in plants in response to diverse abiotic stresses including drought, heat, cold, salt, heavy metals, CO_2 fluctuations, and nutrient deficiencies. Recent advancements in molecular biology and transcriptomics have unveiled their complex regulatory mechanisms, including modulation of gene expression through endogenous target mimicry (eTM), chromatin modification, antisense transcription, and epigenetic regulation. Though limited, studies show that lncRNAs play roles in gene expression, epigenetic regulation, and alternative splicing under elevated CO_2 conditions. Collectively, these findings highlight the versatility and importance of lncRNAs as critical components in the regulation of plant responses to environmental stresses. While many stress-responsive lncRNAs have been identified, their precise roles and mechanisms in regulating gene expression remain to be fully elucidated. Advanced tools like CRISPR/Cas9 can be used to validate their functions in vivo. Combining genomics, transcriptomics, and proteomics may provide deeper insights into the lncRNA-mediated signaling networks under specific stresses and reveal novel stress-regulatory pathways. Investigating the role of lncRNAs in plants subjected to combined abiotic stresses, such as drought and salinity or heat and nutrient deficiency, will help understand their multi-stress adaptive responses. Engineering stress-responsive lncRNAs through biotechnological approaches can enhance crop resilience to environmental challenges, contributing to sustainable agriculture and food security. Future research should explore the roles of lncRNAs in chromatin remodeling, histone modification, and methylation patterns under stress conditions, as these processes significantly impact gene expression. Expanding studies beyond model organisms (e.g. *Arabidopsis*) to diverse crop species and wild relatives will uncover conserved and species-specific lncRNAs, enriching stress-adaptive knowledge. Understanding the role of lncRNAs under changing environmental conditions, such as rising temperatures and fluctuating CO_2 levels, will aid in developing climate-resilient crops. High-throughput sequencing, single-cell RNA-seq, and machine learning approaches may facilitate the identification, prediction, and functional analysis of novel lncRNAs under abiotic stress. By addressing these future directions, lncRNA research can provide innovative strategies to improve plant stress tolerance, ensuring agricultural productivity under ever-changing environmental conditions.

References

Ali H, Khan E, Ilahi I (2019) Environmental chemistry and ecotoxicology of hazardous heavy metals: environmental persistence, toxicity, and bioaccumulation. J Chem 6730305:1

Amor BB et al (2009) Novel long non-protein coding RNAs involved in Arabidopsis differentiation and stress responses. Genome Res 19(1):57–69

Ariel F et al (2015) Battles and hijacks: noncoding transcription in plants. Trends Plant Sci 20(6):362–371

Bastow R et al (2004) Vernalization requires epigenetic silencing of FLC by histone methylation. Nature 427(6970):164–167

Bhatia G, Sharma S, Upadhyay SK, Singh K (2019) Long non-coding RNAs coordinate developmental transitions and other key biological processes in grapevine. Sci Rep 9:3552. https://doi.org/10.1038/s41598-019-38989-7

Bohmdorfer G, Sethuraman S, Rowley MJ, Krzyszton M, Rothi MH, Bouzit L, Wierzbicki AT (2016) Long non-coding RNA produced by RNA polymerase V determines boundaries of heterochromatin. elife 5:e19092. https://doi.org/10.7554/eLife.19092

Bond A, Row P, Dudley E (2011) Post-translation modification of proteins; methodologies and applications in plant sciences. Phytochemistry 72(10):975–996

Bridges MC, Daulagala AC, Kourtidis A (2021) LNCcation: LncRNA localization and function. J Cell Biol 220:e202009045. https://doi.org/10.1083/jcb.202009045

Calixto CPG, Tzioutziou NA, James AB, Hornyik C, Guo W, Zhang R, Nimmo HG, Brown JWS (2019) Cold-dependent expression and alternative splicing of Arabidopsis long non-coding RNAs. Front Plant Sci 10:235. https://doi.org/10.3389/fpls.2019.00235

Cao H, Wahlestedt C, Kapranov P (2018) Strategies to annotate and characterize long noncoding RNAs: advantages and pitfalls. Trends Genet 34:704–721. https://doi.org/10.1016/j.tig.2018.06.002

Carlevaro-Fita J, Johnson R (2019) Global positioning system: understanding long noncoding RNAs through subcellular localization. Mol Cell 73:869–883. https://doi.org/10.1016/j.molcel.2019.02.008

Chekanova JA (2015) Long non-coding RNAs and their functions in plants. Curr Opin Plant Biol 27:207–216

Chen D et al (2018a) Architecture of gene regulatory networks controlling flower development in *Arabidopsis thaliana*. Nat Commun 9(1):4534

Chen L, Shi S, Jiang N, Khanzada H, Wassan GM, Zhu C et al (2018b) Genome-wide analysis of long non-coding RNAs affecting roots development at an early stage in the rice response to cadmium stress. BMC Genomics 19:460. https://doi.org/10.1186/s12864-018-4807-6

Chen Li, Z, Qian-Hao, K, Kerstin (2020) Long non-coding RNAs in plants: emerging modulators of gene activity in development and stress responses. Planta 2020 Oct 24;252(5):92. https://doi.org/10.1007/s00425-020-03480-5

Chung PJ et al (2016) Transcriptome profiling of drought responsive noncoding RNAs and their target genes in rice. BMC Genomics 17(1):1–12

Csorba T et al (2014) Antisense COOLAIR mediates the coordinated switching of chromatin states at FLC during vernalization. Proc Natl Acad Sci 111(45):16160–16165

Ding Z et al (2019) Strand-specific RNA-seq based identification and functional prediction of drought-responsive lncRNAs in cassava. BMC Genomics 20(1):1–13

Dykes IM, Emanueli C (2017) Transcriptional and post-transcriptional gene regulation by long non-coding RNA. Genomics Proteomics Bioinformatics 15(3):177–186

Feng SJ et al (2016) Characterization of long non-coding RNAs involved in cadmium toxic response in Brassica napus. RSC Adv 6(85):82157–82173

Fesenko I, Shabalina SA, Mamaeva A, Knyazev A, Glushkevich A, Lyapina I, Ziganshin R, Kovalchuk S, Kharlampieva D, Lazarev V et al (2021) A vast pool of lineage-specific microproteins encoded by long non-coding RNAs in plants. Nucleic Acids Res 49:10328–10346. https://doi.org/10.1093/nar/gkab816

Ford Doolittle W (2013) Is junk DNA bunk? A critique of ENCODE. Proc Natl Acad Sci USA 110(14):5294–5300

Franco-Zorrilla JM et al (2007) Target mimicry provides a new mechanism for regulation of microRNA activity. Nat Genet 39(8):1033–1037

Henriques R, Wang H, Liu J, Boix M, Huang L-F, Chua N-H (2017) The antiphasic regulatory module comprising CDF5 and its antisense RNA FLORE links the circadian clock to photoperiodic flowering. New Phytol 216(3):854–867

Heo JB, Sung S (2011) Vernalization-mediated epigenetic silencing by a long intronic noncoding RNA. Science 331(6013):76–79

Huanca-Mamani W et al (2018) Long non-coding RNAs responsive to salt and boron stress in the hyper-arid Lluteno maize from Atacama Desert. Genes 9(3):170

Huang R et al (2019) A potential role for epigenetic processes in the acclimation response to elevated CO2 in the model diatom *Phaeodactylum tricornutum*. Front Microbiol 9:3342

Itabashi E et al (2018) Epigenetic regulation of agronomical traits in Brassicaceae. Plant Cell Rep 37:87–101

Janni M et al (2020) Molecular and genetic bases of heat stress responses in crop plants and breeding for increased resilience and productivity. J Exp Bot 71(13):3780–3802

Kim ED, Xiong Y, Pyo Y, Kim DH, Kang BH, Sung S (2017) Spatio-temporal analysis of coding and long noncoding transcripts during maize endosperm development. Sci Rep 7:3838. https://doi.org/10.1038/s41598-017-03878-4

Kindgren P et al (2018) Transcriptional read-through of the long non-coding RNA SVALKA governs plant cold acclimation. Nat Commun 9(1):4561

Li S, Yamada M, Han X, Ohler U, Benfey PN (2016) High-resolution expression map of the Arabidopsis root reveals alternative splicing and lincRNA regulation. Dev Cell 39:508–522. https://doi.org/10.1016/j.devcel.2016.10.012

Li S et al (2017) Genome-wide identification and functional prediction of cold and/or drought-responsive lncRNAs in cassava. Sci Rep 7(1):45981

Li Y, Li X, Yang J, He Y (2020) Natural antisense transcripts of MIR398 genes suppress microR398 processing and attenuate plant thermotolerance. Nat Commun 11(1):5351

Liu J et al (2012) Genome-wide analysis uncovers regulation of long intergenic noncoding RNAs in Arabidopsis. Plant Cell 24(11):4333–4345

Ma L, Bajic VB, Zhang Z (2013) On the classification of long non-coding RNAs. RNA Biol 10(6):924–933

Marquardt S, Raitskin O, Wu Z, Liu F, Sun Q, Dean C (2014) Functional consequences of splicing of the antisense transcript COOLAIR on FLC transcription. Mol Cell 54:156–165. https://doi.org/10.1016/j.molcel.2014.03.026

Niu F et al (2021) Large DNA fragment deletion in lncRNA77580 regulates neighboring gene expression in soybean (Glycine max). Funct Plant Biol 48(11):1139–1147

Pang J et al (2019) Spatio-temporal transcriptional dynamics of maize long non-coding RNAs responsive to drought stress. Genes 10(2):138

Piao S et al (2019) Plant phenology and global climate change: current progresses and challenges. Glob Chang Biol 25(6):1922–1940

Qi X et al (2013) Genome-wide annotation of genes and noncoding RNAs of foxtail millet in response to simulated drought stress by deep sequencing. Plant Mol Biol 83:459–473

Qin T, Zhao HY, Cui P, Albesher N, Xiong LM (2017) A nucleus-localized long non-coding RNA enhances drought and salt stress tolerance. Plant Physiol 175:1321–1336. https://doi.org/10.1104/pp.17.00574

Qingguo et al (2018) The PILNCR1-miR399 Regulatory Module Is Important for Low Phosphate Tolerance in Maize. Plant Physiology177(4):pp.00034.2018 https://doi.org/10.1104/pp.18.00034

Ramírez Gonzales L, Shi L, Bergonzi SB, Oortwijn M, Franco-Zorrilla JM, Solano-Tavira R, Visser RGF, Abelenda JA, Bachem CWB (2021) Potato CYCLING DOF FACTOR 1 and its lncRNA counterpart StFLORE link tuber development and drought response. Plant J 105(4):855–869

Ramirez-Parra E, Desvoyes B, Gutierrez C (2005) Balance between cell division and differentiation during plant development. Int J Dev Biol 49(5–6):467–477

Roulé T, Christ A, Hussain N, Huang Y, Hartmann C, Benhamed M, Gutierrez-Marcos J, Ariel F, Crespi M, Blein T (2022) The lncRNA MARS modulates the epigenetic reprogramming of the marneral cluster in response to ABA. Mol Plant 15(5):840–856

Sanchita G et al (2020) Updates on plant long non-coding RNAs (lncRNAs): the regulatory components. Plant Cell Tiss Org Cult (PCTOC) 140:259–269

Shea DJ et al (2019a) Long noncoding RNAs in *Brassica rapa* L. following vernalization. Sci Rep 9(1):9302

Shea DJ, Nishida N, Takada S, Itabashi E, Takahashi S, Akter A et al (2019b) Long noncoding RNAs in *Brassica rapa* L following vernalization. Sci Rep 9:9302. https://doi.org/10.1038/s41598-019-45650-w

Shuai P et al (2014) Genome-wide identification and functional prediction of novel and drought-responsive lincRNAs in Populus trichocarpa. J Exp Bot 65(17):4975–4983

Song X, Hu J, Wu T, Yang Q, Feng X, Lin H, Feng S, Cui C, Yu Y, Zhou R et al (2021) Comparative analysis of long noncoding RNAs in angiosperms and characterization of long noncoding RNAs in response to heat stress in Chinese cabbage. Hortic Res 8:48. https://doi.org/10.1038/s41438-021-00484-4

St Laurent G, Wahlestedt C, Kapranov P (2015) The landscape of long noncoding RNA classification. Trends Genet 31:239–251. https://doi.org/10.1016/j.tig.2015.03.007

Swiezewski S et al (2009) Cold-induced silencing by long antisense transcripts of an Arabidopsis Polycomb target. Nature 462(7274):799–802

Tang Z et al (2019) Deciphering the non-coding RNA-level response to arsenic stress in rice (*Oryza sativa*). Plant Signal Behav 14(9):1629268

Wang T-Z et al (2015) Identification and characterization of long non-coding RNAs involved in osmotic and salt stress in *Medicago truncatula* using genome-wide high-throughput sequencing. BMC Plant Biol 15(1):1–13

Wang P et al (2019) Identification and functional prediction of cold-related long non-coding RNA (lncRNA) in grapevine. Sci Rep 9(1):6638

Wei L et al (2022) Identification and expression analysis of genome-wide long noncoding RNA responsive CO2 fluctuated environment in marine microalga Nannochloropsisoceanica. Mar Pollut Bull 176:113419

Whittaker C, Dean C (2017) The FLC locus: a platform for discoveries in epigenetics and adaptation. Annu Rev Cell Dev Biol 33:555–575

Wierzbicki AT, Blevins T, Swiezewski S (2021) Long noncoding RNAs in plants. Annu Rev Plant Biol 72:245–271. https://doi.org/10.1146/annurev-arplant-093020-035446

Wood CC et al (2006) The *Arabidopsis thaliana* vernalization response requires a polycomb-like protein complex that also includes VERNALIZATION INSENSITIVE 3. Proc Natl Acad Sci 103(39):14631–14636

Xin M et al (2011) Identification and characterization of wheat long non-protein coding RNAs responsive to powdery mildew infection and heat stress by using microarray analysis and SBS sequencing. BMC Plant Biol 11:1–13

Yatusevich R, Fedak H, Ciesielski A, Krzyczmonik K, Kulik A, Dobrowolska G, Swiezewski S (2017) Antisense transcription represses Arabidopsis seed dormancy QTL DOG1 to regulate drought tolerance. EMBO Rep 18(12):2186–2196

Zhang YC, Liao JY, Li ZY, Yu Y, Zhang JP, Li QF, Qu LH, Shu WS, Chen YQ (2014a) Genome-wide screening and functional analysis identify a large number of long noncoding RNAs involved in the sexual reproduction of rice. Genome Biol 15:512. https://doi.org/10.1186/s13059-014-0512-1

Zhang W et al (2014b) Identification of maize long non-coding RNAs responsive to drought stress. PLoS One 9(6):e98958

Zhang C et al (2018) Long non-coding RNAs of switchgrass (Panicum virgatum L.) in multiple dehydration stresses. BMC Plant Biol 18:1–15

Zhang J et al (2019) Crop improvement through temperature resilience. Annu Rev Plant Biol 70:753–780

Zhao X et al (2018) Global identification of Arabidopsis lncRNAs reveals the regulation of MAF4 by a natural antisense RNA. Nat Commun 9(1):5056

Zhou ZS, Song JB, Yang ZM (2012) Genome-wide identification of Brassica napus microRNAs and their targets in response to cadmium. J Exp Bot 63(12):4597–4613

Chapter 5
Epigenetic Regulation of Abiotic Stress Responses

Sirmast Faiz, Nikhil LNU, Marya Bibi, Nitish Joshi, Imran Khan, Mohammad Hassan Ukasha, Shah Saud, Shah Fahad, Bibek Adhikari, Asmatullah, Muhammad Sarmad, and Taufiq Nawaz

Abstract Plants, as sessile organisms, are constantly exposed to various abiotic stresses such as drought, salinity, heat, cold, heavy metal toxicity, and nutrient deficiencies. These environmental factors negatively impact plant growth, productivity, and survival. To cope with such challenges, plants have developed intricate regulatory mechanisms, including epigenetic modifications, that allow for dynamic gene expression changes without altering DNA sequences. Epigenetic regulation, which includes DNA methylation, histone modifications, non-coding RNAs, and chromatin remodeling, plays a crucial role in stress response, adaptation, and memory formation in plants.

A significant aspect of epigenetic regulation is its role in stress memory and transgenerational inheritance. Studies demonstrate that DNA methylation and histone modifications contribute to inherited stress tolerance. This knowledge has major implications for crop improvement strategies, including epigenetic breeding, CRISPR-based epigenome editing, and priming techniques that enhance stress

S. Faiz · Asmatullah
Dr Panjwani Center for Molecular Medicine and Drug Research, International Center of Chemical and Biological Sciences, University of Karachi, Karachi, Pakistan

N. LNU · M. Bibi · N. Joshi
Department of Agronomy, Horticulture & Plant Science, South Dakota State University, Brookings, SD, USA

I. Khan
Center for Life Science Ventures, Cornell University, Ithaca, NY, USA

M. H. Ukasha
Department of Public Health, Cornell University, Ithaca, NY, USA

S. Saud
College of Life Science, Linyi University, Linyi, Shandong, China

S. Fahad · B. Adhikari · T. Nawaz (✉)
Department of Biology/Microbiology, South Dakota State University, Brookings, SD, USA
e-mail: Taufiq.nawaz@jacks.sdstate.edu

M. Sarmad
Huazhong Agriculture University, Wuhan, Hubei Province, People's Republic of China

S. Tiwari et al. (eds.), *Decoding Non-Coding RNA*, Concepts and Strategies in Plant Sciences, https://doi.org/10.1007/978-981-96-9859-2_5

resilience in plants. With climate change intensifying abiotic stresses, understanding epigenetic mechanisms offers promising solutions for sustainable agriculture. Advances in epigenomic tools such as bisulfite sequencing, ChIP-seq, and CRISPR-dCas9 are paving the way for targeted crop enhancements. However, challenges remain in identifying stable epigenetic markers and integrating epigenetic modifications into breeding programs. Future research should focus on multi-omics approaches and regulatory considerations to harness epigenetic potential for stress-resilient crop development.

Keywords Abiotic stress · Epigenetic regulation · DNA methylation · Non-coding RNAs · CRISPR-based epigenome editing · Climate change adaptation · Multi-omics approaches

5.1 Introduction

Plants are immobile organisms that face a constant array of environmental challenges, including drought, salinity, heat stress, exposure to heavy metals, and ultraviolet radiation. These stresses test their resilience and adaptability as they remain rooted in place. These environmental stresses affect physiological functions and metabolic activities which result in poor growth along with lower productivity and even mortality (Jing et al. 2024). Drought stress affects photosynthesis and reduces nutrient absorption, and salinity stress creates both ion disruption and osmotic stress within the plants. The cumulative effect of these stresses presents a significant challenge to world agricultural and food security systems (Atta et al. 2023). Plants have developed specialized genetic regulatory systems to cope with environmental changes, ensuring their survival in changing conditions (Nawaz et al. 2025). Plants detect environmental signals which activate a chain reaction of signal transmission networks that facilitate transcription factor accumulation to activate gene expression leading to environmental stress adaptation (Singhal et al. 2023). Epigenetic regulation also serves as an important gene control mechanism during stressful situations through DNA-based and histone-based covalent modifications that modify chromatin but do not alter DNA sequences (Abdulraheem et al. 2024; Jin et al. 2024a, b). Epigenetics encompasses the study of heritable changes in gene function that occur without alterations in the underlying DNA sequence. Such modifications control the function and activity of genes alongside their critical roles in biological processes. Plants require epigenetic mechanisms for proper development while using those mechanisms to respond to environmental triggers and adapt to changes in their surroundings (Wu et al. 2023). The dynamic response mechanisms of gene expression to abiotic stress consist of epigenetic modifications that include DNA methylation and histone modifications combined with non-coding RNAs. DNA methylation enables gene regulation through the addition of methyl groups to cytosine bases, although the final gene expression outcome depends on the cellular environment (Akhter et al. 2021). Acetylation along with methylation of histones modifies chromatin packaging which determines both gene availability and transcriptional activity. Post-transcriptional gene expression control occurs through non-coding RNAs that include both microRNAs (miRNAs) and long non-coding RNAs (lncRNAs) which also direct DNA methylation processes. Plants use these epigenetic mechanisms to rapidly change their gene expression profiles which enables them to become stress tolerant while adapting to their environment (Statello et al. 2021). The significance of epigenetic modifications in stress adaptation lies in their role in establishing stress memory. Plants retain stress memory from previous exposure, leading to a primed state that enhances their ability to respond more rapidly and effectively to similar stress conditions. Living organisms maintain genetically marked patterns that control gene expression as a form of stress memory (Nawaz et al. 2025). The process of transgenerational epigenetic inheritance allows specific epigenetic changes to transfer across generations that enhance offspring resistance toward environmental stressors encountered by their parents (Abdulraheem et al. 2024). This chapter explains how epigenetic regulation operates as a vital mechanism for plant responses to environmental stress. The first part of this discussion will explain epigenetic fundamental principles followed by an analysis of abiotic stress impacts on plant physiology and a discussion about how epigenetic modifications help plants adapt to environmental stressors. Recent research findings help us explore the complete role of epigenetic mechanisms which promote plant resilience. Our research will assess applications of acquired findings to establish crop improvement strategies that aim to boost agricultural sustainability under environmental pressures.

5.2 Fundamentals of Epigenetic Mechanisms

Plants rely on epigenetic mechanisms as the essential biological framework for detecting and responding to abiotic stresses and their adaptations. These mechanisms dynamically control how genes express themselves through changes to DNA access without modifying the DNA sequence thus enabling flexible and reversible stress responses. The essential epigenetic processes involved in stress-responsive gene regulation include DNA methylation together with histone modifications and non-coding RNAs and chromatin remodeling.

5.2.1 *Role of DNA Methylation in Regulating Gene Expression Under Stress*

DNA methylation occurs when scientists attach a methyl group to the 5-carbon position of cytosine bases, yet this modification primarily occurs in CpG sites and also appears in CHG and CHH sequences (H represents A, T, or C) in plant genomes. The DNA methyltransferases (DNMTs) such as MET1 for CpG maintenance and CMT3 for CHG maintenance and DRM2 for de novo methylation function together in this process (Zang et al. 2023). The expression levels of genes depend on methylation patterns which typically lead to gene silencing when DNA methylation occurs in promoter regions. The effects of gene body methylation vary because transcription changes based on both the gene structure and regulatory mechanisms within its environment (Kumar and Mohapatra 2021).

5.2.1.1 Dynamic Changes Under Stress Conditions

The expression of stress-responsive genes depends on how abiotic stress affects DNA methylation patterns. Rice plants (*Oryza sativa*) experience increased expression of the transcription factor OsDREB2A when drought stress induces hypomethylation at its promoter region. The transcription factor actively turns on genes controlling Osmo protectant biosynthesis and stomatal conductance while helping plants conserve water (Huiming Zhang et al. 2022). Researchers explored how maize (*Zea mays*) experiences drought-induced methylation changes that affect aquaporin gene activation for enhanced membrane water transport resulting in drought resistance (Wang et al. 2023a, b, c, d). Under salt stress conditions, *Arabidopsis thaliana* plants undergo promoter demethylation at the RD29A gene which controls osmotic regulation and ion homeostasis (Sako et al. 2012). The promotional effect of demethylation boosts transcription processes that allow plants to better handle salt stress.

5.2.1.2 Mechanisms Influencing Stress-Responsive Genes

DNA demethylation functions as an opposite process to DNA methylation through the activity of enzymes including ROS1 (Repressor of Silencing 1), DME (DEMETER), and DML (DEMETER-like proteins). These enzymes function to break down methyl groups from cytosine residues which specifically target loci responsible for stress response (Rai et al. 2024). The enzyme ROS1 from *Arabidopsis thaliana* mediates transposable element demethylation together with neighboring stress-responsive genes when plants experience cold stress, thus blocking the expansion of silencing marks yet enabling gene functionality (Feng et al. 2023a, b). Under drought conditions, DME works to activate ABA-responsive genes by demethylating their DNA sequences. The regulatory mechanism helps plants execute stomatal closure as a water conservation strategy (Muhammad Aslam et al. 2022).

5.2.1.3 Case Studies

Heat Stress in Tomato

Heat stress triggers distinct methylation patterns in *Solanum lycopersicum* (tomato) which leads to suppression of unnecessary metabolic pathways. The transfer of resources to essential processes, particularly heat shock proteins (HSPs) production, allows plants to defend cellular structures against thermal damage. During prolonged heat stress, the cellular survival rates are increased by methylation-based regulatory mechanisms (Singh et al. 2021).

Heavy Metal Stress in Indian Mustard

Cadmium toxicity exposure in *Brassica juncea* (Indian mustard) results in hypermethylation of genes that control secondary metabolism. The process of hypermethylation blocks the growth-related genes from functioning properly while steering metabolic resources toward detoxification mechanisms. The epigenetic regulation through enhanced phytochelatins and glutathione production allows the plant to trap cadmium ions while protecting against oxidative damage (Al-Khayri et al. 2023).

Cold Stress in Wheat

Wheat plants (*Triticum aestivum*) develop hypomethylation at DNA regions controlling the cold-responsive transcription factors including CBF (C-repeat binding factor) genes under cold stress conditions. Through activation of pathways, these transcription factors produce antifreeze proteins and osmoprotectants such as trehalose so plants can better resist freezing temperatures (Zhou et al. 2023a, b, c). DNA methylation functions as an essential mechanism for controlling stress-responsive genes through changes in transcriptional activity. The mechanism helps plants adjust to different abiotic stress environments including drought and salinity conditions as well as heat stress and toxic heavy metals. The development of epigenomic sequencing tools like whole-genome bisulfite sequencing reveals important stress regulation processes while providing potential genetic strategies to enhance crop resilience.

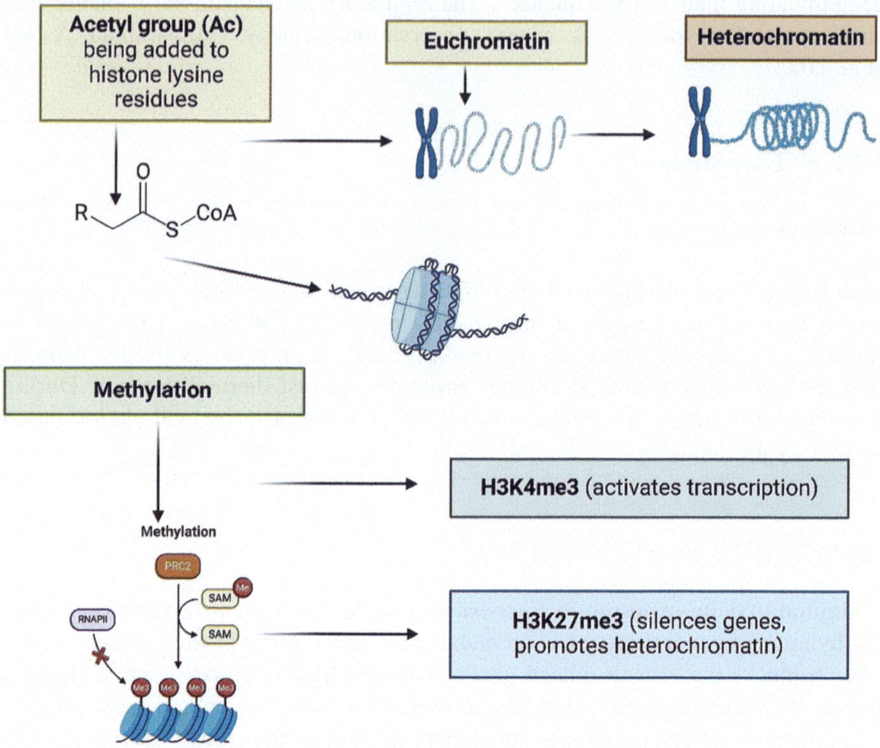

Fig. 5.1 Types of histone modifications and their impact on chromatin structure and gene expression. Acetylation promotes euchromatin formation and gene activation, while methylation, phosphorylation, ubiquitination, and sumoylation contribute to dynamic chromatin regulation, influencing transcriptional control and cellular responses

5.2.2 Histone Modifications

The basic mechanisms behind epigenetics involve histone modifications which change chromatin structures to control DNA accessibility for gene expression (Fig. 5.1). Different modifications affect histone protein amino-terminal tails which function as transcription activation or repression signals based on their characteristics and environmental settings. Plant responses to abiotic stresses heavily depend on these chemical changes which serve to dynamically regulate stress-responsive genes (Zhao et al. 2019).

5.2.2.1 Types of Histone Modifications

Acetylation

The addition of acetyl groups to histone lysine residues within tails occurs through HATs' enzymatic activities. Through histone acetylation, the neutralizing effect on positive charges reduces DNA-intrinsic negative charge interactions which create an accessible euchromatic chromatin environment. The relaxed chromatin structure grants RNA polymerase along with transcription factors direct access to DNA molecules for transcription to take place. The activities of HDACs result in chromatin condensation that forms heterochromatin which suppresses transcriptional activity. The expression of genes under stress conditions heavily depends on how well histone acetyltransferase activities work against histone deacetylase activities Wang et al. 2023a, b, c, d).

Methylation

The histone methyltransferase enzymes modify lysine and arginine residues through sequential methylation from one to three methyl groups. Transcriptional control happens through different methylation sites and modification levels which control this process. For example:

- H3K4me3: associated with transcriptional activation, particularly at gene promoters
- H3K27me3: linked to transcriptional repression and heterochromatin formation The activity of histone demethylase enzymes enables dynamic gene control through the reversal of methylation effects (Zhou et al. 2023a, b, c).

Phosphorylation

The enzyme kinases apply phosphate groups to specific amino acid residues found in serine or threonine and tyrosine. A study by (Feng et al. 2023a, b) shows how this reaction produces relaxed chromatin which helps DNA repair mechanisms alongside stress signaling functions.

Ubiquitination and Sumoylation

The attachment of ubiquitin proteins alongside SUMO (small ubiquitin-like modifier) analogs modifies histones during their binding process. The binding process of ubiquitin proteins and SUMO (small ubiquitin-like modifier) analogs modifies histones. The attachment of ubiquitin to histones leads to gene activation and starts proteolysis pathways to degrade proteins, yet sumoylation primarily functions to silence genes (Morrell and Sadanandom 2019).

5.2.2.2 Impact on Chromatin Structure and Transcription

Certain proteins modify histone molecules which function as signaling molecules that determine chromatin states. Expression activation occurs in euchromatin after acetylation along with the H3K4me3 modifications make DNA more accessible for transcription (Bannister and Kouzarides 2011). Through the activities of H3K27me3 along with deacetylation, heterochromatin transitions into a dense state which prevents transcriptional machinery from functioning, thus silencing genes (Shen et al. 2021). Multiple chromatin modifications operate together to generate specific histone signals that decode environmental stimuli for direct genetic responses to stress conditions.

5.2.2.3 Examples in Abiotic Stress Responses

Histone modifications are key regulators of gene expression in plants under abiotic stress, acting through changes in chromatin structure and accessibility. To adapt to environmental challenges such as drought, salinity, cold, and heat, plants dynamically modify histone marks to control stress-responsive gene expression (Fig. 5.2). Below are a few examples of histone modifications associated with various abiotic stresses.

Cold Stress

The cold treatment affects H3K27me3 levels at cold-responsive genes such as CBF1 in *Arabidopsis thaliana*. Repressive marks reduce through cold exposure to activate genes that promote cold tolerance which allows plants to produce defensive proteins and metabolites (Zhou et al. 2023a, b, c).

Drought Stress

The research on *Triticum aestivum* wheat plants reveals that drought stress causes an elevation of promoter marks H3K4me3 marks on genes which respond to drought conditions. Under water deficit conditions, the modified histone state brings about increased gene expression of functionally beneficial genes that conserve water while maintaining cellular equilibrium (Zhang et al. 2024a, b).

Salt Stress

Stress caused by salt in rice plants of the *Oryza sativa* species prompts acetylation mechanisms at gene sites that control ion transporters to establish regulatory systems for maintaining ion balance and osmotic regulation (Wang et al. 2023a, b, c, d).

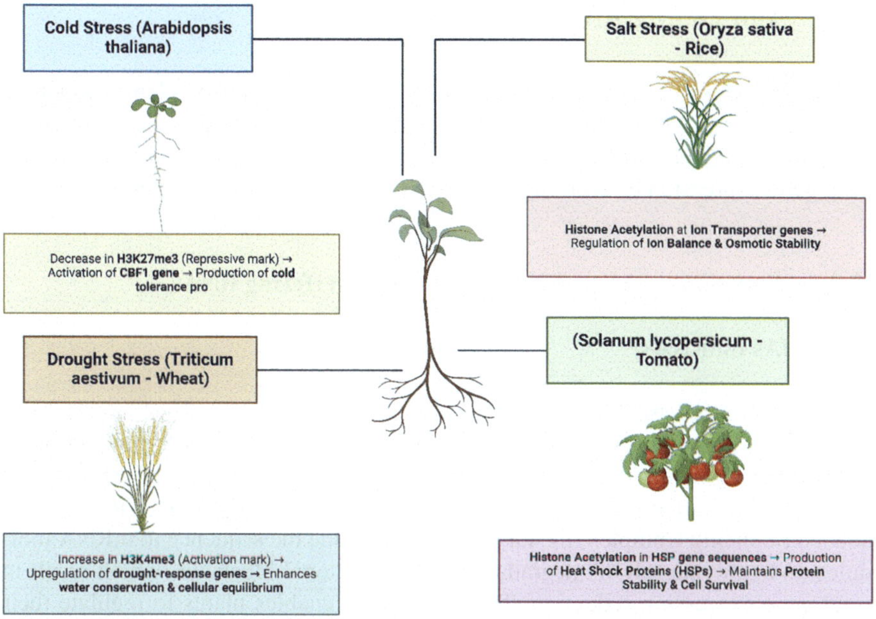

Fig. 5.2 Histone modifications in abiotic stress responses: plants adapt to environmental stresses by modifying histone marks to regulate gene expression. Cold and drought stress alter H3K27me3 and H3K4me3, respectively, while salt and heat stress induce histone acetylation. These modifications activate stress response genes, aiding survival through osmotic balance, water conservation, and protein stability

Heat Stress

Under heat stress conditions, tomato species (*Solanum lycopersicum*) increase histone acetylation patterns within their heat shock protein (HSP) coding sequences. During heat stress, tomato organisms (*Solanum lycopersicum*) show increased histone acetylation patterns across heat shock protein (HSP) gene sequences because proteins serve protective functions and maintain protein stability at high temperatures. High temperature environments enable proteins to protect essential functions as they stabilize protein stability (Alsamir et al. 2021). The proper response from gene expression in abiotic stress conditions requires effective histone modifications for regulation to function. Through modified chromatin structures, plants establish precise gene control as a fast mechanism to adapt when facing environmental challenges. Modern sequencing technology ChIP-seq together with CRISPR-dCas9 systems provide scientists better methods for studying histone modification to develop improved environmental stress resistance in crops.

5.2.3 *Non-coding RNAs*

Non-coding RNAs act as essential regulators of gene expression in plants when they experience abiotic stress conditions. Research on ncRNAs has focused on microRNAs (miRNAs) and small interfering RNAs (siRNAs) due to their key roles in managing gene expression by working at transcriptional and post-transcriptional stages.

5.2.3.1 Functions of MicroRNAs and Small Interfering RNAs

MicroRNAs (miRNAs)

miRNAs represent small RNA molecules with lengths between 21–24 nucleotides that function post-transcriptionally to manage gene expression. The production of miRNAs through Dicer-like enzymes occurs from precursor RNA transcripts which then integrate into the RNA-induced silencing complex (RISC). Within RNA-Induced Silencing Complex, the target miRNAs guide the sequence to mRNA molecules and generate mRNA degradation or translational blockades which result in target gene expression reduction. The mechanism enables plants to regulate their protein synthesis levels according to environmental triggers (Ma et al. 2023).

Small Interfering RNAs (siRNAs)

The 21–24 nucleotide RNA molecules, known as siRNAs, are derived from double-stranded RNA precursors and perform vital functions in maintaining genome stability. RNA-directed DNA methylation (RdDM) utilizes siRNAs as guides which direct silencing machinery to DNA sequences causing DNA methylation and transcriptional repression at target genes and transposable elements. Genome defense against potential transposon damage requires this essential process to control gene expression (Vaucheret and Voinnet 2024).

5.2.3.2 Regulation of Gene Expression via Chromatin Targeting

The regulatory mechanism of gene expression through ncRNAs depends on their control of chromatin structures and their abilities to access DNA:

miRNAs

The regulatory function of miRNAs goes beyond mRNA degradation because these nucleotide molecules also affect chromatin states indirectly. The miR398 in *Arabidopsis thaliana* targets messenger RNA molecules containing copper/zinc superoxide dismutase (CSD1 and CSD2) sequences which then become degraded. ROS levels receive regulatory control through this mechanism which then leads to modifications in stress response gene expression. Under stressful circumstances, miR398 controls homeostasis of Reactive Oxygen Species which subsequently modulates chromatin structure alongside gene expression patterns (Pawłasek et al. 2024).

siRNAs

SiRNAs within the RdDM pathway guide the silencing complex toward specific DNA sequences to establish cytosine methylation and histone modifications which create condensed chromatin and block transcriptional activity. DNA silencing and genome stability require this mechanism to control transposable elements and repetitive sequences. siRNA-mediated chromatin modifications operate to control stress-responsive genes through changes in their chromatin conditions (Vaucheret and Voinnet 2024).

5.2.3.3 Case Studies

Drought Stress

The expression of homeodomain-leucine zipper (HD-ZIP) transcription factors in *Z. mays* (maize) depends on miR166 regulation for proper drought response function. The plant employs miR166 to control transcription factors which enable adaptation through water conservation leading to improved drought resistance (Li et al. 2020).

Salt Stress

Rice plants (*Oryza sativa*) regulate auxin receptor genes through miR393 which causes their reduced expression. The suppressed auxin signaling pathways lead to increased salt tolerance according to this mechanism. The hormone signaling pathway becomes more adaptable due to miR393-based regulation which enables plants to handle salt stress conditions better (Haldar and Bandyopadhyay 2021). The examples demonstrate how ncRNAs play vital roles in plant stress responses which suggests their potential use in genetic engineering approaches to boost crop resistance.

5.2.3.4 Recent Advances

Recent studies have expanded our understanding of ncRNAs in plant stress responses:

Long Non-coding RNAs (lncRNAs)

The recent research shows that lncRNAs that exceed 200 nucleotides act as key regulators for gene expression during stressful conditions. The molecular scaffold function of lncRNAs enables them to bring proteins together with other RNAs for controlling chromatin arrangement while influencing gene expression patterns. Research continues to explore how particular lncRNAs control genes essential for drought along with salinity response although their specific regulatory processes remain unclear (Jin et al. 2024a, b).

Interaction Networks

The regulatory networks controlling gene expression in response to environmental stress develop through interactions between miRNAs, siRNAs, and lncRNAs. Investigating these regulatory networks remains essential to create stress tolerance improvement strategies for agricultural crops (Yu et al. 2019). The plant stress response machinery depends heavily on two non-coding RNAs known as miRNAs and siRNAs. Through multi-level gene expression control, these elements empower plants to adapt in troublesome environmental conditions. Research activities focused on ncRNA mechanisms and regulatory systems produce promising methods to boost crop resistance along with yields during climate change and stress conditions.

5.2.4 *Chromatin Remodeling*

The fundamental epigenetic mechanism of chromatin remodeling controls gene expression through modifications which affect how chromatin units called nucleosomes position and structure themselves around DNA molecules embedded in histone proteins (Fig. 5.3). Plants need the dynamic process of chromatin remodeling to regulate their genetic information because it determines their accessibility to DNA (Song et al. 2021).

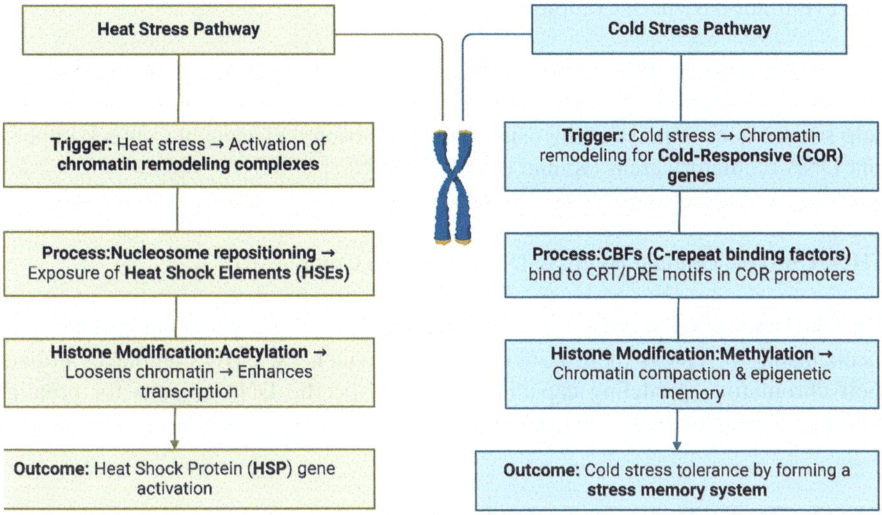

Fig. 5.3 Chromatin remodeling during extreme temperature conditions. Heat stress induces nucleosome repositioning and histone acetylation, exposing HSEs for HSF binding and HSP gene activation. Cold stress promotes CBF binding to COR gene promoters through histone methylation, aiding stress memory and adaptation

5.2.4.1 Mechanisms Altering Nucleosome Positioning

The energy from ATP hydrolysis enables chromatin remodeling complexes to relocate or remove, or restructure nucleosomes which control the accessibility of DNA for transcriptional proteins. The main families of plant chromatin complexes include:

SWI/SNF (Switch/Sucrose Non-fermentable) Complexes

According to scientific research, these nucleosome remodeling complexes use their biological function to reveal promoter regions by sliding or evicting nucleosomes, thus enabling transcription factors and RNA polymerase II binding. In Arabidopsis the SWI/SNF complex serves as an important component that regulates stress-responsive genes through structural manipulation of chromatin to enable transcriptional activation (Bhadouriya et al. 2021).

ISWI (Imitation Switch) Complexes

The main function of ISWI involves chromatin assembly and nucleosome organization which contributes to maintaining proper chromatin structure. These enzymes help silence gene expression by using their nucleosome placement to block important DNA regulatory areas (Kamei et al. 2021).

CHD (Chromo-Domain Helicase DNA-Binding) Complexes

The CHD complexes serve as regulators that activate specific genes but also develop mechanisms to repress their associated DNA sequences. CHD complexes utilize their chromatin remodeling capability to open specific DNA regions for protein binding, dependent on the environment and cellular framework (Guo and He 2024).

5.2.4.2 Role in Transcriptional Regulation Under Stress

The adaptation of plants to abiotic stress conditions depends on chromatin remodeling complexes for modifying gene expression patterns. During drought conditions, the SWI/SNF complex in Arabidopsis enhances the binding of transcription factors with ABA-responsive activities to gene promoters that are drought-sensitive (Wang et al. 2024).

Examples in Abiotic Stress Responses

Heat Stress

Research on *Zea mays* indicates HSP genes activate through chromatin remodeling mechanisms. Heat-stressed plants become thermotolerant when proteins act as molecular chaperones to stabilize and normalize denatured proteins (Thakur and Prasad 2024).

Salt Stress

The chromatin remodeling mechanisms activate genes responsible for ion transport in tomato (*Solanum lycopersicum*) plants to maintain proper ionic balance when exposed to high salt conditions. Activated genes allow plants to seize sodium ions and compartmentalize them which results in better salt tolerance responses (Sharma et al. 2022). Plants use chromatin remodeling to regulate gene expression during their defense against abiotic environmental stresses. Chromatin remodeling complexes position nucleosomes which either reveal or hide DNA regions to establish gene accessibility through transcriptional processes. The regulatory mechanism operates rapidly to activate stress-responsive genes which enable plants to adapt effectively when facing adverse environmental conditions (Fahad et al. 2024).

5.3 Epigenetic Regulation in Abiotic Stress Responses

Plants face multiple abiotic stress conditions which include drought stress alongside salt stress and extreme temperatures along with nutrient deficiencies and heavy metal toxicity as serious threats to their growth and productivity levels. Plants use different biological detection and environmental defense systems to monitor risks that epigenetic modifications help regulate stress-responsive genes. The epigenetic regulators DNA methylation alongside histone modifications and non-coding RNAs (ncRNAs) regulate chromatin structure to modify stress-related gene expression (Ahmad et al. 2022). This is further illustrated in the tables below, where Table 5.1 highlights small RNA-mediated regulation in plants under various stress conditions, while Table 5.2 focuses on epigenetic regulation in response to different environmental stimuli.

Table 5.1 Small RNA-mediated regulation of plants under different stress conditions

Stress type	Plant species	Regulation	Mechanism	References
Drought stress	*Zea mays* (maize)	miR166 regulates HD-ZIP transcription factors	Water conservation leads to improved drought resistance	Li et al. (2020)
Salt stress	*Oryza sativa* (Rice)	miR393 regulates auxin receptor genes, reducing their expression	Enhanced salt tolerance due to altered hormone signaling pathways	Haldar and Bandyopadhyay (2021)
Heat stress	*Arabidopsis thaliana*	miR156 regulates SPL (SQUAMOSA promoter-binding protein-like) genes	Enhanced thermotolerance through gene regulation	Xu et al. (2016)
Cold stress	*Solanum lycopersicum* (tomato)	miR398 regulates copper/zinc superoxide dismutase (CSD) genes	Increased cold tolerance due to oxidative stress management	Li et al. (2022)
Oxidative stress	*Arabidopsis thaliana*	miR398 regulates oxidative stress-related genes like CSDs	Improved defense against reactive oxygen species (ROS)	Sun et al. (2020)
Nitrogen stress	*Glycine max* (soybean)	miR169 regulates NF-Y (nuclear factor Y) transcription factors	Improved nitrogen use efficiency and stress tolerance	Ramesh et al. (2019)
UV stress	*Triticum aestivum* (wheat)	miR398 and miR408 regulate antioxidant defense genes	Enhanced UV tolerance through the regulation of antioxidant pathways	Islam et al. (2022)
Heavy metal stress	*Arabidopsis thaliana*	miR156 regulates heavy metal stress response genes	Enhanced tolerance to heavy metal toxicity	Zhang et al. (2019)

Table 5.2 Epigenetic regulation of plants under different environmental stimuli

Stress type	Plant species	Epigenetic response	Functional impact	References
Heat stress	*Solanum lycopersicum* (Tomato)	Distinct methylation patterns suppress unnecessary pathways	Resource allocation to heat shock proteins (HSPs) for cellular protection	Singh et al. (2021)
Heavy metal stress	*Brassica juncea* (Indian mustard)	Cadmium exposure leads to hypermethylation of growth-related genes	Redirects metabolism to detoxification via phytochelatins and glutathione	Al-Khayri et al. (2023)
Cold stress	*Triticum aestivum* (Wheat)	Hypomethylation at cold-responsive transcription factor (CBF) genes	Enhances production of antifreeze proteins and osmoprotectants	Zhou et al., (2023a, b, c)
Drought stress	*Oryza sativa* (Rice)	DNA methylation changes in drought-responsive genes	Activation of genes involved in osmotic adjustment and root growth	Kim et al. (2020)
Salt stress	*Arabidopsis thaliana* (Arabidopsis)	Histone modifications lead to changes in ion transport genes	Enhances ion homeostasis and salt tolerance	Pang et al. (2010)
Oxidative stress	*Zea mays* (Corn)	Upregulation of antioxidant-related genes through DNA methylation	Increased production of ROS scavengers like superoxide dismutase (SOD)	Yang et al. (2023)
Flooding stress	*Glycine max* (Soybean)	Differential methylation of genes involved in anoxic conditions	Enhanced adaptation to low oxygen levels by activating anaerobic pathways	Yu et al. (2024)
UV-B stress	*Spinacia oleracea* (Spinach)	Changes in histone acetylation and DNA methylation at UV-responsive genes	Enhanced DNA repair mechanisms and production of UV-absorbing compounds	Kasım and Kasım (2017)
Nutrient deficiency	*Cucumis sativus* (Cucumber)	Methylation of nutrient-responsive pathways	Altered root architecture and enhanced nutrient uptake	Rao et al. (2020)
Mechanical stress	*Arabidopsis thaliana* (Arabidopsis)	Altered histone modifications in cell wall biosynthesis genes	Strengthened cell wall integrity and reduced susceptibility to further damage	Wang et al. (2020)

Plants use epigenetic control to develop stress memory across multiple identical stress events, thus improving their defensive capabilities. The regulatory mechanism allows two-way inheritance by transferring stress adaptation properties from generation to generation including both the present and future generations. Understanding these mechanisms enables researchers to develop resilient crops using methods that combine epigenetic breeding with genome editing.

Research shows that plant reactions to abiotic stress conditions heavily rely on epigenetic modifications. DNA methylation patterns allow plants to detect heat stress and to survive dryness plus salty environmental conditions. Environmental stress in plants triggers histone modifications for regulating gene expression levels that determine their adaptability and survival capacity (Chang et al. 2020) (Thakur and Prasad 2024). DNA chemical code changes arising from stress in plants lead to transgenerational epigenetic inheritance over successive generations. This process enables organisms to handle similar environmental challenges more effectively by acquiring proper adaptations for those conditions. Analysis of epigenetic processes helps scientists understand how plants adjust to environmental stress conditions and enables them to develop resilient crop varieties (Fitz-James and Cavalli 2022).

5.3.1 Drought Stress

Plant development and growth and productivity suffer negative effects because of drought stress which functions as a fundamental environmental challenge. Plants employ sophisticated epigenetic control systems that regulate DNA methylation to manage dry conditions by using this mechanism to control how their genetic programs are expressed. The enzymes DNMTs use DNA methylation epigenetic modifications to add methyl groups at CpG dinucleotide sites after cytosine residues for the process (Seem et al. 2024). DNA methylation processes directly impact how chromatin folds along with gene transcription levels while using stress-responsive genes differently based on the situation.

5.3.1.1 Role of DNA Methylation in Drought Stress Response

During drought stress, plants use DNA methylation modifications through hypermethylation processes that silence genes and hypomethylation processes that activate genes which control important drought tolerance pathways. Physiological adaptations become possible through gene-modulating changes which allow for these results:

Osmoprotection and Water Retention

The regulatory functions of DNA methylation operate directly on genes responsible for osmotic adjustment which include elements that produce proline biosynthetic products and manage sugar metabolic activities. When plants experience drought conditions, they build proline levels as osmoprotectant molecules which protect cell water content. The regulation of proline biosynthesis genes through methylation modification including P5CS responsible for encoding Δ^1-pyrroline-5-carboxylate synthetase has been documented for its relationship with plant drought tolerance (Sellamuthu et al. 2024).

Abscisic Acid (ABA) Signaling Pathway

ABA functions as a central plant phytohormone that affects stomatal responses to drought stress while also controlling gene expression that helps plants adapt to stress. The changes in DNA methylation status of ABA-responsive NCED3 (9-cis-epoxycarotenoid dioxygenase) genes lead to drought adaptation. Research shows that ABA signaling gene promoters experience shifting DNA methylation patterns which regulates their expression levels while optimizing the drought response capabilities (Liu et al. 2024).

Drought-Inducible Transcription Factors (TFs)

Multiple transcription factors, including DREB (Dehydration-Responsive Element-Binding) proteins, WRKY, NAC, and MYB, play essential roles in drought stress adaptation. Promoters that control transcription factors undergo DNA methylation alterations which steer their expression activity toward influencing downstream protection genes against stress. Drought tolerance improves through the hypomethylation of DREB2A because it activates transcription (Sadau et al. 2024).

5.3.1.2 DNA Demethylation and Stress Memory

The process of stress memory in plants occurs through epigenetic modifications that remain stable after initial exposures to drought. DNA demethylation operates as a crucial process through demethylases including ROS1, DML2, and DML3 enzymes to sustain gene expression patterns triggered by stress (Yin et al. 2024). Plants use an epigenetic priming mechanism to remember their earlier drought experiences as this improves their quick and strong responses when drought occurs again.

5.3.1.3 Chromatin Remodeling and Histone Modifications in Drought Stress

DNA methylation in plants functions alongside additional epigenetic modifications which include histone methylation as well as acetylation and chromatin remodeling in drought stress adaptation. The transcription of drought-related genes becomes controlled by specific histone modifications where H3K4me3 trimethylation activates genes but H3K9me2 dimethylation silences them. Plants use DNA methylation together with histone modifications to build more precise stress adaptation programs (Rao et al. 2024).

Plants use DNA methylation as an important epigenetic adaptation tool to adapt to drought through dynamic changes in gene methylation and demethylation patterns. The modifications activate vital physiological pathways to achieve an efficient water scarcity response by controlling ABA signaling and osmoregulation functions with transcription factor activities. Plants develop a better capacity to resist future drought stress due to their ability to form epigenetic stress memory mechanisms. The knowledge of these molecular mechanisms leads scientists toward the creation of drought-resistant crops through precise epigenetic alterations.

5.3.2 Salinity Stress

The survival of plants under high salt stress suffers from major setbacks due to improper ion regulation and detrimental osmotic imbalance. Plants have developed advanced defense systems against high salinity through epigenetic modifications which direct essential salt tolerance gene expression. Histone modifications and small RNA-regulated pathways function as vital elements which control both ion transport mechanisms and osmotic pressure balance in saline environments (Balasubramaniam et al. 2023).

5.3.2.1 Histone Modifications in Salt Tolerance Mechanisms

Environmental stress responses to salinity become regulated through histone modifications such as acetylation and methylation and phosphorylation and ubiquitination which control chromatin structure and gene accessibility. Studies have identified some histone modifications that activate or deactivate salt-responsive genetic elements.

The process of histone acetylation primarily occurs through adding acetyl groups to lysine histone residues to create an accessible chromatin state that enhances transcription. Research shows that rice plants (*Oryza sativa*) exhibit increased acetylation of histone proteins within promoter areas of salt-responsive genes which leads to better gene expression during salt stress scenarios. Specific gene loci recruit histone acetyltransferases (HATs) to initiate this modification process when plants experience saline environments (Kumar et al. 2021).

Histone Methylation

The activation or repression of gene expression depends on which specific amino acid residue and how severely it gets methylated. The activation of genes through histone modification occurs when H3K4me3 trimethylation takes place but H3K27me3 trimethylation leads to gene repression. Scientific research on Arabidopsis thaliana plants shows that salt stress triggers reduced H3K27me3 levels at promoters of salt-specific genes which leads to promoter activation. The addition of H3K27me3 at selected sites promotes gene silencing for genes that create problems in high salinity conditions (Wang et al. 2023a, b, c, d).

These histone modifications are dynamic and can be reversed by histone deacetylases (HDACs) and demethylases, allowing plants to fine-tune gene expression in response to fluctuating salinity levels. The precise regulation of these modifications ensures that plants can promptly activate defense mechanisms while conserving energy by repressing unnecessary gene activity (Wang et al. 2023a, b, c, d).

5.3.2.2 Small RNA-Mediated Regulation of Ion Transport and Osmotic Balance

Small RNAs including microRNAs and small interfering RNAs play a vital role as adaptive components during plant stress responses to saline conditions. Non-coding RNAs guide RNA-induced silencing complex (RISC) to mRNA targets that lead to mRNA cleavage or translation interference following transcription.

miR319a

Research has shown that miR319a operates as a vital regulator that helps poplar (Populus species) plants survive salt stress conditions. The enhanced salt tolerance in transgenic poplar plants resulted from overexpressing miR319a through gene modifications that control ion transport and osmotic adjustment mechanisms. MiR319a attaches to TCP transcription factor mRNA sequences and regulates transcription factors that control genes related to stress response mechanisms. The regulatory network demonstrates that miR319a controls both ion equilibrium and osmotic regulation during saline stress exposure (Cheng et al. 2024).

miR398

The plant uses miR398 as a key miRNA to mitigate salinity stress. Natural conditions cause miR398 to silence copper/zinc superoxide dismutase (CSD) gene expression because these genes help plants detoxify reactive oxygen species (ROS). The salt stress triggers a decrease in miR398 expression which allows CSD proteins to accumulate to minimize ROS damage from salt-induced oxidative stress. The adjustment supports cellular redox homeostasis and protects plants from oxidative damage that occurs during exposure to high salinity conditions (Li et al. 2024).

Research demonstrates that siRNAs employ DNA methylation to control specific genomic regions that affect the expression of Na + and K+ transporter genes. Plants employ epigenetic control mechanisms that properly regulate the expression of ion channels and transporters to preserve plant cell osmotic pressure alongside ionic balance during salt stress (Martins and Law 2023).

Plants allow perception and response to salinity stress by utilizing histone modifications along with small RNA-mediated pathways. These epigenetic mechanisms regulate chromatin structure and gene expression to keep plants balanced with their ions and osmotic pressure which results in greater salt stress tolerance.

5.3.3 Temperature Stress

Plants face considerable growth and productivity problems due to extreme temperatures which include both hot and cold environmental conditions (Nawaz et al. 2025). Plants have developed complex mechanisms to detect and react to extreme temperatures through epigenetic regulation which controls gene expression without modifying DNA base sequences. The activation of heat shock proteins (HSPs) and cold-responsive genes and heat shock proteins is controlled by three main epigenetic mechanisms: DNA methylation in combination with histone modifications and chromatin remodeling processes (Sadura and Janeczko 2024).

5.3.3.1 Epigenetic Regulation in Heat and Cold Stress Responses

Plants use epigenetic modification mechanisms to alter gene expression quickly in response to temperature changes. Plants use reversible epigenetic modifications to respond quickly to environmental signals and form stress memories in addition to short-term adaptations.

DNA Methylation

Plants use DNA methyltransferases to place methylation marks on cytosine bases found within CG, CHG, and CHH DNA patterns. Temperature stress causes modifications in DNA methylation patterns that result in the activation or repression of stress-responsive genes. DNA methylation patterns change when plants experience cold stress which controls the expression of cold tolerance genes (Banerjee et al. 2017).

Histone Modifications

Post-translational modifications through acetylation and methylation alongside phosphorylation and ubiquitination control how chromatin structures affect gene expression. Studies during heat stress conditions have shown that enhanced histone acetylation at gene loci activates the expression of HSPs that stabilize proteins and protect them from damage. Conversely, histone deacetylation can repress gene expression, contributing to the fine-tuning of stress responses (Jin et al. 2024a).

5.3.3.2 Activation of Heat Shock Proteins and Cold-Responsive Genes

The activation of HSPs and cold-responsive genes is central to plant survival under extreme temperatures.

Heat Shock Proteins (HSPs)

The HSPs operate as biological chaperones to correctly fold newly synthesized peptides and stop protein misfolds and aggregation while plants encounter heat stress. Under normal conditions the transcription of HSPs is controlled by heat shock transcription factors (HSFs) that bind to heat shock elements (HSEs) present in HSP gene promoters to start the transcriptional process. Histone acetylation creates accessibility for HSFs to reach HSEs and enhances the expression of HSPs (Boopathy et al. 2022).

Cold-Responsive Genes

During cold stress exposure, plants activate genes which enhance their ability to acclimate to this environment. CBFs function as key regulators that activate COR gene expression through binding their promoters at CRT/DRE control regions. The expression of CBFs and COR genes is controlled by epigenetic mechanisms that include histone methylation and chromatin remodeling processes leading to cold temperature adaptation in plants (Banerjee et al. 2017).

5.3.3.3 Chromatin Remodeling During Extreme Temperature Conditions

The process known as chromatin remodeling changes chromosomal structures dynamically to enable transcriptional machinery components to reach DNA where they can turn genes on or off.

Heat Stress

Heat stress triggers chromatin remodeling complexes to shift nucleosome positions and consequently expose HSEs for HSP gene promoter accessibility by HSFs. The histone modifications include acetylation that makes nucleosomes unstable and improves transcriptional initiation potential (Boopathy et al. 2022).

Cold Stress

In cold conditions, chromatin remodeling facilitates the binding of CBFs to CRT/DRE motifs in COR gene promoters. The process involves histone methylation because it controls chromatin compaction and determines the accessibility of genes. The modified epigenetic patterns from cold stress exposures help plants create a memory system which enhances their reaction to later cold temperatures (Banerjee et al. 2017).

The mechanisms of DNA methylation together with histone modifications and chromatin remodeling function as crucial regulators for plant temperature stress responses. Plants make adaptations through these modifications to activate their heat shock proteins and cold-responsive genes which enable them to survive temperature extremes.

5.3.4 Nutrient Deficiency

Plant growth and development face significant reduction when plants lack essential macronutrients such as nitrogen (N), phosphorus (P), and potassium (K). Plants possess advanced sensing and response systems that detect nutrient shortages, while epigenetic regulation significantly controls the expression of genes involved in acquisition and metabolic processes for nutrients (Fig. 5.4; Table 5.3). Gene activity together with chromatin structure control through the mechanisms of DNA methylation and histone modifications and non-coding RNAs responds to changes in nutrient availability (Khan et al. 2023).

5.3.4.1 Epigenetic Mechanisms in Nutrient Uptake and Metabolism

DNA Methylation

The DNA modification includes the attachment of methyl groups to cytosine bases, leading to gene suppression. The DNA methylation patterns of plants modify according to their nutritional state to control genes which enable nutrient extraction and processing. The genes that transport and assimilate nitrogen undergo DNA methylation modifications when plants experience nitrogen deficiency (Zhang et al. 2023).

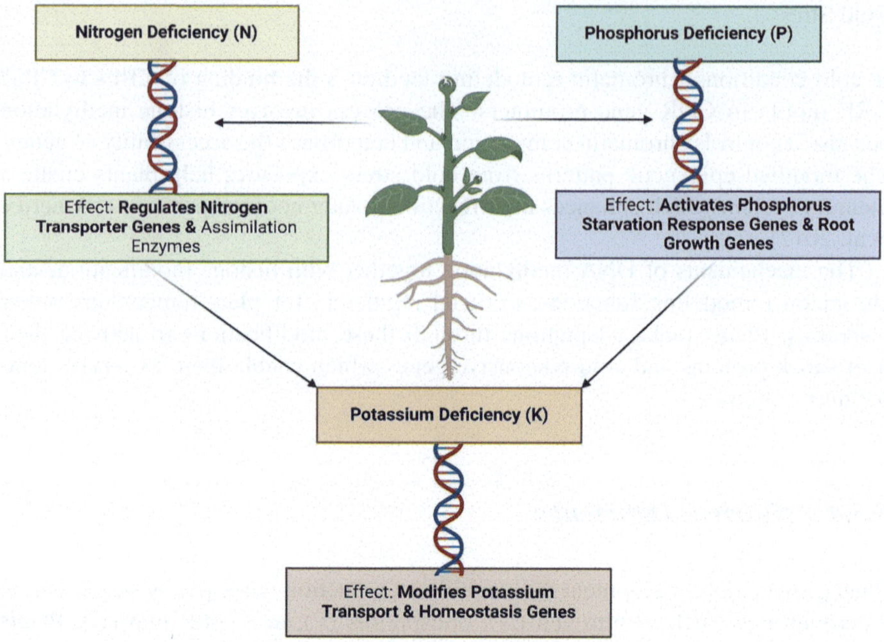

Fig. 5.4 Epigenetic regulation of plant responses to nitrogen, phosphorus, and potassium deficiencies: nutrient limitations trigger DNA methylation and histone modifications, influencing gene expression to optimize nutrient uptake, root development, and cellular homeostasis

Histone Modifications

Histone protein post-translational modifications through acetylation and methylation control the accessibility of chromatin structures and gene expression regulation. Plants modify histones during phosphorus limitation to activate key starvation response genes which improve their phosphorus adaptation (Séré and Martin 2020).

Non-coding RNAs

Small RNAs particularly microRNAs (miRNAs) operate as essential regulators of gene expression during post-transcriptional stages. Plants increase the expression of specific microRNAs which then target gene transcripts related to nutrient uptake and homeostasis for precision control during nutrient stress (Abdulraheem et al. 2024).

Table 5.3 Epigenetic regulation of plants under various nutrients

Nutrient	Epigenetic regulation	Gene	Function	References
N	H3K27me3 enrichment	*AtHNI9*	Involved in the deposition of H3K27me3 at the NRT2.1 locus in response to high N supply	Widiez et al. (2011)
	H3K27me3 enrichment	*NRT2.1*	Undetermined	Bellegarde et al. (2018)
	H3K4me3 enrichment	Detoxification genes	Primary root growth, ROS accumulation	Bellegarde et al. (2019)
	lncRNA expression	*NRT2.4*	Undetermined	Fukuda et al. (2019)
P	H2A.Z incorporation	*AT4, SPX1, BMY1, SRG3, ASK11*	Shorter primary root, enhanced root hair development, reduced pi content	Smith et al. (2010)
	H2A-H2A.Z exchange	*ACP5, RNS1, PHT1*	Reduced pi content	Iglesias et al. (2013)
	Reading H3K4me2 and H3K4me3	*NPC4, SQD2, PS2, ETC1*	Impaired root hair development, shorter primary root, reduced pi content	Chandrika et al. (2013a, b)
	H3K4me3/ H3K27me3 enrichment	*AT4, SPX3*	Shorter primary root, enhanced root hair development	Yen et al. (2017)
	Histone deacetylation	*SPX3, SPX1*	Impaired root hair development, longer epidermal cells, reduced P content	Chen et al. (2015)
	Histone deacetylation	*ALMT1, LPR1, LPR2, PT2*	Altered remodeling of root system architecture	Xu et al. (2019)
	Histone acetylation	*AT4, WRKY6, SBT3.5, RIPK*	Reduced pi content	Wang et al. (2019)
	DNA methylation	*SPX2, SPX1*	Undetermined	Secco et al. (2015)
	H3K4me3	*AtAL6*	Affects transcript maturation and stability of critical genes involved in root hair elongation	Chandrika et al. (2013a, b)
	H2A.Z	*IPK1*	Involved in the transcriptional regulation of some pi starvation-responsive genes	Zhang et al. (2022)
Fe	H4R3me2 enrichment	*FRO2, IRT1, BHLH38, BHLH39, BHLH100, BHLH101*	Higher chlorophyll content	Sasmal et al. (2025)
	H3K27me3 enrichment	*FRO2, IRT1, FIT*	Longer primary root	Park et al. (2019)
	Histone acetylation/ deacetylation	*FRD3, EXO70H2, MLP329, CRK25, BOR1*	Impaired translocation from roots to shoots	Xing et al. (2015)
	H4R3sme2	*AtPRMT5*	Negatively regulates iron homeostasis, via regulation of Ib subgroup bHLH genes	Fan et al. (2014)

5.3.4.2 Focus on Nitrogen, Phosphorus, and Potassium Deficiencies

Nitrogen Deficiency

Amino acid synthesis and nucleotide synthesis require the presence of nitrogen. Plants transform their DNA methylation and histone modifications during nitrogen deficiency which controls nitrogen transporter gene expression in addition to enzymes responsible for nitrogen assimilation to maximize nitrogen efficiency (Zhang et al. 2023).

Phosphorus Deficiency

The molecule phosphorus functions as an essential element for generating ATP and nucleic acids. Under phosphorus limitation conditions, plants activate phosphorus starvation response genes through histone modifications to enhance the development of roots and expression of phosphorus transporters for better phosphorus acquisition (Séré and Martin 2020).

Potassium Deficiency

Plants need potassium for enzyme activation of vital processes and managing water balance in their systems. Cellular functions under low potassium conditions benefit from epigenetic changes which modify the expression of potassium transport and homeostasis genes (Réthoré et al. 2021).

5.3.4.3 Case Studies of Nutrient-Efficient Plants

Rice (*Oryza sativa*)

Research shows that different varieties of rice possess various epigenetic changes which boost their ability to use nitrogen resources efficiently. Plant genes that transport nitrogen become more active because of genetic modifications which enable plants to grow better in environments with low nitrogen content (Zhang et al. 2023).

Arabidopsis thaliana

Studies of *Arabidopsis* demonstrate that phosphorus deficiency response in plants heavily depends on histone modifications. Histone modification variations among mutants affect their phosphorus acquisition efficiency which demonstrates epigenetic regulation plays a key role in nutrient uptake (Séré and Martin 2020).

The study of epigenetic mechanisms creates important knowledge about plant nutrient deficiency responses and offers opportunities for breeding nutrient-efficient crops utilizing genome editing and epigenetic breeding technologies.

5.3.5 Heavy Metal Stress

Plants experience growth limitations due to heavy metal stress which occurs from environmental accumulation of toxic metals including Pb, Cd, and As. Plants utilize complex adaptation systems to counter high metal toxicity through epigenetic modifications that control heavy metal-related gene expression functions, especially metal uptake and transport while executing detoxification processes. DNA methylation and histone modifications together with non-coding RNAs (ncRNAs) function as major epigenetic mechanisms to change chromatin structure alongside gene expression levels when plants experience heavy metal exposure (Kosakivska et al. 2021).

5.3.5.1 Epigenetic Modifications in Response to Heavy Metal Toxicity

DNA Methylation

Methyl groups added to DNA cytosine residues through this process generally produce gene repression. DNA methylation patterns undergo changes when exposed to heavy metals which results in gene expression alterations for metal transport and detoxification processes. Research findings demonstrate that heavy metal exposure activates two opposite DNA modification patterns at specific genomic locations which changes gene expression for metal toxicity protection (Fasani et al. 2023).

Histone Modifications

Histone proteins undergo post-translational transformations through acetylation and methylation processes which determine chromatin accessibility together with gene expression dynamics. Histone modification patterns modified by heavy metal stress result in gene activation or repression which controls metal homeostasis mechanisms. The modifications have two potential outcomes that alter chromatin structure, so they modify transcriptional responses to heavy metals (Fasani et al. 2023).

5.3.5.2 Regulation of Metal Transporters and Detoxification Pathways

Plants employ specific transporters and detoxification pathways to manage heavy metal uptake and mitigate toxicity:

Metal Transporters

Plants utilize three types of proteins including Natural Resistance-Associated Macrophage Proteins (NRAMPs), Heavy Metal ATPases (HMAs), and Zinc/Iron-Regulated Transporter-Like Proteins (ZIPs) for heavy metal uptake and distribution. The expression levels of metal transporters depend on epigenetic modifications which allow plants to regulate heavy metal distribution patterns in different tissues (Fasani et al. 2023).

Detoxification Pathways

Plants detoxify heavy metals through two sequential steps where phytochelatins and metallothioneins combine with heavy metals before transferring them into vacuoles. Plants use gene modifications to control the activity of detoxification agents which leads to stable metal stress responses that prevent damage to the cell (Fasani et al. 2023).

5.3.5.3 Role of Non-coding RNAs in Mitigating Heavy Metal Stress

Non-coding RNAs, including microRNAs (miRNAs) and long non-coding RNAs (lncRNAs), play significant roles in plant responses to heavy metal stress:

MicroRNAs (miRNAs)

MicroRNAs (miRNAs) operate as small RNAs which adjust gene expression by causing mRNA destruction or preventing translation. Under heavy metal stress conditions miR398 regulates oxidative stress response through its modulation of target gene expression (Ma et al. 2023).

Long Non-coding RNAs (lncRNAs)

Scientific evidence suggests that Long Non-Coding RNAs (lncRNAs) function as regulators of gene expression during heavy metal stress. Plants exposed to cadmium reveal certain lncRNAs that seem to control the expression of genes responsible for metal transport and detoxification processes (Feng et al. 2016).

5.3.5.4 Case Studies of Heavy Metal-Tolerant Plants

Research has identified several plant species that exhibit tolerance to heavy metal stress through epigenetic mechanisms:

Arabidopsis thaliana

Scientists demonstrated that heavy metal exposure causes epigenetic changes in Arabidopsis thaliana which affect the expression of genes that detoxify metal substances and manage homeostasis. The plant gains tolerance and adaptation ability toward heavy metal pollution through these genetic modifications (Cong et al. 2019).

Medicago truncatula

The model legume *Medicago truncatula* reveals identification of lncRNA groups which respond to aluminum stress, suggesting these ncRNAs play a role in heavy metal tolerance systems. The research demonstrates that epigenetic controls represent a promising method to improve heavy metal tolerance in leguminous plants (Gui et al. 2022).

Knowledge of these epigenetic processes helps scientists develop techniques to create heavy metal-resistant crop types by using epigenetic manipulation and biotechnological practices.

5.4 Epigenetic Memory and Transgenerational Inheritance

As stationary organisms, plants have evolved complex systems to detect environmental stress situations and respond effectively and even remember previous encounters. Plants establish stress memory through their ability to remember previous stress exposures which permits faster and improved responses to future encounters with similar stress. The essential components of this phenomenon utilize epigenetic modifications which constitute heritable gene expression changes that occur without DNA sequence modifications. Plants maintain inherited modifications in their DNA which allow them to adapt to changing conditions across generations (Rendina González et al. 2018).

5.4.1 Stress Memory in Plants

Plants have a built-in ability to retain knowledge about past environmental situations in order to create more robust responses in future situations. The memory operates on two-time scales: short-term impact on current individual plant behavior as well as long-term effects that transfer across generations. Plants acquire stress memory through epigenetic mechanisms that involve DNA methylation and histone modifications as well as non-coding RNA regulation. A plant exposed to abiotic stresses like drought or salinity undergoes epigenetic modifications which create improved tolerance whenever the same stress occurs again (Iwasaki and Paszkowski 2014).

5.4.2 *Epigenetic Marks Across Generations*

Epigenetic marks represent chemical alterations of DNA or histone proteins which regulate gene expression but do not modify genetic DNA sequences. Key epigenetic marks that determine stress memory include:

5.4.2.1 DNA Methylation

The addition of methyl groups to DNA cytosine residues generates gene suppression through DNA Methylation. DNA methylation changes triggered by stress persist across generations while enabling organisms to pass beneficial traits to their offspring. Studies have demonstrated that organisms subjected to cold events develop DNA methylation patterns which then transmit to their descendants, thus improving their cold tolerance (Ramakrishnan et al. 2022).

5.4.2.2 Histone Modifications

Through histone modifications, the post-translational changes in histone proteins lead to modifications in chromatin structure which determines gene accessibility. Biological modifications can persist during cellular replication as well as between generations which influences how stress-responsive genes function in offspring (Lämke and Bäurle 2017).

5.4.2.3 Non-coding RNAs

Small RNAs of the microRNA and small interfering RNA types serve as regulatory agents which direct the modification of DNA methylation patterns as well as histone modifications to control gene expression. RNA molecules play a role in creating epigenetic states which transmit across generations to produce transgenerational stress memory (Bure et al. 2022).

5.4.3 *Role of Transgenerational Epigenetic Inheritance in Abiotic Stress Adaptation*

The process of transgenerational epigenetic inheritance enables plants to pass environmental knowledge to their offspring which helps them better handle future conditions. Plant populations obtain better adaptive capabilities in uncertain environments through this mechanism. The next generation of plants inherits epigenetic changes from drought-stressed progenitors which develop advanced drought resistance mechanisms. These heritable epigenetic modifications allow populations to adapt quickly through mechanisms that bypass genetic changes, thus offering organisms environmental flexibility during stressful conditions (Hauser et al. 2011).

5.4.4 Examples from Crops and Model Plants

Several studies have documented transgenerational epigenetic inheritance in both model plants and crops:

5.4.4.1 *Arabidopsis thaliana*

Arabidopsis thaliana plants subjected to different abiotic stress factors such as heat and salinity develop genomic instability which results in elevated homologous recombination frequencies throughout the plant population. Environmental stress causes epigenetic changes that transmit across generations because these inherited modifications increase the subsequent generation's stress resistance capabilities (Quadrana and Colot 2016).

5.4.4.2 Rice (*Oryza sativa*)

Epigenetic modifications regulate stress response genes in rice plants (*Oryza sativa*). The DNA methylation patterns which form during drought stress can transmit from parent to offspring and create descendants which show improved drought protection abilities (Lämke and Bäurle 2017).

5.4.4.3 Wheat (*Triticum aestivum*)

Exposure to heat stress in wheat (*Triticum aestivum*) causes DNA methylation pattern alterations that transfer to future generations. The inherited changes help future generations become more tolerant of heat conditions (Xu et al. 2021).

Transgenerational epigenetic inheritance is essential for understanding how plants adapt to abiotic stresses, as demonstrated by these specific examples. Studying epigenetic mechanisms is valuable for developing stress-resilient crops, as it facilitates both epigenetic breeding and biotechnological approaches.

5.5 Role of Epigenetics in Crop Improvement for Stress Resilience

Abiotic stresses that include drought and salinity together with extreme temperatures occur more frequently with growing severity, thus creating substantial difficulties for global agricultural production. Traditional breeding approaches successfully developed stress-tolerant crops, yet they require extended periods of time and do not maximize plant's adaptive capacity. Changing gene expression patterns through heritable epigenetic alterations provides future possibilities for stress-resistant crops because these modifications occur without modifying DNA sequences. Research into epigenetic mechanisms enables scientists to create new crop improvement strategies.

5.5.1 Epigenetic Breeding Strategies

The purpose of epigenetic breeding includes targeted modifications of epigenetic marks to develop beneficial traits in crop plants. Traditional breeding and genetic engineering methods gain additional potential through this approach for developing stress-resilient plant varieties.

5.5.1.1 Use of Epimutants for Stress Tolerance

Plants with epimutant characteristics inherit variations of gene expression through epigenetic modifications rather than DNA sequence alterations. Plant traits like stress tolerance together with flowering time and yield receive different effects from these epigenetic variances (Varotto et al. 2020).

5.5.1.2 Identification and Utilization

More resilient lines for breeding programs emerge from screening natural populations or using artificial methods to generate epimutants which demonstrate stress tolerance. Studies demonstrate that drought-tolerant traits in rice stem from DNA methylation pattern alterations which could be used to develop resilient cultivation lines (Varotto et al. 2020).

5.5.1.3 Stability and Inheritance

The utilization of epimutants requires thorough evaluation of both the stability properties and the heritability traits of epigenetic modifications. Breeding programs should evaluate both the stability and potential return of epigenetic changes which span across different generations (Varotto et al. 2020).

5.5.1.4 CRISPR-Based Epigenome Editing for Improved Stress Responses

Genome editing technologies, particularly CRISPR/Cas systems, now enable researchers to modify the epigenome precisely without altering DNA sequences. Epigenome editing methods provide researchers the ability to control specific gene expression patterns which enhances stress reaction capabilities.

Mechanism

Scientists achieve targeted epigenetic mark modifications in specific genomic regions by linking dCas9 proteins with DNA methyltransferases and histone acetyltransferases epigenetic modifiers. The strategic method targets particular genes to modify their stress response activities in order to enhance tolerance (Zhang et al. 2024a, b).

Applications

Scientists have utilized epigenome editing for investigating modulations of tolerance-related genes. Scientists improve plant adaptability by enhancing expression of stress-responsive genes by directing histone modification to their promoter regions (Zhang et al. 2024a, b).

5.5.1.5 Challenges and Future Prospects in Agricultural Biotechnology

Various barriers exist which need attention before these strategies can achieve their maximum potential:

Epigenetic Variation

The identification of both permanent inheritable epigenetic markers and temporary modifications remains essential for successful breeding efforts. Studying epigenetic landscapes throughout different environmental factors helps identify valuable epimutations (Khan et al. 2021).

Regulatory Considerations

The regulatory system governing crops improved through epigenetic alterations remains in the development phase. The adoption of epigenetic breeding technologies in agriculture needs specific guidelines for implementation purposes (Khan et al. 2021).

Integration with Traditional Breeding

The merger of epigenetic strategies with traditional breeding methods together with genomic selection strengthens the effectiveness of generating stress-tolerant crop varieties. The most effective approach would involve integrating genetic and epigenetic elements (Khan et al. 2021).

Future investigations must focus on explaining how epigenetic mechanisms control stress responses along with creating stable epimutation selection methods and establishing application frameworks for epigenetic modifications in agricultural improvement.

5.5.2 Epigenetic-Based Priming for Stress Tolerance

Plants exposed to initial stimuli through priming will react more swiftly and with enhanced strength when future stress situations occur. The application of epigenetic-based priming utilizes various agents which trigger epigenetic modifications that produce enhanced stress tolerance in plants.

5.5.2.1 Use of Chemical Epigenetic Modulators

Epigenetic modification through chemical agents enables plants to acquire stress tolerance. The DNA methylation inhibitor 5-azacytidine serves as a chemical agent in this process.

Mechanism

The incorporation of 5-azacytidine into DNA replication leads to enzyme inhibition of DNA methyltransferases which results in hypomethylation that might reactivate important stress response genes.

Applications

Research showed that using 5-azacytidine enhances drought resistance in maize plants through the activation of genes that respond to stress. The outcomes of treatment prove unpredictable because investigators must optimize their procedures carefully (Kovalchuk 2023).

5.5.2.2 Field Applications and Practical Implications

Agricultural implementation of epigenetic-based priming strategies needs assessment of various essential elements.

Delivery Methods

Proper delivery systems must be developed for agricultural application of epigenetic modulators because these compounds need to be delivered effectively to crops in the field. Researchers should conduct a full investigation to determine both seed priming techniques and soil and foliar spraying methods alongside finding their best dosage schedules (Ganie et al. 2024).

Environmental Impact

The complete evaluation of environmental consequences remains fundamental for chemical modulator sustainability in agricultural fields. It is vital to measure the rate of breakdown and harmful effects of these agents and their unintended impacts (Ganie et al. 2024).

Crop Specificity

Specific epigenetic priming agents affect different plant species differently due to their varying crop sensitivity. Treatment effectiveness increases when manufacturers develop medicine specifically for particular species and cultivars (Ganie et al. 2024).

The practical use of chemical epigenetic modulators in agriculture remains in early experimental trial phases. The full understanding of long-lasting effects and proper application guidelines, and possible safety issues requires more scientific investigation.

5.6 Tools and Methods for Studying Epigenetic Changes

Modern molecular biological techniques offer essential tools to analyze epigenetic modifications which occur in plants:

5.6.1 Bisulfite Sequencing

Bisulfite Sequencing stands as the premier approach for identifying DNA methylation patterns. The technique transforms unmethylated cytosines into uracil but does not modify methylated cytosines, allowing researchers to observe methylation patterns precisely at single-base resolution (Li et al. 2023).

5.6.2 Chromatin Immunoprecipitation Sequencing (ChIP-Seq)

The ChIP-seq technique uses bound antibodies that recognize transcription factors along with modified histone proteins to identify DNA–protein interactions and histone modifications. This methodology helps researchers decipher changes in chromatin structure as well as regulatory processes (Zhao et al. 2024).

5.6.3 CRISPR-dCas9

Researchers use dCas9 systems that incorporate epigenetic effectors to edit specific DNA loci for studying gene regulation functions of epigenetic marks (Zhou et al. 2024).

These research tools have shaped our understanding of how epigenetic mechanisms function in plant stress responses while creating opportunities to develop resilient agricultural crops.

5.7 Challenges and Future Directions

Studies on epigenetic regulation in plant stress responses have revealed promising methods to improve crop resistance. Despite encouraging findings, additional research is needed to overcome the ongoing gaps and challenges which exist in this field.

5.7.1 Gaps in Research on Epigenetic Regulation Under Abiotic Stress

Scientific discovery of epigenetic mechanisms has increased substantially, but various knowledge gaps need further exploration.

5.7.1.1 Mechanistic Understanding

Scientists currently lack a thorough understanding of molecular mechanisms through which precise epigenetic modifications affect stress-responsive genes. The research field needs to explore how DNA methylation combines with histone modifications and non-coding RNAs to regulate gene expression when plants experience stress (Chang et al. 2020).

5.7.1.2 Stress Combinations

Plants in their natural environments endure simultaneous exposure to various stress factors. The limited investigation of combined abiotic stress effects on the epigenome requires further study to advance resistance capabilities for crops in real farming environments (Shriti et al. 2024; Rahman et al. 2025).

5.7.1.3 Long-Term and Transgenerational Effects

The mechanisms behind transgenerational epigenetic inheritance reveal mixed results due to insufficient understanding of how such epigenetic marks endure across extended generations. The duration of epigenetic modifications together with their effects on offspring development should be studied to advance practical crop enhancement applications (Abdulraheem et al. 2024).

5.7.2 Challenges in Identifying Stable Epigenetic Markers for Stress Resilience

Multiple challenges exist in identifying trustworthy epigenetic markers related to stress tolerance.

5.7.2.1 Epigenetic Variability

The identification of stable markers becomes complex because epigenetic modifications demonstrate high dynamic behavior together with environmental influences. Determining the difference between stress-induced transient modifications and heritable epigenetics stands as a critical requirement (Cadet 2016).

5.7.2.2 Context-Dependent Effects

The effects of epigenetic modifications depend on three factors which include the plant species along with its developmental stage as well as the exact type of stress it experiences. This context dependency necessitates extensive validation across different conditions.

5.7.2.3 Technical Limitations

Technical challenges in current epigenetic analyses occur due to high costs along with the inability of bisulfite sequencing to detect the complete epigenome. Higher throughput along with cost-efficient technology needs improvement (Gottschalk et al. 2020).

5.7.3 Potential Integration with Multi-omics Approaches

Multiple omics approaches linked with epigenomics generate extensive information about plant stress responses.

5.7.3.1 Genomics and Transcriptomics

Using epigenetics alongside genomic and transcriptomic methodologies permits researchers to view how stress affects gene expression networks through epigenetic modifications. DNA methylation patterns and gene expression changes together allow researchers to identify important regulatory genes which control stress responses (Saeed et al. 2022).

5.7.3.2 Proteomics and Metabolomics

The analysis of proteomic and metabolomic data reveals the protein abundance impacts and metabolic pathway changes resulting from epigenetic modifications while stress occurs (Vo et al. 2021).

5.7.3.3 Systems Biology

Through systems biology, researchers can establish comprehensive models by integrating multi-omics data to describe how stress adaptations develop across plant genomes, epigenomes, transcriptomes, proteomes, and metabolomes (Muhie et al. 2023).

5.7.4 Ethical and Regulatory Aspects of Epigenetic Modifications in Crops

The application of epigenetic modifications in crop improvement raises important ethical and regulatory considerations:

5.7.4.1 Unintended Consequences

The evaluation process requires special focus on potential undesired effects generated by epigenetic modifications affecting both plant growth patterns and the relationship between plants and their ecosystem environment. Risk assessment requires complete evaluations to detect possible impacts (Tao et al. 2023).

5.7.4.2 Regulatory Frameworks

The regulatory systems in place presently evaluate genetic modifications in crops without examining epigenetic modifications. A standardized oversight system needs establishment to regulate the approval process of epigenetically modified crops that use specialized genetic processes while offering reversible results (Huang et al. 2022).

5.7.4.3 Public Perception

Public recognition of epigenetically modified crops depends on providing detailed production information and convincing illustrations showing both security and advantages. Stakeholder dialog consisting of consumers and farmers, and policymakers must succeed in resolving concerns through informed decision-making (Shin et al. 2022).

Research solutions for these challenges depend on multi-disciplinary studies and technological advancements, and thorough assessments of social aspects and ethical issues. The resolution of the main obstacles permits epigenetic modification programs to produce resilient crop varieties which survive intensifying abiotic stress patterns because of global warming.

5.8 Conclusion

Research investigations on epigenetic elements unlocked significant advancements in understanding plant stress responses to environmental factors. The integration of DNA methylation alongside non-coding RNAs and histone modifications facilitates the regulation of gene expressions that occur during stressful conditions. Plants undergo immediate adaptation through epigenetic modifications, but these modifications also build stress memory that allows the plants to resist environmental threats in future generations.

The value of epigenetics for strengthening plant resistance stands major during climate change events. The development of crop varieties that withstand multiple abiotic stresses becomes viable through the implementation of epigenetic mechanisms when environmental conditions show increasing unpredictability. The innovative approach to improve crop stress tolerance utilizes two epigenetic breeding methods that combine epimutant development and CRISPR-based epigenome editing. Plants can acquire better adaptive capabilities for adverse situations through the implementation of epigenetic-based priming methods that use chemical modulators for preparation.

Scientists need to tackle present research challenges by discovering durable epigenetic indicators for stress resistance together with investigating extended impacts of epigenetic modification methods. Expanding epigenetic research by using multiple omics methods that combine genomic and transcriptomic, and proteomic data will enable better plant stress understanding. A responsible application of epigenetic modifications in agriculture requires thorough evaluations of ethical laws and regulations.

The improvement of epigenetic science will allow us to use this mechanism optimally for building sustainable farming systems that protect our food supply during global climate shifts.

References

Abdulraheem MI, Xiong Y, Moshood AY, Cadenas-Pliego G, Zhang H, Hu J (2024) Mechanisms of plant epigenetic regulation in response to plant stress: Recent discoveries and implications. Front Plant Sci 13(2):163

Ahmad HM, Fiaz S, Hafeez S, Zahra S, Shah AN, Gul B et al (2022) Plant growth-promoting rhizobacteria eliminate the effect of drought stress in plants: a review. Front Plant Sci 13:875774

Akhter Z, Bi Z, Ali K, Sun C, Fiaz S, Haider FU, Bai J (2021) In response to abiotic stress, DNA methylation confers epigenetic changes in plants. Cells 10(6):1096

Al-Khayri JM, Banadka A, Rashmi R, Nagella P, Alessa FM, Almaghasla MI (2023) Cadmium toxicity in medicinal plants: an overview of the tolerance strategies, biotechnological and omics approaches to alleviate metal stress. Front Plant Sci 13:1047410

Alsamir M, Mahmood T, Trethowan R, Ahmad N (2021) An overview of heat stress in tomato (*Solanum lycopersicum L*). Acta Physiol Plant 28(3):1654–1663

Atta K, Mondal S, Gorai S, Singh AP, Kumari A, Ghosh T et al (2023) Impacts of salinity stress on crop plants: improving salt tolerance through genetic and molecular dissection. Front Plant Sci 14:1241736

Balasubramaniam T, Shen G, Esmaeili N, Zhang H (2023) Plants' response mechanisms to salinity stress. Plant Physiol Biochem 12(12):2253

Banerjee A, Wani SH, Roychoudhury A (2017) Epigenetic control of plant cold responses. Front Plant Sci 8:1643

Bannister AJ, Kouzarides T (2011) Regulation of chromatin by histone modifications. Cell Res 21(3):381–395

Bellegarde F, Herbert L, Séré D, Caillieux E, Boucherez J, Fizames C et al (2018) Polycomb repressive complex 2 attenuates the very high expression of the Arabidopsis gene NRT2. 1. Sci Rep 8(1):7905

Bellegarde F, Maghiaoui A, Boucherez J, Krouk G, Lejay L, Bach L et al (2019) The chromatin factor HNI9 and ELONGATED HYPOCOTYL5 maintain ROS homeostasis under high nitrogen provision. Plant Physiol 180(1):582–592

Bhadouriya SL, Mehrotra S, Basantani MK, Loake GJ, Mehrotra R (2021) Role of chromatin architecture in plant stress responses: an update. Front Genet 11:603380

Boopathy LRA, Jacob-Tomas S, Alecki C, Vera M (2022) Mechanisms tailoring the expression of heat shock proteins to proteostasis challenges. J Biol Chem 298(5):101796

Bure IV, Nemtsova MV, Kuznetsova EB (2022) Histone modifications and non-coding RNAs: mutual epigenetic regulation and role in pathogenesis. Int J Mol Sci 23(10):5801

Cadet JL (2016) Epigenetics of stress, addiction, and resilience: therapeutic implications. Dialogues Clin Neurosci 53:545–560

Chandrika NNP, Sundaravelpandian K, Schmidt W (2013a) A PHD in histone language: on the role of histone methylation in plant responses to phosphate deficiency. Plant Signal Behav 8(6):e24381

Chandrika NNP, Sundaravelpandian K, Yu SM, Schmidt W (2013b) ALFIN-LIKE 6 is involved in root hair elongation during phosphate deficiency in Arabidopsis. New Phytol 198(3):709–720

Chang YN, Zhu C, Jiang J, Zhang H, Zhu JK, Duan CG (2020) Epigenetic regulation in plant abiotic stress responses. Mol Plant 62(5):563–580

Chen CY, Wu K, Schmidt W (2015) The histone deacetylase HDA19 controls root cell elongation and modulates a subset of phosphate starvation responses in Arabidopsis. Sci Rep 5(1):15708

Cheng Y, Wang Q, Yang L, Li Q, Yan X (2024) MiR319a-mediated salt stress response in poplar. Plant Biotechnol J:uhae157

Cong W, Miao Y, Xu L, Zhang Y, Yuan C, Wang J et al (2019) Transgenerational memory of gene expression changes induced by heavy metal stress in rice (*Oryza sativa*L.). BMC Plant Biol 19:1–14

Fahad, S., Adnan, M., Munir, I., Lal, R., Nawaz, T., & Saud, S. (Eds.). (2024). Challenges and solutions of climate impact on agriculture. Elsevier. Chicago

Fan H, Zhang Z, Wang N, Cui Y, Sun H, Liu Y et al (2014) SKB1/PRMT 5-mediated histone H 4 R 3 dimethylation of I b subgroup bHLH genes negatively regulates iron homeostasis in Arabidopsis thaliana. Plant J 77(2):209–221

Fasani E, Giannelli G, Varotto S, Visioli G, Bellin D, Furini A, DalCorso G (2023) Epigenetic control of plant response to heavy metals. Plants 12(18):3195

Feng SJ, Zhang XD, Liu XS, Tan SK, Chu SS, Meng JG et al (2016) Characterization of long non-coding RNAs involved in cadmium toxic response in Brassica napus. Sci Rep 6(85):82157–82173

Feng D, Gao Q, Liu J, Tang J, Hua Z, Sun X (2023a) Categories of exogenous substances and their effect on alleviation of plant salt stress. Chemosphere 142:126656

Feng S, Jiang X, Wang R, Tan H, Zhong L, Cheng Y et al (2023b) Histone H3K4 methyltransferase DcATX1 promotes ethylene induced petal senescence in carnation. Plant J 192(1):546–564

Fitz-James MH, Cavalli G (2022) Molecular mechanisms of transgenerational epigenetic inheritance. Nat Rev Genet 23(6):325–341

Fukuda M, Nishida S, Kakei Y, Shimada Y, Fujiwara T (2019) Genome-wide analysis of long intergenic noncoding RNAs responding to low-nutrient conditions in Arabidopsis thaliana: possible involvement of trans-acting siRNA3 in response to low nitrogen. Plant Cell Physiol 60(9):1961–1973

Ganie SA, McMulkin N, Devoto A (2024) The role of priming and memory in rice environmental stress adaptation: current knowledge and perspectives. J Exp Bot 47(5):1895–1915

Gottschalk MG, Domschke K, Schiele MA (2020) Epigenetics underlying susceptibility and resilience relating to daily life stress, work stress, and socioeconomic status. Front Psychol 11:163

Gui Q, Yang Z, Chen C, Yang F, Wang S, Dong R (2022) Identification and characterization of long noncoding RNAs involved in the aluminum stress response in Medicago truncatula via genome-wide analysis. Front Plant Sci 13:1017869

Guo J, He X-J (2024) Composition and function of plant chromatin remodeling complexes. Annu Rev Plant Biol 81:102613

Haldar S, Bandyopadhyay S (2021) Co-ordinated regulation of miRNA and their target genes by CREs during salt stress in *Oryza sativa* (Rice). Gene Expr Patterns 28:100323

Hauser M-T, Aufsatz W, Jonak C, Luschnig C (2011) Transgenerational epigenetic inheritance in plants. FEBS Lett 1809(8):459–468

Huang Y, Liu Y, Liu C, Birchler JA, Han F (2022) Prospects and challenges of epigenomics in crop improvement. Plant J 44(3):251–257

Iglesias J, Trigueros M, Rojas-Triana M, Fernández M, Albar JP, Bustos R et al (2013) Proteomics identifies ubiquitin–proteasome targets and new roles for chromatin-remodeling in the Arabidopsis response to phosphate starvation. J Proteome 94:1–22

Islam W, Waheed A, Idrees A, Rashid J, Zeng F (2022) Role of plant microRNAs and their corresponding pathways in fluctuating light conditions. Biochimica et Biophysica Acta (BBA) - molecular. Cell Res 1870(2):119304. https://doi.org/10.1016/j.bbamcr.2022.119304

Iwasaki M, Paszkowski J (2014) Epigenetic memory in plants. EMBO J 33(18):1987–1998

Jin Q, Chachar M, Ali A, Chachar Z, Zhang P, Riaz A et al (2024a) Epigenetic regulation for heat stress adaptation in plants: new horizons for crop improvement under climate change. Front Plant Sci 14(9):2105

Jin X, Wang Z, Li X, Ai Q, Wong DCJ, Zhang F et al (2024b) Current perspectives of lncRNAs in abiotic and biotic stress tolerance in plants. Front Plant Sci 14:1334620

Jing Z, Liu N, Zhang Z, Hou X (2024) Research progress on plant responses to stress combinations in the context of climate change. Plant Signal Behav 13(4):469

Kamei M, Ameri AJ, Ferraro AR, Bar-Peled Y, Zhao F, Ethridge CL et al (2021) Imitation switch is required for normal chromatin structure and gene repression in PRC2 target domains. Proc Natl Acad Sci 118(4):e2010003118

Kasım MU, Kasım R (2017) Yellowing of fresh-cut spinach (Spinacia oleracea L.) leaves delayed by UV-B applications. Info Proc Agri 4(3):214–219. https://doi.org/10.1016/j.inpa.2017.05.006

Khan N, Ray RL, Sargani GR, Ihtisham M, Khayyam M, Ismail S (2021) Current progress and future prospects of agriculture technology: gateway to sustainable agriculture. J Agric Food Res 13(9):4883

Khan F, Siddique AB, Shabala S, Zhou M, Zhao C (2023) Phosphorus plays key roles in regulating plants' physiological responses to abiotic stresses. Front Plant Sci 12(15):2861

Kim Y, Chung YS, Lee E, Tripathi P, Heo S, Kim K (2020) Root response to drought stress in Rice (*Oryza sativa L.*). Int J Mol Sci 21(4):1513. https://doi.org/10.3390/ijms21041513

Kosakivska IV, Babenko LM, Romanenko KO, Korotka IY, Potters G (2021) Molecular mechanisms of plant adaptive responses to heavy metals stress. Plant Physiol Biochem 45(2):258–272

Kovalchuk I (2023) Role of epigenetic factors in response to stress and establishment of somatic memory of stress exposure in plants. Int J Mol Sci 12(21):3667

Kumar S, Mohapatra T (2021) Dynamics of DNA methylation and its functions in plant growth and development. Front Genet 12:596236

Kumar V, Thakur JK, Prasad M (2021) Histone acetylation dynamics regulating plant development and stress responses. Plant Mol Biol Report 78:4467–4486

Lämke J, Bäurle I (2017) Epigenetic and chromatin-based mechanisms in environmental stress adaptation and stress memory in plants. Curr Opin Plant Biol 18:1–11

Li N, Yang T, Guo Z, Wang Q, Chai M, Wu M et al (2020) Maize microRNA166 inactivation confers plant development and abiotic stress resistance. Plant J 21(24):9506

Li J, Song Q, Zuo Z, Liu L (2022) MicroRNA398: a master regulator of plant development and stress responses. Int J Mol Sci 23(18):10803. https://doi.org/10.3390/ijms231810803

Li J, Han F, Yuan T, Li W, Li Y, Wu HX et al (2023) The methylation landscape of giga-genome and the epigenetic timer of age in Chinese pine. New Phytol 14(1):1947

Li Z, Yang J, Zou J-J, Cai X, Zeng X, Xing W (2024) A systematic review on the role of miRNAs in plant response to stresses under the changing climatic conditions. Plant Sci 14:100674

Liu Y, Zhao C, Tang X, Wang L, Guo R (2024) Transcriptomic and metabolomic insights into ABA-related genes in Cerasus humilis under drought stress. Int J Mol Sci 25(14):7635

Ma B, Wang S, Wu W, Shan P, Chen Y, Meng J et al (2023) Mechanisms of circRNA/lncRNA-miRNA interactions and applications in disease and drug research. Pharmacol Res 162:114672

Martins LM, Law JA (2023) Moving targets: mechanisms regulating siRNA production and DNA methylation during plant development. Curr Opin Plant Biol 75:102435

Morrell R, Sadanandom A (2019) Dealing with stress: a review of plant SUMO proteases. Front Plant Sci 10:1122

Muhammad Aslam M, Waseem M, Jakada BH, Okal EJ, Lei Z, Saqib HSA et al (2022) Mechanisms of abscisic acid-mediated drought stress responses in plants. Int J Mol Sci 23(3):1084

Muhie S, Gautam A, Misganaw B, Yang R, Mellon SH, Hoke A et al (2023) Integrated analysis of proteomics, epigenomics and metabolomics data revealed divergent pathway activation patterns in the recent versus chronic post-traumatic stress disorder. Mol Psychiatry 113:303–316

Nawaz, T., Fahad, S., Nawaz, T., Saud, S., Rahman, T. U., Adnan, M., ... & Hassan, S. (2025). Climate change– resilient crop production. In Challenges and Solutions of Climate Impact on Agriculture (pp. 329-357). Academic Press.

Nawaz, T., Nelson, D., Saleem, A., Bibi, M., Joshi, N., Fahad, S., ... & Khan, I. (2025). Genetic Engineering for Crop Biofortification. Crop Biofortification: Biotechnological Approaches for Achieving Nutritional Security Under Changing Climate, 263–294

Nawaz, T., Fahad, S., Joshi, N., Rahman, T. U., Khan, M. N. R., Hassan, S., & Saud, S. (2025). Legume Production and Problems in the Mediterranean Region. In Marker-Assisted Breeding in Legumes for Drought Tolerance (pp. 41-75). Springer, Singapore

Pang Q, Chen S, Dai S, Chen Y, Wang Y, Yan X (2010) Comparative proteomics of salt tolerance in Arabidopsis thaliana and Thellungiella halophila. J Proteome Res 9(5):2584–2599. https://doi.org/10.1021/pr100034f

Park EY, Tsuyuki KM, Hu F, Lee J, Jeong J (2019) PRC2-mediated H3K27me3 contributes to transcriptional regulation of FIT-dependent iron deficiency response. Front Plant Sci 10:627

Pawłasek N, Sokołowska A, Koter M, Oracz K (2024) The interaction between miR165/166 and miR160 regulates Arabidopsis thaliana seed size, weight, and number in a ROS-dependent manner. Planta 260(3):72

Quadrana L, Colot V (2016) Plant transgenerational epigenetics. Annu Rev Genet 50(1):467–491

Rai P, Kumari P, Gaur V (2024) Erasing Methylation Marks on DNA by Plant-Specific DEMETER Family DNA Glycosylases. Int J Mol Sci:1–17

Ramakrishnan M, Zhang Z, Mullasseri S, Kalendar R, Ahmad Z, Sharma A et al (2022) Epigenetic stress memory: A new approach to study cold and heat stress responses in plants. Front Plant Sci 13:1075279

Ramesh SV, Govindasamy V, Rajesh MK, Sabana AA, Praveen S (2019) Stress-responsive miRNAome of Glycine max (L.) Merrill: molecular insights and way forward. Planta 249(5):1267–1284. https://doi.org/10.1007/s00425-019-03114-5

Rao S, Balyan S, Jha S, Bansal C, Das JR, Gupta A, Mathur S (2020) Orchestration of MicroRNAs and transcription factors in the regulation of plant abiotic stress response. In: Springer eBooks, pp 251–277. https://doi.org/10.1007/978-981-15-9380-2_8

Rao X, Yang S, Lü S, Yang PJP (2024) DNA methylation dynamics in response to drought stress in crops. Front Plant Sci 13(14):1977
Rendina González AP, Preite V, Verhoeven KJ, Latzel V (2018) Transgenerational effects and epigenetic memory in the clonal plant Trifolium repens. Front Plant Sci 9:1677
Réthoré E, Jing L, Ali N, Yvin J-C, Pluchon S, Hosseini SA (2021) K deprivation modulates the primary metabolites and increases putrescine concentration in Brassica napus. Front Plant Sci 12:681895
Sadau SB, Liu Z, Ninkuu V, Guan L, Sun X (2024) DREB transcription factors are crucial regulators of abiotic stress responses in Gossypium spp. Plant Cell Physiol 11:100350
Sadura I, Janeczko A (2024) Are heat shock proteins important in low-temperature-stressed plants? A Minireview. Plants 14(6):1296
Saeed F, Chaudhry UK, Bakhsh A, Raza A, Saeed Y, Bohra A, Varshney RK (2022) Moving beyond DNA sequence to improve plant stress responses. Front Plant Sci 13:874648
Sako K, Maki Y, Kanai T, Kato E, Maekawa S, Yasuda S et al (2012) Arabidopsis RPT2a, 19S proteasome subunit, regulates gene silencing via DNA methylation. PLoS One 7(5):e37086
Sasmal P, Som S, Patali R, Nag U, Nath P, Singh S (2025) Insights into epigenetic modulation of iron homeostasis in plants. Plant Mol Biol Report:1–11
Secco D, Wang C, Shou H, Schultz MD, Chiarenza S, Nussaume L et al (2015) Stress induced gene expression drives transient DNA methylation changes at adjacent repetitive elements. elife 4:e09343
Seem K, Kaur S, Selvan T, Kumar S (2024) Plant response to drought stress: epigenomic perspective. In: Current omics advancement in plant abiotic stress biology. Elsevier, pp 323–341
Sellamuthu G, Tarafdar A, Jasrotia RS, Chaudhary M, Vishwakarma H, Padaria JC (2024) Introgression of Δ 1-pyrroline-5-carboxylate synthetase (PgP5CS) confers enhanced resistance to abiotic stresses in transgenic tobacco. Physiol Plant:1–17
Séré D, Martin A (2020) Epigenetic regulation: another layer in plant nutrition. J Exp Bot 15(1):1–6
Sharma M, Kumar P, Verma V, Sharma R, Bhargava B, Irfan MJ (2022) Understanding plant stress memory response for abiotic stress resilience: Molecular insights and prospects. Plant Physiol Biochem 179:10–24
Shen Q, Lin Y, Li Y, Wang G (2021) Dynamics of H3K27me3 modification on plant adaptation to environmental cues. Cells 10(6):1165
Shin H, Choi WL, Lim JY, Huh JH (2022) Epigenome editing: targeted manipulation of epigenetic modifications in plants. Plant Biotechnol J:1–9
Shriti S, Bhar A, Roy A (2024) Unveiling the role of epigenetic mechanisms and redox signaling in alleviating multiple abiotic stress in plants. Front Plant Sci 15:1456414
Singhal, R. K., Fahad, S., Kumar, P., Choyal, P., Javed, T., Jinger, D., ... & Nawaz, T. (2023). Beneficial elements: New Players in improving nutrient use efficiency and abiotic stress tolerance. Plant Growth Regulation, 100(2), 237-265
Singh PK, Miller G, Faigenboim A, Lieberman-Lazarovich M (2021) The tomato ddm1b mutant shows decreased sensitivity to heat stress accompanied by transcriptional alterations. Int J Mol Sci 12(9):1337
Smith AP, Jain A, Deal RB, Nagarajan VK, Poling MD, Raghothama KG, Meagher RB (2010) Histone H2A. Z regulates the expression of several classes of phosphate starvation response genes but not as a transcriptional activator. Plant Physiol 152(1):217–225
Song ZT, Liu JX, Han JJ (2021) Chromatin remodeling factors regulate environmental stress responses in plants. Front Plant Sci 63(3):438–450
Statello L, Guo C-J, Chen L-L, Huarte M (2021) Gene regulation by long non-coding RNAs and its biological functions. Nat Rev Mol Cell Biol 22(2):96–118
Sun Z, Shu L, Zhang W, Wang Z (2020) Cca-miR398 increases copper sulfate stress sensitivity via the regulation of CSD mRNA transcription levels in transgenic Arabidopsis thaliana. Peer J 8:e9105. https://doi.org/10.7717/peerj.9105
Tao Z, Yan F, Hahn M, Ma Z (2023) Regulatory roles of epigenetic modifications in plant-phytopathogen interactions. 1(1):6
Thakur R, Prasad P (2024) Epigenetic regulation of abiotic stress responses in plants. Plant Sci 1868:130661

Varotto S, Tani E, Abraham E, Krugman T, Kapazoglou A, Melzer R et al (2020) Epigenetics: possible applications in climate-smart crop breeding. Front Plant Sci 71(17):5223–5236

Vaucheret H, Voinnet O (2024) The plant siRNA landscape. Curr Opin Plant Biol 36(2):246–275

Vo KTX, Rahman MM, Rahman MM, Trinh KTT, Kim ST, Jeon J-S (2021) Proteomics and metabolomics studies on the biotic stress responses of rice: an update. Int J Mol Sci 14:1–16

Wang T, Xing J, Liu Z, Zheng M, Yao Y, Hu Z et al (2019) Histone acetyltransferase GCN5-mediated regulation of long non-coding RNA At4 contributes to phosphate starvation response in Arabidopsis. J Exp Bot 70(21):6337–6348

Wang X, Wang D, Xu W, Kong L, Ye X, Zhuang Q, Fan D, Luo K (2020) Histone methyltransferase ATX1 dynamically regulates fiber secondary cell wall biosynthesis in Arabidopsis inflorescence stem. Nucleic Acids Res 49(1):190–205. https://doi.org/10.1093/nar/gkaa1191

Wang B, Wang X, Wang Z, Zhu K, Wu W (2023a) Comparative metagenomic analysis reveals rhizosphere microbial community composition and functions help protect grapevines against salt stress. Front Plant Sci 14:1102547

Wang M, Wang Z, Guo M, Qu L, Biere A (2023b) Effects of arbuscular mycorrhizal fungi on plant growth and herbivore infestation depend on availability of soil water and nutrients. Front Plant Sci 14:1101932

Wang M, Zhong Z, Gallego-Bartolomé J, Feng S, Shih Y-H, Liu M et al (2023c) Arabidopsis TRB proteins function in H3K4me3 demethylation by recruiting JMJ14. Nat Commun 14(1):1736

Wang W, Liu H, Wang F, Liu X, Sun Y, Zhao J et al (2023d) N4-acetylation of cytidine in mRNA plays essential roles in plants. Nat Commun 35(10):3739–3756

Wang F, Li C-H, Liu Y, He L-F, Li P, Guo J-X et al (2024) Plant responses to abiotic stress regulated by histone acetylation. Front Plant Sci 15:1404977

Widiez T, El Kafafi ES, Girin T, Berr A, Ruffel S, Krouk G et al (2011) High nitrogen insensitive 9 (HNI9)-mediated systemic repression of root NO3− uptake is associated with changes in histone methylation. Proc Natl Acad Sci 108(32):13329–13334

Wu Y-L, Lin Z-J, Li C-C, Lin X, Shan S-K, Guo B et al (2023) Epigenetic regulation in metabolic diseases: mechanisms and advances in clinical study. Genes & Diseases 8(1):98

Xing J, Wang T, Liu Z, Xu J, Yao Y, Hu Z et al (2015) GENERAL CONTROL NONREPRESSED PROTEIN5-mediated histone acetylation of FERRIC REDUCTASE DEFECTIVE3 contributes to iron homeostasis in Arabidopsis. Plant Physiol 168(4):1309–1320

Xu M, Hu T, Zhao J, Park M, Earley KW, Wu G, Yang L, Poethig RS (2016) Developmental functions of miR156-regulated SQUAMOSA PROMOTER BINDING PROTEIN-LIKE (SPL) genes in Arabidopsis thaliana. PLoS Genet 12(8):e1006263. https://doi.org/10.1371/journal.pgen.1006263

Xu JM, Wang ZQ, Wang JY, Li PF, Jin JF, Chen WW, Fan W, Kochian LV, Zheng SJ, Yang JL (2019) Low phosphate represses histone deacetylase complex1 to regulate root system architecture remodeling in Arabidopsis. New Phytol 225:1732

Xu P, Chen H, Hu J, Cai W (2021) Potential evidence for transgenerational epigenetic memory in Arabidopsis thaliana following spaceflight. NPJ Microgravity 4(1):835

Yang X, Ren J, Lin X, Yang Z, Deng X, Ke Q (2023) Melatonin alleviates chromium toxicity in maize by modulation of cell wall polysaccharides biosynthesis, glutathione metabolism, and antioxidant capacity. Int J Mol Sci 24(4):3816. https://doi.org/10.3390/ijms24043816

Yen MR, Suen DF, Hsu FM, Tsai YH, Fu H, Schmidt W, Chen PY (2017) Deubiquitinating enzyme OTU5 contributes to DNA methylation patterns and is critical for phosphate nutrition signals. Plant Physiol 175(4):1826–1838

Yin M, Wang S, Wang Y, Wei R, Liang Y, Zuo L et al (2024) Impact of abiotic stress on rice and the role of DNA methylation in stress response mechanisms. Plants 13(19):2700

Yu Y, Zhang Y, Chen X, Chen Y (2019) Plant noncoding RNAs: hidden players in development and stress responses. Curr Opin Plant Biol 35(1):407–431

Yu X, An J, Liang J, Yang W, Zeng Z, Zhang M, Wu H, Liu S, Cao X (2024) Comparative analysis of two soybean cultivars revealed tolerance mechanisms underlying soybean adaptation to flooding. Curr Issues Mol Biol 46(11):12442–12456. https://doi.org/10.3390/cimb46110739

Zang Y, Xie L, Su J, Luo Z, Jia X, Ma X (2023) Advances in DNA methylation and demethylation in medicinal plants: a review. J Ethnopharmacol 50(9):7783–7796

Zhang L, Ding H, Jiang H, Wang H, Chen K, Duan J, Feng S, Wu G (2019) Regulation of cadmium tolerance and accumulation by miR156 in Arabidopsis. Chemosphere 242:125168. https://doi.org/10.1016/j.chemosphere.2019.125168

Zhang H, Zhu J, Gong Z, Zhu J-K (2022) Abiotic stress responses in plants. Annu Rev Plant Biol 23(2):104–119

Zhang H, Zhang X, Xiao JJP (2023) Epigenetic regulation of nitrogen signaling and adaptation in plants. Plants 12(14):2725

Zhang J, Wang L, Xue M, Gan L, Pei X (2024a) MicroRNA171a regulates plant development and enhances drought stress tolerance. Plant Sci 219:105613

Zhang P, He Y, Huang SJA (2024b) Unlocking epigenetic breeding potential in tomato and potato. Trends Plant Sci:1–12

Zhao T, Zhan Z, Jiang D (2019) Histone modifications and their regulatory roles in plant development and environmental memory. J Integr Plant Biol 46(10):467–476

Zhao X, Xue Z, Liu Y, Huang Z, Sun Y, Wu C et al (2024) ChIP-seq revealed the role of tomato SNAC4 in response to biological and abiotic stresses and mediating auxin signaling response. Plant Sci 209:112696

Zhou C, Xiong H, Fu M, Guo H, Zhao L, Xie Y et al (2023a) Genetic mapping and identification of Rht8-B1 that regulates plant height in wheat. J Integr Plant Biol 23(1):333

Zhou L, Gao G, Li X, Wang W, Tian S, Qin G (2023b) The pivotal ripening gene SlDML2 participates in regulating disease resistance in tomato. Plant Commun 21(11):2291–2306

Zhou M, Li Y, Yan Y, Gao L, He C, Wang J et al (2023c) Proteome and phosphoproteome analysis of 2, 4-epibrassinolide-mediated cold stress response in cucumber seedlings. Front Plant Sci 14:1104036

Zhou S, Li Y, Wu Q, Gong C (2024) Nanotechnology-based CRISPR/Cas9 delivery system for genome editing in cancer treatment. Adv Sci 3(1):e70

Chapter 6
Mobile Genetic Elements and Abiotic Stress Tolerance

Shikha Verma and Pankaj Kumar Verma

Abstract In the face of escalating climate change, crop plants are increasingly challenged by diverse abiotic stresses, threatening global food security. Mobile genetic elements (MGEs) have emerged as critical players in shaping plant genomes and modulating stress responses. This chapter explores the multifaceted contributions of MGEs, including transposable elements and retrotransposons, to abiotic stress adaptation in crops. We discuss their mechanisms of transposition, their impact on genome structure and gene expression, and their role in generating genetic variation for stress-responsive traits. Through case studies, we examine specific MGEs activated under stress conditions, influencing gene regulation and epigenetic modifications. Furthermore, we highlight the potential of MGEs in epigenetic regulation, stress memory, and transgenerational inheritance of stress tolerance. Finally, we explore the application of MGEs as molecular tools for targeted genome editing and breeding stress-resilient crop varieties. By elucidating the dynamic role of MGEs in abiotic stress tolerance, this chapter aims to provide insights into their evolutionary and functional significance, ultimately contributing to the development of genetically resilient crops for sustainable agriculture.

Keywords Mobile genetic elements · Retrotransposons · DNA transposons · Abiotic stress adaptation · Gene activation

S. Verma
French Associates Institute for Agriculture and Biotechnology of Drylands, Ben-Gurion University of the Negev, Be'er Sheva, Israel

P. K. Verma (✉)
Department of Plant and Soil Sciences, Texas Tech University, Lubbock, TX, USA

S. Tiwari et al. (eds.), *Decoding Non-Coding RNA*, Concepts and Strategies in Plant Sciences, https://doi.org/10.1007/978-981-96-9859-2_6

6.1 Introduction

In the current climate change scenario, continually changing environmental factors significantly affect crop growth and productivity (Jovović et al. 2025). These environmental stressors adversely impact plant growth, development, and productivity, posing a significant challenge to global food security (Malhi et al. 2021). To overcome the loss caused by environmental stress, it is essential to understand the underlying molecular and genetic mechanisms for plant responses to abiotic stress, which is critical for developing stress-tolerant varieties. In recent years, in addition to gene expression analysis, mobile genetic elements (MGEs) have emerged as key players in shaping plant genomes and modulating stress responses (Glazko et al. 2022; Hwang et al. 2024).

Mobile genetic elements (MGEs) are diverse genetic sequences capable of transposing or relocating within a genome (Hwang et al. 2024). They include transposable elements (TEs), retrotransposons, insertion sequences, and certain classes of non-coding RNAs that influence genome plasticity and gene regulation (Jordan 2009). MGEs contribute significantly to genome evolution by generating genetic variation, restructuring chromosomal regions, and controlling gene expression. In crops, MGEs have been implicated in the evolution of stress-responsive genes, epigenetic modifications, and adaptive traits that enhance survival under harsh environmental conditions (Shapiro 2013; Shapiro 2014; Shapiro 2002).

The role of MGEs in abiotic stress tolerance is increasingly being recognized through transcriptomic, epigenomic, and functional genomic studies (Hamdan et al. 2022). Many stress-responsive transposable elements are activated under adverse environmental conditions, leading to genome-wide regulatory changes. For instance, specific long terminal repeat (LTR) retrotransposons are transcriptionally activated under drought and heat stress, influencing the expression of nearby genes (Alzohairy et al. 2014; Deneweth et al. 2022; Grandbastien 2015; Vangelisti et al. 2019). Similarly, DNA transposons can create novel gene regulatory networks by inserting into promoter or enhancer regions, modulating gene activity in response to stress cues (Negi et al. 2016; Malfait 2024).

This chapter provides an in-depth exploration of the contributions of MGEs to abiotic stress adaptation in crops. It discusses the classification and characteristics of MGEs, their mechanisms of transposition, and their impact on genome structure and gene expression. We further examine case studies of key MGEs involved in abiotic stress responses across various crop species. The role of MGEs in epigenetic regulation, stress memory, and transgenerational inheritance of stress tolerance is also highlighted. Finally, we explore the potential applications of MGEs in modern crop breeding, including their use as molecular tools for targeted genome modifications and stress-resilient trait development.

By elucidating the dynamic role of MGEs in abiotic stress tolerance, this chapter aims to provide new insights into these elements' evolutionary and functional significance in plant genomes. Understanding their regulatory mechanisms will contribute to developing genetically resilient crops capable of withstanding environmental fluctuations, ultimately supporting global agricultural sustainability.

6.2 Mobile Genetic Elements: An Overview

Mobile genetic elements are segments of DNA that can change their position within the genome. They are generally categorized into Class I elements, also known as retrotransposons (Eickbush and Malik 2007), and Class II Elements (DNA Transposons).

Class I Elements (Retrotransposons) are a major class of transposable elements (TEs) that proliferate via a copy-and-paste mechanism involving an RNA intermediate (Ramakrishnan et al. 2022). Retrotransposons significantly shape plant genomes by contributing to genetic diversity, genome expansion, and gene expression regulation (Mita and Boeke 2016; Papolu et al. 2022). Retrotransposons are classified into two main groups: those with Long Terminal Repeats (LTRs) and those without. LTR retrotransposons, such as the Gypsy and Copia superfamilies, are characterized by the presence of identical repeat sequences at both ends. In contrast, non-LTR retrotransposons, including LINEs (Long Interspersed Nuclear Elements) and SINEs (Short Interspersed Nuclear Elements), do not contain these terminal repeats (Eickbush and Malik 2007). Retrotransposons drive genome size variation among plant species, as seen in maize and wheat, and their amplification is balanced by host suppression mechanisms such as DNA methylation and RNA interference (Lee and Kim 2014; Galindo-González et al. 2018). They also influence gene expression through insertional mutagenesis, promoter activation, and epigenetic modifications, with some becoming transcriptionally active under environmental stress, leading to novel gene regulatory networks that enhance plant adaptation (Papolu et al. 2022; Deneweth et al. 2022; Hirsch and Springer 2017). Furthermore, retrotransposons are widely used in plant biotechnology, particularly in functional genomics and marker-assisted breeding, where transposon-based tagging strategies help identify genes associated with key agronomic traits, and retrotransposon insertion polymorphisms serve as molecular markers for genetic diversity studies (Kalendar et al. 2018). Given their significant impact on genome evolution, gene regulation, and adaptation, retrotransposons are essential components of plant genomes, and their study offers valuable insights into plant genomics with potential applications in crop improvement and genetic research.

Class II transposable elements, or DNA transposons, are sequences that can move within plant genomes via a "cut-and-paste" mechanism. This process involves the element being excised from one genomic location and inserted into another, facilitated by the enzyme transposase (Feschotte and Pritham 2007). These elements are characterized by terminal inverted repeats (TIRs) at their ends, which are recognized by transposase during transposition, and the creation of target site duplications (TSDs) upon integration (Ling and Cordaux 2010). In plants, Class II elements are categorized into several superfamilies based on the structure and sequence of their transposases and TIRs (Han et al. 2013). Notable superfamilies include the Tc1/Mariner, hAT, Mutator, PIF/Harbinger, and CACTA (Han et al. 2013). Miniature Inverted-repeat Transposable Elements (MITEs) are a distinct group of non-autonomous DNA transposons. Although they do not encode their own transposase enzymes, they can still move within the genome by utilizing transposases produced by related autonomous transposable elements (Lu et al. 2012). MITEs are typically short, possess TIRs, and are prevalent in plant genomes, often inserting near or within genes, thereby influencing gene expression and contributing to genetic diversity (Lu et al. 2012). The activity of DNA transposons can induce mutations, drive genome evolution, and play roles in plant adaptation by creating genetic variability (Bennetzen and Wang 2014; Ramakrishnan et al. 2021). Despite their mobility, the activity of these elements is carefully controlled by the host genome through epigenetic mechanisms, including DNA methylation and histone modifications, to preserve genomic integrity (Galindo-González et al. 2018; Lisch 2009). Understanding the behavior and regulation of Class II transposable elements is crucial for insights into plant genome dynamics and for potential applications in plant breeding and genetic engineering.

Mobile genetic elements (MGEs) move within the genome through distinct mechanisms, primarily classified into class I (retrotransposons) and class II (DNA transposons). These transposition mechanisms are crucial in genome evolution, gene expression regulation, and plant adaptation to abiotic stresses.

6.2.1 Class I: Retrotransposons (Copy-and-Paste Mechanism)

Retrotransposons, also known as RNA transposons, transpose via an RNA intermediate and follow a "copy-and-paste" mechanism (Pandita and Pandita 2025). These elements are highly abundant in plant genomes, often comprising a significant portion of the total DNA content, particularly in crops with large genomes, such as wheat, maize, and rice. Retrotransposons are first transcribed into RNA by RNA polymerase II or RNA polymerase III, depending on the element type (Luqman-Fatah et al. 2024). The RNA produced from retrotransposons is reverse-transcribed into complementary DNA (cDNA) by the enzyme reverse transcriptase, which is typically encoded within the retrotransposon sequence itself. Once synthesized, the cDNA is integrated into a new genomic location with the help of integrase, an enzyme also encoded by the retrotransposon (Luqman-Fatah et al. 2024). Unlike DNA transposons, retrotransposons do not excise from their original position; instead, they create multiple copies, leading to genome expansion.

6.2.2 Class II: DNA Transposons (Cut-and-Paste Mechanism)

DNA transposons move through a cut-and-paste mechanism, in which they are excised from one site in the genome and inserted into another. This mechanism is catalyzed by the enzyme transposase, which is encoded by the transposon itself. The transposase enzyme identifies and binds to specific terminal inverted repeat (TIR) sequences located at the ends of the transposon. It cleaves the transposon from its original location, creating a double-stranded DNA break. After excision, the transposon is inserted into a new site within the genome, usually guided by the transposase's recognition of a specific target sequence. DNA repair mechanisms fill gaps left by the excision process, sometimes leading to mutations or chromosomal rearrangements that can contribute to stress tolerance (Luqman-Fatah et al. 2024).

6.3 Epigenetic Regulation of Transposition

Since uncontrolled transposition can be harmful, plants have evolved epigenetic mechanisms to regulate MGE activity, including DNA Methylation, such as Cytosine methylation in transposable elements, which represses their transcription (Lisch 2009). Histone Modifications such as heterochromatin formation through histone modifications (e.g., H3K9me2) suppress transposon mobility (Pachamuthu and Borges 2023). Small RNAs (siRNAs and miRNAs): RNA interference (RNAi) pathways silence transposons at the transcriptional or post-transcriptional level. Under abiotic stress, these regulatory mechanisms can be relaxed, allowing selective activation of MGEs that contribute to stress adaptation (Underwood et al. 2017; Pachamuthu and Borges 2023; Chuong et al. 2017).

6.4 Influence of Retrotransposons on Gene Regulation and Environmental Stress Response

6.4.1 From Jumping Genes to Gene Regulators

Long Terminal Repeat (LTR) retrotransposons often serve as regulatory components within the genome, acting as promoters, enhancers, or silencers for adjacent genes. By interacting with transcription factors, they can modulate gene activity and contribute to the regulation of important plant traits (Table 6.1) (Grandbastien 2015; Papolu et al. 2021; Papolu et al. 2022). For instance, in Japanese plum (*Prunus salicina*), the insertion of an LTR retrotransposon in the promoter region of the PsMYB10.2 gene influences anthocyanin production, thereby playing a key role in determining the pigmentation of the fruit flesh (Fiol et al. 2022). Likewise, in *Citrus sinensis* (Sicilian blood orange), the insertion of a Copia-like retrotransposon in the promoter region of the Ruby gene enhances anthocyanin accumulation, particularly in response to low-temperature conditions (Butelli et al. 2012; Butelli et al. 2017; Butelli et al. 2019). In *Daucus carota* (carrot), the DcMYB7 gene, responsible for purple root pigmentation, is transcriptionally inactivated in non-purple cultivars due to tandem duplication and transposon insertions (Xu et al. 2019). Transposable elements also drive phenotypic variation by disrupting gene function through insertional mutagenesis and serving as natural markers in transposon tagging approaches. In flowering plants like Nicotiana tabacum, Ipomoea, and Petunia, TEs influence

Table 6.1 Stress-responsive transposons in plants

Stress	TE	Group	Plant species	References
Heat	*Athila*	Gypsy	*Arabidopsis thaliana*	Buchmann et al. (2009)
Heat, high-intensity light	ONSEN/ Atcopia78	Copia	*Arabidopsis thaliana*	Tittel-Elmer et al. (2010) and Matsunaga et al. (2012)
Cold	*Tcs1, Tcs2*	Copia	*Citrus sinensis*	Butelli et al. (2012)
Salt, osmotic stress, cold, heat, abscisic acid	*At2G06045*	Copia	*Arabidopsis thaliana*	Zeller et al. (2009)
Nutrition starvation, salt, cytokinin, abscisic acid	*AtCopeg1*	Copia	*Arabidopsis thaliana*	Duan et al. (2008)
UV light	*Reme1*	Copia	*Cucumis melo*	Ramallo et al. (2008)
Wounding, salt	*CLCoy1*	Copia	*Citrus limon*	De Felice et al. (2009)
Wounding, auxin	*CIRE1*	Copia	*Citrus sinensis*	Rico-Cabanas and Martínez-Izquierdo (2007)
Wounding	*HRET1*	Gypsy	*Hibiscus syriacus*	Jeung et al. (2005)
Senescence	*BAGY1*	Gypsy	*Hordeum vulgare*	Ay et al. (2008)
Abscisic acid	*BARE-1*	Copia	*Hordeum vulgare*	Suoniemi et al. (1996)
Salicylic acid	*OARE1*	Copia	*Hordeum vulgare*	Kimura et al. (2001)

anthocyanin-related pigmentation, leading to diverse flower colors (Pandita and Pandita 2023).

6.4.2 Role of MGEs in Environmental Stress Response

Mobile genetic elements (MGEs), especially Long Terminal Repeat (LTR) and non-LTR retrotransposons, are involved in various aspects of plant stress responses. In *Phaeodactylum tricornutum*, LTR retrotransposons are activated under nitrate deprivation and in the presence of reactive aldehydes, indicating their responsiveness to environmental stressors (Maumus et al. 2009). In *Capsicum baccatum*, the activation of Athila-type LTR retrotransposons has been linked to the expansion of disease-resistance gene families, suggesting a role in enhancing the plant's defense mechanisms (Kim et al. 2017; de Assis et al. 2020). Similarly, in *Arabidopsis thaliana* and *Vigna angularis*, treatment with the DNA demethylation agent zebularine leads to the augmentation of Ty1/copia-like ONSEN retrotransposons, highlighting their role in stress adaptation (Boonjing et al. 2020). Similarly, in the resurrection plant *Craterostigma plantagineum*, dehydration triggers the activation of the Craterostigma Desiccation-Tolerant (CDT-1) retrotransposon, which contributes to the plant's ability to tolerate desiccation (Boonjing et al. 2020).

Similarly, in transgenic *Arabidopsis thaliana*, the introduction of the Osmotic and Alkaline Resistance 1 (OAR1) gene from *Boea hygrometrica* improves tolerance to osmotic and alkaline stresses by preserving photochemical efficiency, maintaining cell membrane integrity, and regulating the expression of stress-responsive genes (Zhao et al. 2014; Zhao Yan et al. 2014). In addition to stress responses, retrotransposons also influence plant metabolic traits. In maize, they affect centromere structure and mutagenesis (DeFraia and Slotkin 2014; Gao et al. 2016). In tomatoes, differences in volatile ester profiles among fruit color variants are associated with retrotransposon insertions near esterase family genes, which influence the activity of these enzymes (Goulet et al. 2012). In addition, the LTR retrotransposons, PHRE1 and PHRE2 in *Phyllostachys edulis* (Moso bamboo) and *Arabidopsis thaliana* are activated under high-temperature stress, with their 5′ LTR regions serving as promoters that regulate transposition events (Papolu et al. 2021; Papolu et al. 2022). Non-LTR retrotransposons are equally influential, contributing to gene regulation, genome evolution, stress response, flowering time, and gene duplication. They serve as molecular markers in genetic studies and can be harnessed for transgene introduction to improve drought resistance, pest resistance, and crop yield (Hassan et al. 2024; Han 2010; Hunter 2020).

6.5 Role of Transposition in Specific Abiotic Stress Tolerance

Mobile genetic elements (MGEs) contribute to stress tolerance by modulating gene expression, generating genetic diversity, and influencing epigenetic regulation. The activation of MGEs under stress conditions leads to the mobilization of transposable elements (TEs), creating novel gene regulatory networks that enhance stress resilience. Below, we discuss the role of MGEs in specific abiotic stress responses in crop plants. Mobile genetic elements contribute significantly to abiotic stress tolerance in crops by influencing gene regulation, generating genetic diversity, and shaping epigenetic landscapes (Gul 2025). Their stress-induced activation enables plants to adapt to environmental challenges by modulating drought, salinity, temperature, heavy metal, and oxidative stress responses. Exploiting MGEs in crop improvement programs through genomic selection, transposon-based gene editing, and epigenetic modification holds promise for developing stress-resilient crop varieties.

6.5.1 MGEs in Drought Stress Tolerance

Plants respond to drought by activating stress-responsive genes, some of which are influenced by transposable elements (Lanciano and Mirouze 2018). MGEs contribute to drought adaptation by altering gene expression, creating novel regulatory elements, and enhancing water-use efficiency (Negi et al. 2016). Several retrotransposons have been shown to activate under drought conditions, leading to increased transcription of stress-related genes (Alzohairy et al. 2014). Moreover, mobile genetic elements (MGEs) can influence the expression of adjacent genes by acting through cis-regulatory elements within their sequences or by modifying DNA and chromatin methylation patterns. Moreover, MGEs can modulate the expression of neighboring genes either through cis-regulatory elements within their sequences or by altering DNA and chromatin methylation patterns. For instance, the insertion of a Miniature Inverted-Repeat Transposable Element (MITE) in the promoter region of the ZmNAC111 gene has been associated with enhanced drought tolerance in maize (Mao et al. 2015). This 82-bp MITE insertion is linked to decreased expression of ZmNAC111 in maize, and when transferred into Arabidopsis, it silences ZmNAC111 expression via the RNA-directed DNA methylation (RdDM) pathway. Functional studies demonstrate that overexpression of ZmNAC111 enhances drought tolerance, improves water-use efficiency, and upregulates stress-responsive genes in both transgenic maize and Arabidopsis seedlings (Mao et al. 2015). Similarly, transposable elements contribute to the regulation of drought-related traits through complex gene networks. Research on the Drought-Related Environment-Specific Super eQTL Hotspot on Chromosome 8 (DRESH8) and its target gene, ZmMYBR38, revealed a TE-derived inverted repeat (TE-IR)-mediated small RNA regulatory system (Sun et al. 2023). Genome-wide analyses indicate that TE-IRs are associated with drought response and agronomic traits that were selectively expanded during maize domestication. These findings underscore the importance of TE-IRs in post-transcriptional regulation, highlighting potential targets for breeding stress-resilient yet high-yielding crops (Sun et al. 2023). A study by Thieme et al. (2022) proposed that heat-induced transposition enhances drought tolerance in *Arabidopsis thaliana* through a key mutation, involving an exonic insertion of the ONSEN retrotransposon in the ribose-5-phosphate isomerase 2 (RPI2) gene. This insertion led to a loss-of-function mutation, resulting in reduced photosynthesis and lower water consumption a beneficial trait under drought conditions. This study offers valuable empirical evidence that underscores the adaptive potential of stress-responsive transposable elements (TEs) in eukaryotes. By illuminating the intricate interactions between TEs and their host genomes, our findings emphasize the importance of TE-mediated loss-of-function mutations in stress adaptation, especially in the context of global climate change.

6.5.2 MGEs in Salinity Stress Tolerance

Salt stress disrupts cellular homeostasis and causes osmotic imbalance in plants. MGEs contribute to salinity tolerance by modifying stress-responsive genes, enhancing antioxidant defense systems, and regulating ion transport mechanisms. For instance, MGE-induced expression of AtHKT1;1 enhances Na^+ exclusion and salinity tolerance in *Arabidopsis* (Venkataraman et al. 2021). Plett et al. (2010) used a GAL4 GFP enhancer trap rice line where AtHKT1;1 was specifically expressed in the root cortex. Compared with the background GAL4 GFP line, transgenic rice plants expressing AtHKT1;1 exhibited significantly improved salinity tolerance. This represents an important advancement in developing abiotic stress tolerance in crop plants through targeted modifications in mineral transport (Plett et al. 2010). Moreover, DNA methylation alterations in TE regions can activate genes involved in salinity tolerance, such as SOS1 (Salt Overly Sensitive 1) in rice (Li and Guo 2023).

6.5.3 MGEs in Temperature Stress Tolerance

Extreme temperatures cause protein denaturation, membrane instability, and oxidative damage in plants. MGEs contribute to heat and cold stress adaptation by regulating temperature-responsive genes and influencing chromatin remodeling. For example, the activation of *ONSEN*, an LTR-*Copia*-type retrotransposon in *Arabidopsis thaliana*, is associated with high-temperature tolerance (Cavrak et al. 2014).

6.6 Biotechnological Applications of MGEs in Crop Improvement

Mobile genetic elements (MGEs), including transposable elements (TEs) and endogenous retroviruses, play a significant role in plant genome evolution and stress adaptation. Their ability to mobilize and restructure genomes has been harnessed in biotechnology for improving crop resilience to abiotic stresses, enhancing genetic diversity, and facilitating targeted genome editing. The biotechnological applications of MGEs in crop improvement encompass transposon tagging, gene expression regulation, genome engineering, stress-resilient breeding, and epigenome modifications. This section explores various ways MGEs can be leveraged for crop improvement through advanced molecular and genomic tools.

6.6.1 Transposon Tagging for Gene Discovery

Transposon tagging is a powerful tool for functional genomics, enabling the identification of genes involved in abiotic stress tolerance. Transposons can insert themselves into genes, causing mutations that reveal gene function. The Ac/Ds transposon system has been used to identify genes related to drought and salt stress. For example, the ac/ds transposon tagging system widely used for gene expression profiling in abiotic stress in rice (Rengasamy 2009; Nugroho et al. 2011; Jiang et al. 2018; Greco et al. 2003), tomato (Carter et al. 2013), wheat (Sigmaz et al. 2015). Another system Tos17 used for the identification of genes involved in ABA signaling and osmotic stress responses (Li et al. 2021; Kumar and Rani 2023). En/Spm System is used to generate mutations in genes associated with cold stress resistance (Marsch-Martínez and Pereira 2011).

6.6.2 Transposons as Delivery Vectors

Transposons serve as efficient vectors for gene transfer in crops, providing an alternative to Agrobacterium-mediated transformation. These mobile genetic elements (MGEs) move DNA within and between genomes using only two components, the transposon's DNA sequence and its transposase enzyme. Several transposon systems are widely used in research, including Sleeping Beauty, piggyBac, and Tol2, each with mutagenesis, transgenesis, and gene transfer applications. The recent discovery of RNA-guided transposons enables CRISPR-based precision for inserting large DNA fragments (>100 kB) into specific genomic sites. The piggyBac transposon, originally from the cabbage looper moth, is notable for its seamless excision, making it ideal for precise genome editing. Studies have demonstrated its efficacy in integrating stress-responsive genes in rice and tobacco (Nishizawa-Yokoi and Toki 2023). The piggyBac targets TTAA sequences and excises without leaving footprints. It can mobilize DNA fragments greater than 100 kB across various organisms and preferentially integrates at transcription start sites and regulatory elements. The hyperactive variant, hyPB, increases transposition efficiency tenfold in mammalian cells. Tol2, discovered in medaka fish, integrates at TNA(C/G)TTATAA(G/C)TNA sites and can deliver DNA fragments up to 200 kB, favoring regulatory elements (Koga et al. 2003). A truncated version, miniTol2, enhances transposition efficiency threefold. Sleeping Beauty, reconstructed from inactive fish transposons, integrates at TA dinucleotides and has a high cargo capacity exceeding 100 kB, though efficiency decreases with larger DNA inserts (Ivics et al. 1997). The SB100X variant improves efficiency 100-fold, with hySB100x adding another 30% boost (Jin et al. 2011).

DNA transposons' reliance on specific target sequences is a limitation, leading to random insertions. The INTEGRATE system, derived from Vibrio cholerae, overcomes this by using a Type I-F CRISPR-Cas system for precise DNA targeting. This system includes a CRISPR RNA (crRNA) that guides the process, a QCascade complex for DNA binding, a transposase complex for integration, and donor DNA (mini-transposon) inserted approximately 50 bp downstream of the target sequence (Vo et al. 2021). This approach achieves nearly 100% efficiency for inserting up to 10 kB DNA fragments in bacterial genomes.

Transposon insertions generate genetic polymorphisms valuable for stress-resistance breeding. Retrotransposon-based markers have been used in drought-tolerant wheat and rice (Kalendar et al. 2011), while IRAP and REMAP techniques facilitate genotyping of plant genomes (Kalendar and Schulman 2006). Miniature Inverted Repeat Transposable Elements (MITEs) in maize have been linked to salt tolerance (Mao et al. 2015), and LTR retrotransposons enhance heat and drought resistance in *Arabidopsis* and barley (Thieme et al. 2022; Paul et al. 2023). In rice, transposon-generated variants have been associated with cold resistance (Zhang et al. 2024). These transposon-based strategies contribute to improving crop resilience and advancing agricultural productivity.

6.7 Future Perspectives and Challenges

While mobile genetic elements (MGEs) offer exciting potential for crop improvement, their application presents several challenges that require further research. One key challenge is controlling unintended transposition, ensuring that MGE activation does not induce genome instability. Additionally, developing precise regulatory mechanisms to manage stress-induced MGE activation is critical to avoid undesirable effects. Biosafety concerns and public acceptance of MGE-based genome modifications also remain important issues to address. As MGEs are harnessed for crop improvement, it will be essential to integrate AI-based genome analysis to predict beneficial MGE insertions that enhance crop resilience. Looking forward, exploring the potential of MGEs for multi-stress tolerance is crucial. This includes engineering crops that can endure combined drought, heat, and salinity stress. Moreover, investigating the role of MGEs in horizontal gene transfer in crops could open new avenues for introducing novel traits through natural genetic exchange mechanisms.

6.8 Conclusion

The biotechnological potential of MGEs in crop improvement is vast, spanning gene discovery, genome editing, stress-adaptive breeding, and synthetic biology. Harnessing these elements offers a promising strategy for developing resilient crops suited to changing climate conditions. With genome engineering and epigenetic control advancements, MGEs can be leveraged as precision tools for sustainable agricultural development. Future research in MGE-based biotechnology will pave the way for innovative crop improvement strategies that enhance food security and environmental resilience. Mobile genetic elements significantly contribute to plant abiotic stress tolerance by generating genetic diversity, modulating gene expression, and facilitating epigenetic adaptations. Harnessing MGEs through modern biotechnological tools can provide novel strategies for developing stress-resilient crops in the face of climate change and environmental challenges. The dynamic nature of MGEs and their role in plant stress adaptation underscore their importance in evolutionary biology and crop improvement. Understanding the mechanisms governing MGE activity and their interactions with the genome will pave the way for novel biotechnological applications aimed at enhancing abiotic stress tolerance in crops.

Further research into TE regulation, epigenetic modifications, and biotechnological applications will help harness the full potential of MGEs for sustainable agriculture and food security.

References

Alzohairy AM, Sabir JS, Gyulai G, Younis RA, Jansen RK, Bahieldin A (2014) Environmental stress activation of plant long-terminal repeat retrotransposons. Funct Plant Biol 41:557–567

Ay N, Clauss K, Barth O, Humbeck K (2008) Identification and characterization of novel senescence-associated genes from barley (Hordeum vulgare) primary leaves. Plant Biol 10:121–135

Bennetzen JL, Wang H (2014) The contributions of transposable elements to the structure, function, and evolution of plant genomes. Annu Rev Plant Biol 65:505–530

Boonjing P, Masuta Y, Nozawa K, Kato A, Ito H (2020) The effect of zebularine on the heat-activated retrotransposon ONSEN in Arabidopsis thaliana and Vigna angularis. Genes Genet Syst 95:165–172

Buchmann RC, Asad S, Wolf JN, Mohannath G, Bisaro DM (2009) Geminivirus AL2 and L2 proteins suppress transcriptional gene silencing and cause genome-wide reductions in cytosine methylation. J Virol 83:5005–5013

Butelli E, Licciardello C, Zhang Y, Liu J, Mackay S, Bailey P et al (2012) Retrotransposons control fruit-specific, cold-dependent accumulation of anthocyanins in blood oranges. Plant Cell 24:1242–1255

Butelli E, Garcia-Lor A, Licciardello C, Las Casas G, Hill L, Recupero GR et al (2017) Changes in anthocyanin production during domestication of citrus. Plant Physiol 173:2225–2242

Butelli E, Licciardello C, Ramadugu C, Durand-Hulak M, Celant A, Recupero GR et al (2019) Noemi controls production of flavonoid pigments and fruit acidity and illustrates the domestication routes of modern citrus varieties. Curr Biol 29:158–164. e152

Carter JD, Pereira A, Dickerman AW, Veilleux RE (2013) An active ac/ds transposon system for activation tagging in tomato cultivar M82 using clonal propagation. Plant Physiol 162:145–156

Cavrak VV, Lettner N, Jamge S, Kosarewicz A, Bayer LM, Mittelsten Scheid O (2014) How a retrotransposon exploits the plant's heat stress response for its activation. PLoS Genet 10:e1004115

Chuong EB, Elde NC, Feschotte C (2017) Regulatory activities of transposable elements: from conflicts to benefits. Nat Rev Genet 18:71–86

de Assis R, Baba VY, Cintra LA, Gonçalves LSA, Rodrigues R, Vanzela ALL (2020) Genome relationships and LTR-retrotransposon diversity in three cultivated capsicum L.(Solanaceae) species. BMC Genomics 21:1–14

De Felice B, Wilson RR, Argenziano C, Kafantaris I, Conicella C (2009) A transcriptionally active copia-like retroelement in Citrus Limon. Cell Mol Biol Lett 14:289–304

DeFraia C, Slotkin RK (2014) Analysis of retrotransposon activity in plants. In: Plant epigenetics and epigenomics: methods and protocols, pp 195–210

Deneweth J, Van de Peer Y, Vermeirssen V (2022) Nearby transposable elements impact plant stress gene regulatory networks: a meta-analysis in a thaliana and S lycopersicum. BMC Genomics 23:18

Duan K, Ding X, Zhang Q, Zhu H, Pan A, Huang J (2008) AtCopeg1, the unique gene originated from AtCopia95 retrotransposon family, is sensitive to external hormones and abiotic stresses. Plant Cell Rep 27:1065–1073

Eickbush TH, Malik HS (2007) Origins and evolution of retrotransposons. Mobile DNA ii:1111–1144

Feschotte C, Pritham EJ (2007) DNA transposons and the evolution of eukaryotic genomes. Annu Rev Genet 41:331–368

Fiol A, García S, Dujak C, Pacheco I, Infante R, Aranzana MJ (2022) An LTR retrotransposon in the promoter of a PsMYB10. 2 gene associated with the regulation of fruit flesh color in Japanese plum. Hortic Res 9:uhac206
Galindo-González L, Sarmiento F, Quimbaya MA (2018) Shaping plant adaptability, genome structure and gene expression through transposable element epigenetic control: focus on methylation. Agronomy 8:180
Gao D, Li Y, Kim KD, Abernathy B, Jackson SA (2016) Landscape and evolutionary dynamics of terminal repeat retrotransposons in miniature in plant genomes. Genome Biol 17:1–17
Glazko V, Kosovsky GY, Glazko T (2022) The sources of genome variability as domestication drivers. Sel'skokhozyaistvennaya Biologiya [Agri Biol] 57:832–851
Goulet C, Mageroy MH, Lam NB, Floystad A, Tieman DM, Klee HJ (2012) Role of an esterase in flavor volatile variation within the tomato clade. Proc Natl Acad Sci 109:19009–19014
Grandbastien M-A (2015) LTR retrotransposons, handy hitchhikers of plant regulation and stress response. Biochim Biophys Acta Gene Regul Mech 1849:403–416
Greco R, Ouwerkerk PB, De Kam RJ, Sallaud C, Favalli C, Colombo L et al (2003) Transpositional behaviour of an Ac/Ds system for reverse genetics in rice. Theor Appl Genet 108:10–24
Gul A (2025) Retrotransposons and plant stress responses. In: Plant Retrotransposons. CRC Press, pp 244–265
Hamdan MF, Karlson CKS, Teoh EY, Lau S-E, Tan BC (2022) Genome editing for sustainable crop improvement and mitigation of biotic and abiotic stresses. Plan Theory 11:2625
Han JS (2010) Non-long terminal repeat (non-LTR) retrotransposons: mechanisms, recent developments, and unanswered questions. Mob DNA 1:15
Han Y, Qin S, Wessler SR (2013) Comparison of class 2 transposable elements at superfamily resolution reveals conserved and distinct features in cereal grass genomes. BMC Genomics 14:1–10
Hassan AH, Mokhtar MM, El Allali A (2024) Transposable elements: multifunctional players in the plant genome. Front Plant Sci 14:1330127
Hirsch CD, Springer NM (2017) Transposable element influences on gene expression in plants. Biochim Biophys Acta, Gene Regul Mech 1860:157–165
Hunter RG (2020) Stress, adaptation, and the deep genome: why transposons matter. Integr Comp Biol 60:1495–1505
Hwang J, Ye D-Y, Jung GY, Jang S (2024) Mobile genetic element-based gene editing and genome engineering: recent advances and applications. Biotechnol Adv 108343
Ivics Z, Hackett PB, Plasterk RH, Izsvák Z (1997) Molecular reconstruction of sleeping beauty, a Tc1-like transposon from fish, and its transposition in human cells. Cell 91:501–510
Jeung J-U, Cho SK, Lee SJ, Shin JS (2005) Characterization of Ty3-gypsy-like elements in Hibiscus syriacus. Mol Cells 19:318–327
Jiang S-Y, Ma A, Vanitha J, Xie L, Ramachandran S (2018) Functional characterization of transposon-tagged abiotic stress-responsive rice genes and their molecular polymorphisms among various stress-tolerant genotypes. Mol Breed 38:1–15
Jin Z, Maiti S, Huls H, Singh H, Olivares S, Mátés L et al (2011) The hyperactive sleeping beauty transposase SB100X improves the genetic modification of T cells to express a chimeric antigen receptor. Gene Ther 18:849–856
Jordan K (2009) Transposable elements, chromatin and gene regulation. In: Genomic impact of eukaryotic transposable elements, vol 10
Jovović Z, Velimirović A, Yaman N (2025) Climate and crop production crisis. In: Agriculture and water management under climate change. Springer, pp 1–28
Kalendar R, Schulman AH (2006) IRAP and REMAP for retrotransposon-based genotyping and fingerprinting. Nat Protoc 1:2478–2484
Kalendar R, Flavell AJ, Ellis THN, Sjakste T, Moisy C, Schulman AH (2011) Analysis of plant diversity with retrotransposon-based molecular markers. Heredity 106:520–530
Kalendar R, Amenov A, Daniyarov A (2018) Use of retrotransposon-derived genetic markers to analyse genomic variability in plants. Funct Plant Biol 46:15–29

Kim S, Park J, Yeom S-I, Kim Y-M, Seo E, Kim K-T et al (2017) New reference genome sequences of hot pepper reveal the massive evolution of plant disease-resistance genes by retroduplication. Genome Biol 18:1–11

Kimura Y, Tosa Y, Shimada S, Sogo R, Kusaba M, Sunaga T et al (2001) OARE-1, a Ty1-copia retrotransposon in oat activated by abiotic and biotic stresses. Plant Cell Physiol 42:1345–1354

Koga A, Iida A, Kamiya M, Hayashi R, Hori H, Ishikawa Y et al (2003) The medaka fish Tol2 transposable element can undergo excision in human and mouse cells. J Hum Genet 48:231–235

Kumar M, Rani K (2023) Epigenomics in stress tolerance of plants under the climate change. Mol Biol Rep 50:6201–6216

Lanciano S, Mirouze M (2018) Transposable elements: all mobile, all different, some stress responsive, some adaptive? Curr Opin Genet Dev 49:106–114

Lee S-I, Kim N-S (2014) Transposable elements and genome size variations in plants. Genome Inform 12:87

Li Y, Guo D (2023) Transcriptome and dna methylome analysis of two contrasting rice genotypes under salt stress during germination. Int J Mol Sci 24:3978

Li X, Yu B, Wu Q, Min Q, Zeng R, Xie Z et al (2021) OsMADS23 phosphorylated by SAPK9 confers drought and salt tolerance by regulating ABA biosynthesis in rice. PLoS Genet 17:e1009699

Ling A, Cordaux R (2010) Insertion sequence inversions mediated by ectopic recombination between terminal inverted repeats. PLoS One 5:e15654

Lisch D (2009) Epigenetic regulation of transposable elements in plants. Annu Rev Plant Biol 60:43–66

Lu C, Chen J, Zhang Y, Hu Q, Su W, Kuang H (2012) Miniature inverted–repeat transposable elements (MITEs) have been accumulated through amplification bursts and play important roles in gene expression and species diversity in Oryza sativa. Mol Biol Evol 29:1005–1017

Luqman-Fatah A, Nishimori K, Amano S, Fumoto Y, Miyoshi T (2024) Retrotransposon life cycle and its impacts on cellular responses. RNA Biol 21:1048–1064

Malfait J (2024) The role of Epromoters in the regulation of stress response genes. Aix-Marseile Université

Malhi GS, Kaur M, Kaushik P (2021) Impact of climate change on agriculture and its mitigation strategies: a review. Sustain For 13:1318

Mao H, Wang H, Liu S, Li Z, Yang X, Yan J et al (2015) A transposable element in a NAC gene is associated with drought tolerance in maize seedlings. Nat Commun 6:8326

Marsch-Martínez N, Pereira A (2011) Activation tagging with En/Spm-I/dSpm transposons in Arabidopsis. Methods and Protocols, Plant Reverse Genetics, pp 91–105

Matsunaga W, Kobayashi A, Kato A, Ito H (2012) The effects of heat induction and the siRNA biogenesis pathway on the transgenerational transposition of ONSEN, a copia-like retrotransposon in Arabidopsis thaliana. Plant Cell Physiol 53:824–833

Maumus F, Allen AE, Mhiri C, Hu H, Jabbari K, Vardi A et al (2009) Potential impact of stress activated retrotransposons on genome evolution in a marine diatom. BMC Genomics 10:1–19

Mita P, Boeke JD (2016) How retrotransposons shape genome regulation. Curr Opin Genet Dev 37:90–100

Negi P, Rai AN, Suprasanna P (2016) Moving through the stressed genome: emerging regulatory roles for transposons in plant stress response. Front Plant Sci 7:1448

Nishizawa-Yokoi A, Toki S (2023) Precise genetic engineering with piggyBac transposon in plants. Plant Biotechnol (Tokyo) 40:255–262

Nugroho S, Zannati A, Situmorang A, Windiastri VE, Widyajayantie D, Pantouw CF et al (2011) Phenotypic screening of ds transposon and activation-tag insertional mutant Rice population for drought and salinity tolerant related traits. In: Annales bogorienses, pp 21–25

Pachamuthu K, Borges F (2023) Epigenetic control of transposons during plant reproduction: from meiosis to hybrid seeds. Curr Opin Plant Biol 75:102419

Pandita D, Pandita A (2023) Plant transposable elements: biology and bio technology. CRC Press

Pandita D, Pandita A (2025) Retrotransposons: a 360 degree overview. Plant Retrotransp 1–26

Papolu PK, Ramakrishnan M, Wei Q, Vinod KK, Zou L-H, Yrjala K et al (2021) Long terminal repeats (LTR) and transcription factors regulate PHRE1 and PHRE2 activity in Moso bamboo under heat stress. BMC Plant Biol 21:585

Papolu PK, Ramakrishnan M, Mullasseri S, Kalendar R, Wei Q, Zou LH et al (2022) Retrotransposons: how the continuous evolutionary front shapes plant genomes for response to heat stress. Front Plant Sci 13:1064847

Paul M, Tanskanen J, Jääskeläinen M, Chang W, Dalal A, Moshelion M et al (2023) Drought and recovery in barley: key gene networks and retrotransposon response. Front Plant Sci 14:1193284

Plett D, Safwat G, Gilliham M, Skrumsager Møller I, Roy S, Shirley N et al (2010) Improved salinity tolerance of rice through cell type-specific expression of AtHKT1; 1. PLoS One 5:e12571

Ramakrishnan M, Satish L, Kalendar R, Narayanan M, Kandasamy S, Sharma A et al (2021) The dynamism of transposon methylation for plant development and stress adaptation. Int J Mol Sci 22:11387

Ramakrishnan M, Satish L, Sharma A, Kurungara Vinod K, Emamverdian A, Zhou M et al (2022) Transposable elements in plants: recent advancements, tools and prospects. Plant Mol Biol Report 40:628–645

Ramallo E, Kalendar R, Schulman AH, Martínez-Izquierdo JA (2008) Reme1, a Copia retrotransposon in melon, is transcriptionally induced by UV light. Plant Mol Biol 66:137–150

Rengasamy R (2009) Functional genomics in rice. In: Oryza Sativa L (ed) Using Ac/Ds transposon tagging system and gene expression profiling. scholarbank.nus.edu.sg

Rico-Cabanas L, Martínez-Izquierdo JA (2007) CIRE1, a novel transcriptionally active Ty1-copia retrotransposon from Citrus sinensis. Mol Gen Genomics 277:365–377

Shapiro JA (2002) Genome organization and reorganization in evolution: formatting for computation and function. Ann N Y Acad Sci 981:111–134

Shapiro JA (2013) How life changes itself: The Read–Write (RW) genome. Phys Life Rev 10:287–323

Shapiro JA (2014) Epigenetic control of mobile DNA as an interface between experience and genome change. Front Genet 5:87

Sigmaz B, Agar G, Arslan E, Aydin M, Taspinar MS (2015) The role of putrescine against the long terminal repeat (LTR) retrotransposon polymorphisms induced by salinity stress in Triticum aestivum. Acta Physiol Plant 37:1–9

Sun X, Xiang Y, Dou N, Zhang H, Pei S, Franco AV et al (2023) The role of transposon inverted repeats in balancing drought tolerance and yield-related traits in maize. Nat Biotechnol 41:120–127

Suoniemi A, Anamthawat-Jónsson K, Arna T, Schulman AH (1996) Retrotransposon BARE-1 is a major, dispersed component of the barley (Hordeum vulgare L.) genome. Plant Mol Biol 30:1321–1329

Thieme M, Brêchet A, Bourgeois Y, Keller B, Bucher E, Roulin AC (2022) Experimentally heat-induced transposition increases drought tolerance in Arabidopsis thaliana. New Phytol 236:182–194

Tittel-Elmer M, Bucher E, Broger L, Mathieu O, Paszkowski J, Vaillant I (2010) Stress-induced activation of heterochromatic transcription. PLoS Genet 6:e1001175

Underwood CJ, Henderson IR, Martienssen RA (2017) Genetic and epigenetic variation of transposable elements in Arabidopsis. Curr Opin Plant Biol 36:135–141

Vangelisti A, Usai G, Mascagni F, Natali L, Giordani T, Cavallini A (2019) A whole genome analysis of long-terminal-repeat retrotransposon transcription in leaves of Populus trichocarpa L. subjected to different stresses. Caryologia 72:69–79

Venkataraman G, Shabala S, Véry A-A, Hariharan GN, Somasundaram S, Pulipati S et al (2021) To exclude or to accumulate? Revealing the role of the sodium HKT1; 5 transporter in plant adaptive responses to varying soil salinity. Plant Physiol Biochem 169:333–342

Vo PLH, Ronda C, Klompe SE, Chen EE, Acree C, Wang HH et al (2021) CRISPR RNA-guided integrases for high-efficiency, multiplexed bacterial genome engineering. Nat Biotechnol 39:480–489

Xu Z-S, Yang Q-Q, Feng K, Xiong A-S (2019) Changing carrot color: insertions in DcMYB7 alter the regulation of anthocyanin biosynthesis and modification. Plant Physiol 181:195–207

Zeller G, Henz SR, Widmer CK, Sachsenberg T, Rätsch G, Weigel D et al (2009) Stress-induced changes in the Arabidopsis thaliana transcriptome analyzed using whole-genome tiling arrays. Plant J 58:1068–1082

Zhang C, Wang H, Tian X, Lin X, Han Y, Han Z et al (2024) A transposon insertion in the promoter of OsUBC12 enhances cold tolerance during japonica rice germination. Nat Commun 15:2211

Zhao Yan ZY, Shen ChunYing SC, Xu Tao XT, Chen ShiXuan CS, Li MeiJing LM, Deng Xin DX (2014) A 30-kb genomic DNA fragment of resurrection plant Boea hygrometrica enhances Arabidopsis osmotic and alkaline tolerance. J. Food Agric. Environ. 12, 210–215.

Zhao Y, Xu T, Shen C-Y, Xu G-H, Chen S-X, Song L-Z et al (2014) Identification of a retroelement from the resurrection plant Boea hygrometrica that confers osmotic and alkaline tolerance in Arabidopsis thaliana. PLoS One 9:e98098

Chapter 7
Genome-Wide Profiling of Abiotic Stress-Responsive Non-coding RNAs

Varucha Misra and A. K. Mall

Abstract Environmental fluctuations consistently challenge plant development and productivity by triggering a range of abiotic stressors, including water stress, soil salinization, temperature extremes, and oxidative damage. In response, plants have evolved intricate survival strategies that encompass early stress detection, dynamic epigenetic remodelling, and precise regulation of genetic activity during both RNAsynthesis and protein production stages. These mechanisms collectively enhance the plant's ability to exhibit phenotypic flexibility and adapt effectively to challenging external conditions. Recent advancements in molecular research—particularly those utilizing high-throughput transcriptomic techniques and next-generation sequencing—have considerably expanded our insights into the genomic mechanism underlying plant stress adaptation. Within this framework, regulatory RNA molecules lacking protein-coding potential, including microRNAs (miRNAs), small interfering RNAs (siRNAs), and long non-coding RNAs (lncRNAs), have gained prominence as key modulators orchestrating gene activity in response to environmental stressors. This chapter delves into the regulatory functions of ncRNAs during environmental stress adaptation, reviews advanced methodologies for their genome-wide discovery, and discusses how cutting-edge genome-editing platforms like associated proteins (Clustered Regularly Interspaced Short Palindromic Repeats (CRISPR)/Cas) systems can be harnessed to develop stress-resilient plant varieties.

Keywords Abiotic stress · ncRNA · associated proteins(Clustered Regularly Interspaced Short Palindromic Repeats (CRISPR)/Cas) · High-throughput technologies

---

V. Misra (✉) · A. K. Mall
ICAR-Indian Institute of Sugarcane Research, Lucknow, UP, India
e-mail: ashutoshkumarmall@gmail.com

S. Tiwari et al. (eds.), *Decoding Non-Coding RNA*, Concepts and Strategies in Plant Sciences, https://doi.org/10.1007/978-981-96-9859-2_7

7.1 Introduction

The escalating unpredictability of global climate patterns—marked by temperature extremes, prolonged droughts, and irregular rainfall—has emerged as a major obstacle to sustaining agricultural output and ensuring global food security (Rahman et al. 2024; Muhammad et al. 2024; Tarolli et al. 2024; Altieri et al. 2024; Vincze et al. 2024; FAO 2023). To withstand these environmental pressures, plants deploy a broad spectrum of regulatory pathways that modulate their physiological and molecular responses (Traubenik et al. 2024). Survival under such conditions is supported by sophisticated networks that control gene and protein activity at multiple regulatory stages—including DNA level transcription, RNA processing, protein synthesis, and post-synthesis modification (Ingram and Bartels 1996). Recent discoveries have highlighted the essential roles of non-coding RNAs (ncRNAs) in managing plant resilience to abiotic stresses (Traubenik et al. 2024; Imaduwage and Hewadikaram 2024; Jin et al. 2024). These RNA molecules, which include microRNAs (miRNAs), small interfering RNAs (siRNAs), and long non-coding RNAs (lncRNAs), exert control over genes that govern essential physiological functions such as embryonic development, tissue differentiation, flowering, and senescence (Correa et al. 2018; Cordeiro et al. 2022). Their involvement extends to modulating stress perception, signalling cascades, and downstream physiological responses under adverse conditions like drought, high salinity, and temperature extremes (Traubenik et al. 2024; Yadav et al. 2024; Imaduwage and Hewadikaram 2024; Jin et al. 2024). Understanding how ncRNAs regulate plant adaptation opens up new possibilities for enhancing crop performance. By uncovering their mechanistic roles, researchers can design targeted interventions to increase harvest efficiency, adaptive capacity, agronomic traits, and overall agricultural value in the context of environmental adversity.

This chapter delves into the functional significance of regulatory RNAs in plant abiotic stress responses, outlines cutting-edge technologies used for their discovery, and explores how genome engineering approaches such as Clustered Regularly Interspaced Short Palindromic Repeats (CRISPR) engineer climate-resilient crops.

7.2 Non-coding RNAs in Plants: An Overview

Although genomic DNA functions as the foundational template for RNAsynthesis, its contribution to protein production is largely indirect. Despite the fact that nearly 90% of the genome is capable of being transcribed into RNA, just a limited portion give rise to protein-coding messenger RNAs (mRNAs). This discrepancy initially led to the belief that the remainder of the transcribed sequences—termed non-coding RNAs (ncRNAs)—were unimportant remnants, often dismissed as "junk DNA" or "genomic dark matter" (Kuska 1998). However, following the development of next-generation sequencing (NGS) platforms and expansive transcriptome profiling, it has become evident that a significant portion of the eukaryotic genome produces functional RNAs molecules that are not translated into polypeptides. These ncRNAs are categorized based on their length: small ncRNAs (sRNAs) measure 18 to 30 nucleotides, medium-sized ncRNAs span 31 to 200 nucleotides, and long ncRNAs (lncRNAs) exceed 200 nucleotides (Wang et al. 2017). In plant systems, the small RNA group predominantly consists of microRNAs (miRNAs) and small interfering RNAs (siRNAs), each with distinct origins and biological roles. miRNAs—usually 21 to 24 nucleotides in size—are generated from imperfect hairpin-structured precursors and are among the most studied ncRNAs due to their influence on plant growth, organ formation, hormone signalling, and environmental stress management (D'Ario et al. 2017; Tang and Chu 2017). In contrast, siRNAs originate from long, perfectly paired double-stranded RNA precursors and often rely on RNA-dependent RNA polymerases for their amplification (Kidner and Martienssen 2004). Another emerging class of ncRNAs includes circular RNAs (circRNAs), formed through reverse splicing mechanisms events that create a covalently linked circular structure by connecting the RNA's 3′ and 5′ ends (Patop et al. 2019). lncRNAs have garnered considerable focus for the involvement in gene regulation through mechanisms such as transcriptional repression, alternative splicing modulation, and translational control (Zhu and Wang 2012). In accordance with their genomic placement and alignment with protein-coding loci, lncRNAs can be grouped into several types: intergenic ncRNAs located between the genes (lincRNAs), expressed from non-coding stretches located between protein-coding sequences; intronic ncRNAs (incRNAs), arising from intron-derived segments; and natural antisense transcripts (NATs), which are synthesized from the complementary strand of coding region (Mattrick and Rinn 2015). Additional forms are transcribed near promoter regions, gene ends, enhancer sequences, or splice junctions. As a result of the progress in modern RNA profiling platforms such as microarrays and high-throughput sequencing, the functional landscape of ncRNAs has become increasingly clear across various life forms, including plants, animals, microbes, and viruses (Chen et al. 2022; Kang et al. 2021). Numerous investigations have underscored the essential roles of ncRNAs in key developmental processes in plants—such as meristem identity, leaf architecture, and floral transition (Knauer et al. 2013; Csorba et al. 2014; Yang et al. 2019a). Moreover, mounting evidence suggests that ncRNAs are integral to how plants perceive and adapt to environmental stressors, playing central roles in gene expression regulation during both abiotic and biotic challenges (Hou et al. 2019; Jabnoune et al. 2013; Li et al. 2018; Kinoshita et al. 2012; Song et al. 2019; Zhao et al. 2018).

7.3 Exploring the ncRNA Involvement in Plant Adaptation to Harsh Environmental Conditions

In response to adverse environmental stresses, plants activate a range of adaptive strategies to sustain development and ensure survival. Emerging research has underscored the significant role of ncRNAs in these adaptive responses, revealing their distinct expression profiles under various stress scenarios (Jha et al. 2020; Vakilian 2020). These ncRNAs contribute fundamentally to stress adaptation, either by influencing gene regulatory networks or by engaging directly in molecular pathways that mitigate stress impacts (Helliwell et al. 2011).

7.3.1 miRNAs for Abiotic Stress Resilience in Plants

miRNAs are increasingly acknowledged as pivotal molecules in plants, especially in managing plant responses to non-living stress factors such as water-deficit stress. Their broad impact on gene expression makes them valuable targets for enhancing stress resilience and crop productivity (Pandita 2019). During environmental stress, specific miRNAs are upregulated and can inhibit genetic elements that negatively affect stress tolerance—for instance, genes that repress the activation of stress-responsive pathways (Sunkar et al. 2007). Conversely, when miRNAs are down-regulated, their target genes—often positive regulators of stress response—can accumulate and support adaptive mechanisms. miRNA expression patterns are highly dynamic and vary depending on the type and severity of the abiotic stress encountered. Some miRNAs are consistently responsive to a broad range of environmental cues. In *Arabidopsis*, for instance, miR398 expression is suppressed under oxidative stress, salinity, Abscisic Acid (ABA) treatment, and elevated copper levels. Meanwhile, miR169a and miR169c are downregulated during drought and Abscisic Acid (ABA) exposure (Sunkar et al. 2012). In contrast, miR393 is activated under conditions like salinity, low temperatures, drought, Abscisic Acid (ABA) treatment, and UV-B radiation. Notably, miR169 in *Arabidopsis* is activated by salt and cold but repressed by drought and Abscisic Acid (ABA), underscoring the complex and sometimes contradictory regulatory nature of miRNAs. Studies in crops such as maize and rice further reveal that stress conditions may differentially influence miRNA expression, not only between species but also among different tissues such as roots and shoots (Ding et al. 2009; Kantar et al. 2010). These differences may stem from environmental variations or the intensity of stress experienced by individual tissues. Moreover, a subset of miRNAs is evolutionarily retained among diverse plant lineages, highlighting their indispensable roles in stress adaptation. Drought stress influences the transcription of numerous genes and metabolic components such as dehydrins, vacuolar acid invertase, Glutathione S- transferases (GST), LEA (Late Embryogenesis Abundant), RAB (Ras related proteins), COR (One Carbon (1C) metabolism), Rubisco, helicases, proline, and sugars (Nezhadahmadi et al. 2013). miRNAs likely act as upstream regulators of these drought-responsive elements. For example, conserved miRNAs are involved in orchestrating responses to drought by fine-tuning gene expression networks (Islam et al. 2022). When exposed to water deficit, miRNAs may exhibit altered expression patterns, which in turn regulate genes linked to hormone signalling, ROS detoxification, and metabolism (Wirsiy and Casir 2024). In drought-exposed rice seedlings, 17 miRNAs were suppressed, while 16 were activated (Zhang et al. 2016). Similarly, in maize, drought triggered the activation of 8 and suppression of 13 miRNAs in leaves; in roots, 7 miRNAs were induced and 7 were suppressed (Liu et al. 2019). Elevated expression of miR156, 160, 162, 164, and 171 was associated with the downregulation of key target genes. For example, miR160 targets ARF16 and ARF17, both involved in auxin-mediated root development. Suppression of these targets helps limit growth, enhancing drought tolerance. Additionally, miR169 downregulation increases NF-YA accumulation, which in turn activates nitrate transporters like *AtNRT2.1* and *AtNRT1.1* (Zhao et al. 2011). In rice, drought induces miR169g and miR393, where miR393 suppresses TIR1, an auxin receptor involved in plant growth regulation (Shukla et al. 2008). In *Medicago truncatula*, miR398a/b and miR408 are upregulated in both shoots and roots under drought, while miR169 is repressed in roots. However, miR169 does not show antagonistic regulation of its target *MtHAP2-1*, suggesting a limited role in drought response in this legume (Trindade et al. 2010). The increased expression of miR398a/b and miR408 corresponds to reduced levels of their targets like Cu/Zn superoxide dismutases and cytochrome c oxidase, supporting drought acclimation and copper balance. Differential miRNA responses across plant species under drought stress have been frequently reported. For example, miR159 is reduced in *Arabidopsis* and maize but increased in rice and wheat. miR399 is elevated in maize and wheat but decreased in rice. Likewise, miR168, miR397, and miR188 show increased levels in *Arabidopsis* but reduced expression in maize (Liu et al. 2019; Akdogan et al. 2016). miR166, miR172, miR827, and miR1432 are often suppressed under drought in crops like maize, rice, and wheat (Zhou et al. 2010). Despite some inconsistencies, the conserved nature of these miRNAs highlights their fundamental biological roles. For instance, miR171 shows varying expression in drought-stressed maize versus drought-stressed rice (Cheah et al. 2015; Akdogan et al. 2016), suggesting species-specific regulatory functions. The distinct expression within plant tissues is an important factor for plant. In peach, roots showed a higher abundance of miRNAs than leaves under drought, indicating greater root sensitivity (Eldem et al. 2012). In maize, drought repressed miR166 within foliar tissue while enhancing its levels in the roots (Liu et al. 2019). Conversely, in wheat, several miRNAs were activated in foliar tissues but suppressed in root zone (Akdogan et al. 2016). Similarly, in barley, miRNAs like hvu-miR156a, hvu-miR166, and hvu-miR408 were induced in leaves, while only hvu-miR166 was suppressed in roots during drought (Kantar et al. 2010). The method of inducing drought stress also alters miRNA profiles. For instance, polyethylene glycol (PEG)-mediated drought stress does not fully mimic natural field conditions, often leading to distinct expression patterns. In maize, miR398 was upregulated in response to PEG-induced osmotic stress, yet suppressed during water deficit (Li et al. 2013; Wei et al. 2009). Similar discrepancies were observed for miR159 and miR166 (Wei et al. 2009; Wang et al. 2014a, b). Under moderate water deficit, miR156 is often suppressed, while severe water deficit induces it. PEG treatments sometimes reversed these effects depending on exposure duration (Li et al. 2013). Likewise, miR399 was induced under field water deficit but suppressed by PEG (Li et al. 2013), highlighting that stress simulation methods can influence miRNA behaviour. Furthermore, miR393, miR160, and miR167 are upregulated under both water deficit and salt stress (Sunkar et al. 2012). miR169, a key player in water-deficit and salt stress responses, is repressed via an Abscisic Acid (ABA)-mediated pathway to upregulate its target *NFYA5*, which boosts the expression of stress-adaptive genes (Li et al. 2008). Interestingly, some miRNAs show dual behaviour under similar stress in different species. For instance, miR398a/b in *Medicago truncatula* has been reported to be both induced and suppressed under drought (Ferdous et al. 2015). miR156, miR164, and miR171 also display contrasting regulation patterns in maize and rice (Cheah et al. 2015; Liu et al. 2019; Zhou et al. 2010). Overall, the variability in miRNA expression across species, tissues, and treatment conditions underscores the multifaceted regulation orchestrated by miRNAs in response to plant stressors. Further in-depth studies are necessary to decode these regulatory networks and harness them for engineering drought-resilient crops.

Under heat stress, plant microRNAs (miRNAs) play pivotal roles in enhancing thermotolerance by modulating key developmental processes such as the transition from vegetative development to the flowering stage and the formation of floral structures (Zhao et al. 2016). These heat-responsive miRNAs and their corresponding targets form intricate regulatory circuits that oversee vital physiological mechanisms, including protein folding, antioxidant defence systems, preservation of photosynthetic efficiency, and protection of reproductive tissues. They are also involved in regulating miRNA synthesis itself (Ding et al. 2020). Among the miRNAs affected by heat, the miR169 group showed a predominant upregulation. Conversely, miRNAs like miR159, miR156, and miR160 were consistently found to be downregulated under elevated temperatures. In tomato, the miR167, miR396, and miR482 families were notably enriched in stamens, while miR159, miR166, and miR482 were active in pistils when exposed to 35 °C. Despite their tissue-specific expression, none of the miRNAs were prominently upregulated in both floral organs, suggesting that prolonged heat exposure may lead to a general decline in miRNA expression. Functional validation through techniques such as Gene Ontology (GO) and Kyoto Encyclopedia of Genes and Genomes (KEGG) pathway analysis, quantitative real-time PCR (qRT-PCR), and RNA ligase–mediated rapid amplification of cDNA ends (RLM-5′RACE) has revealed that miR393 targets TIR1, and miR160 regulates SlARF10/16, both key players in auxin signalling. Meanwhile, miR398 and miR397 participate in ROS-driven growth regulation by targeting genes like *S1CSD1* and laccases (LACs). miR156, through its interaction with SPL transcription factors, serves as a master regulator under heat stress conditions, influencing various developmental and physiological responses (Pan et al. 2017). In heat-tolerant wild tomato varieties exposed to moderate heat (33 °C), miR156 and miR396 were activated, while miR168 expression declined, underscoring the stress-specific and genotype-dependent nature of miRNA regulation.

At elevated temperatures around 40 °C, the expression of certain microRNAs such as miR319 and miR398 is significantly diminished, indicating that extreme heat conditions can suppress miRNA activity (Zhou et al. 2016). Experimental manipulation using synthetic miR160 mimics or inhibitors, which block the natural activity of miR160, led to weakened heat stress resistance in plants. In contrast, boosting miR160 levels through genetic modification improved thermotolerance in *Arabidopsis*, highlighting its protective role. This miRNA influences auxin-responsive regulators ARF10, ARF16, and ARF17, thereby modulating the production of heat-inducible proteins and contributing to plant survival during elevated temperatures (Lin et al. 2018). Similarly, the enhancement of miR156 expression has been associated with improved heat stress resilience in alfalfa (Arshad et al. 2020). In Chinese cabbage, exposure to 38 °C resulted in the detection of multiple miRNAs such as miR156, miR159, miR168, miR171, and miR1885 (Ahmed et al. 2019). Conversely, wheat plants under heat stress conditions showed a reduction in levels of miR156, Tae-miR818, miR169, miR528, and miR398 (Ragupathy et al. 2016). Among these, miR528 and miR9662 were found to be responsive to heat and linked to antioxidant pathways and mitochondrial stress–related proteins (Ravichandran et al. 2019). Terminal heat exposure in wheat has pronounced effects on both grain quality and yield. miRNA-based SSR markers such as miR159c and miR165b have emerged as useful tools for identifying heat-resilient wheat genotypes, offering potential applications in breeding programs for improved heat tolerance (Sihag et al. 2021). Furthermore, in Populus, suppression of miR168 led to increased expression of its target AGO1, indicating a fine-tuned regulatory mechanism under thermal stress (Chen et al. 2012). Numerous miRNA-target regulatory modules have been implicated in heat response across various plant species. Notable examples include miR169-SBP, miR159-MYB, miR156-SBP/SPL, miR172-AP2/ERF, miR164-NAC, miR166-HD-ZIP, miR396-GRF, and miR5381-SAC2. However, due to interspecies variation in stress treatments and responses, miRNA regulatory patterns under heat stress remain largely species specific and difficult to generalize (Zhang et al. 2019a, b). Among all, miR156 is recognized as a central regulator of heat tolerance, frequently upregulated in stressful conditions. Its suppression of SPL transcription factors at the mRNA level is key to managing stress responses (Stief et al. 2014). In bananas, miR156, together with miR535, targets SPL genes and is downregulated during heat exposure, indicating a coordinated role in thermal stress management (Zhu et al. 2019). In cotton, various miRNAs take part in different stages of anther development during heat stress. For instance, miR2949, miR167, and miR160 influence early sporogenous cell formation; miR156 and miR172 are involved during meiosis; miR156 also plays a role during microspore release; and miR393 along with miR3476 functions in pollen maturation. These regulatory RNAs contribute to the plant's reproductive development under thermal stress, in both heat-sensitive and heat-tolerant cotton cultivars (Chen et al. 2020).

miRNAs serve as vital regulators in plants' adaptive responses to chilling stress, functioning across a broad spectrum of plant species. According to Megha et al. (2018), these small RNAs orchestrate gene expression by modulating specific genetic pathways that facilitate tolerance to low temperatures. Various miRNAs, such as miR319, miR394, miR398, and miR156, are regarded as central components in the plant response to chilling conditions. In important crops like rice, sugarcane, and cassava, miR319 has demonstrated a role in enhancing cold resistance by downregulating TCP (teosinte branched1/cycloidea/proliferating cell factor) regulatory proteins, which participate in developmental and growth related pathways that must be reprogrammed during stress. The highly conserved miR394 is implicated in cold stress acclimation in both monocot and dicot species, with a well-established function in *Arabidopsis*. In rice, miR1320 overexpression has been linked to enhanced cold stress resilience. This miRNA targets the AP2/ERF family transcription factor OsERF096, which influences cold response through the jasmonic acid (JA)-dependent signalling cascade (Sun et al. 2022). Likewise, boosting the levels of miR156, miR319, and miR528 in rice has shown a positive effect on cold tolerance (Huo et al. 2022). miR319, in particular, confers chilling resistance by regulating the expression of OsPCF6 and OsTCP21, both members of the TCP (teosinte branched1/cycloidea/proliferating cell factor) transcription factor family. Suppression of these genes has been correlated with improved cold stress adaptation in rice (Wang et al. 2014a, b). This expanding field of study highlights the intricate function of miRNA-mediated gene regulation in enhancing plant resilience to cold-induced challenges.

In *Arabidopsis thaliana*, the expression of miR398 is notably inhibited under conditions such as oxidative stress, salinity, abscisic acid (Abscisic Acid (ABA)) exposure, and elevated copper ion (Cu^{2+}) levels. Under oxidative stress, the suppression of miR398 permits the upregulation of its targets—two key Cu/Zn superoxide dismutases, CSD1 and CSD2. These enzymes are essential for neutralizing ROS, thereby connecting miR398 activity to the plant's oxidative stress response mechanism (Bonnet et al. 2004; Jones-Rhoades and Bartel 2004). Moreover, genetically modified *Arabidopsis* lines engineered to be unresponsive to miR398—thereby accumulating more CSD2 protein—exhibit enhanced tolerance to oxidative challenges. These include environmental stressors such as intense light exposure and toxic heavy metals, highlighting the protective role of elevated superoxide dismutase activity when miR398 suppression is alleviated (Sunkar et al. 2006).

miRNAs serve as key modulators in plant adaptation to salt stress, particularly by managing the balance of ROS and safeguarding cellular integrity. This role has been observed in several species, including *Arabidopsis thaliana*, cotton, *Phaseolus vulgaris*, and European aspen, where miRNAs contribute to structural resilience and improved stress tolerance (Xie et al. 2015; Naya et al. 2014; Jia et al. 2009). In *Arabidopsis*, miR398 suppresses the expression of cytosolic and plastid Cu/Zn superoxide dismutases (CSD1 and CSD2), while in moss, miR1073 targets similar antioxidant enzymes (Higashi et al. 2013), and in cowpea, miR408 regulates translation-related proteins (Paul et al. 2011). These regulatory mechanisms enable better membrane stability through precise ROS control. Beyond ROS detoxification, miRNAs play a part in protein degradation, cellular signalling, and metabolic regulation under salinity stress. For example, in *Populus euphratica*, miR894 modulates the *UBE2H* gene involved in proteolysis, while miR2867 supports genome maintenance by targeting replication factor C (RFC). Calcium signalling, essential for stress response, is influenced by miR5077 and miR2619, which regulate PLC and PPP3C, respectively. Additionally, miR159 modulates the MAPK pathway by targeting *ATM*, a gene linked to programmed cell death under salt conditions (Yang et al. 2015; Sachin and Birendra Prasad 2015). The miR156 group also plays a pivotal role in salinity tolerance through its *interactions* with SPL regulatory proteins such as AtSPL10 and AtSPL11, promoting developmental transitions and seedling establishment (Liu et al. 2008; Nodine and Bartel 2010). In alfalfa, increased levels of miR156 have been associated with improved seed germination and early plant development during salt exposure (Muhammad et al. 2017). Likewise, in rice and *Arabidopsis*, miR393 enhances salinity resistance by targeting TIR1, leading to downstream modulation of salt-responsive genes like *P5CDH* and *SR05* (Gao et al. 2011). Maize demonstrates elevated miR159 levels under saline conditions, which suppress MYB transcription factors within the Abscisic Acid (ABA) pathway, thereby enhancing seed germination and overall salt tolerance (Ding et al. 2009; Xue et al. 2017). In rice, miR169 is upregulated during salt stress, targeting NF-YA, which reduces Abscisic Acid (ABA) sensitivity and boosts tolerance when expressed in *Arabidopsis* (Sunkar and Zhu 2004). In *Spathiphyllum* roots under salt stress, conserved and novel miRNAs such as PC-3p-153871_47 and PC-3p-613_8603 are upregulated, whereas PC-5p-838_6216 is suppressed, indicating diverse miRNA responses (Nai et al. 2023). In *Chrysanthemum*, miR398 targets CSD2 and copper chaperone *CCS* genes, both involved in ROS detoxification. Overexpressing a salt-resistant variant of Os-CSD2 in rice enhances salinity tolerance (Covarrubias and Reyes 2010; Chen et al. 2015). However, temporal expression varies—miR398 is repressed within 12 h in *Chrysanthemum*, while in *Arabidopsis*, the effect appears after 48 h (Zhao et al. 2019). In mustard, CSD1 and CSD2 negatively regulate miR398, underscoring species-specific regulatory dynamics. Another notable regulator is miR168, which targets AGO1, a crucial protein in the RNA-induced silencing complex. In *Chrysanthemum* roots, miR168 is markedly upregulated under salt stress, suppressing AGO1 expression. This leads to altered miRNA processing and broader activation of stress response pathways (Beauclair et al. 2010; Lu et al. 2010). Several other miRNAs have also been implicated in salinity adaptation. In African cultivated rice (*Oryza glaberrima*), miR399i, miR396-5p, miR159b, miR398b, and others influence diverse metabolic activities tied to salt response. In the halophytic wild rice *Oryza coarctata*, OcmiR12477 is a key regulator of salt resilience (Mondal et al. 2018; Parmar et al. 2020). Additionally, in Vicia faba Abscisic Acid (ABA), miR393 and miR171 are activated under salt exposure, emphasizing the conserved and divergent miRNA regulatory frameworks across plant taxa (Alzahrani et al. 2019).

In maize, the regulatory functions of miRNAs under shaded environments have been well documented. According to Yuan et al. (2016), miR319 and miR166 play roles in downregulating TCP (TEOSINTE BRANCHED1/CYCLOIDEA/PROLIFERATING CELL FACTOR) transcription factors, which facilitate enhanced shade adaptation. In contrast, exposure to low-light conditions elevates the expression of miR164, which targets and reduces transcripts encoding NAC-domain proteins, ultimately leading to diminished ear development and lower crop productivity. In *Arabidopsis*, Xie et al. (2017) found that suppression of MIR156 expression intensified the shade avoidance response. Similarly, in rice, Panigrahy et al. (2022) demonstrated that miR172b, through regulation of OsbHLH153, detrimentally affects pollen viability under shaded environments. Although shade-responsive miRNA mechanisms have been extensively studied in crops like maize, rice, and Arabidopsis, research in potatoes is still emerging. Several conserved miRNA families have been identified in potato (*Solanum tuberosum*) that are involved in key biological processes including transcriptional regulation, signal transduction, and hormone-mediated development. These include miR156, miR164, miR166, miR167, miR172, miR171, miR390, miR394, miR395, miR398, miR399, miR414, miR530, miR778, and miR829 (Zhang et al. 2009a, b, c; Zhang et al. 2013; Lakhotia et al. 2014; Rey-Burusco et al. 2019). A substantial subset of miRNAs is recognized for its involvement in countering microbial threats and managing abiotic challenges. However, the specific involvement of these small RNAs in modulating potato responses to shade or low light remains an underexplored area of study, warranting further investigation.

Under low-oxygen (hypoxic) environments, where plants shift their energy production from oxygen-dependent pathways towards fermentation-based processes (Bailey-Serres and Voesenek 2008), miRNAs have been recognized as key components in managing stress adaptation. In *Arabidopsis*, water-deficit conditions trigger an increase in the expression of miR393, miR397b, and miR402, whereas miR319c and miR389a exhibit reduced expression levels (Sunkar and Zhu 2004). Similarly, nutrient limitation, particularly nitrogen deficiency, elicits miRNA-mediated responses. In both *Arabidopsis* and maize, nitrogen scarcity leads to the upregulation of multiple miRNAs such as miR169, miR171, miR395, miR397, miR398, miR399, miR408, and miR827 (Nguyen et al. 2015; Zhao et al. 2013). Additional studies report that miR160, miR162, miR395, and miR827 also show increased expression when plants are subjected to nitrogen-limited environments (Zhou et al. 2010; Akdogan et al. 2016; Liu et al. 2008). These findings underscore the diverse and dynamic roles of miRNAs in helping plants adjust to changing oxygen and nutrient conditions.

miRNAs are instrumental in post-transcriptionally controlling gene activity during exposure to mercury (Hg) toxicity. According to Zhou et al. (2008), several miRNAs—including miR171, miR319, miR393, and miR529—are significantly upregulated in response to Hg stress, while others like miR166 also show elevated expression. However, miR319, miR160, and miR395 exhibit minimal or no changes in their expression levels under the same conditions. A particularly noteworthy observation is the suppression of miR398 during heavy metal exposure and nutrient deprivation, indicating its vital function in managing ROS during such stresses (Ding et al. 2011). In relation to heavy metal stress, miR528 acts as a central modulator of copper-binding proteins (CBPs), which facilitate the transport of excess metals such as zinc and copper from the endoplasmic reticulum (ER), thereby influencing auxin homeostasis. This miRNA aids in mitigating arsenic toxicity in *Oryza sativa*, even in the absence of the IAR1 protein (Ferdous et al. 2017). Research on *Brassica juncea* roots exposed to arsenic has uncovered a variety of miRNAs involved in processes linked to development, nutrient uptake, and hormonal signalling, all of which play essential roles in stress mitigation (Srivastava et al. 2013). Moreover, external application of indole-3-acetic acid (IAA) and jasmonic acid (JA) has been shown to modulate miR167, miR319, and miR854 expression, thereby promoting adaptive developmental changes under arsenic induced exposure (Hu et al. 2015).

miRNAs are also responsive to physical forces such as tension and compression experienced within xylem tissues. During tensile stress, miR408 expression is significantly increased, whereas several others—including miR156, miR162, miR164, miR475, miR480, and miR481—are notably suppressed. Conversely, under compressive stress, there is a marked reduction in the levels of miR160 and miR172, while miR168 becomes more active, particularly in tissues exposed to tension (Lu et al. 2005). In the context of mechanical injury, such as wounding, miRNA expression also shifts. In *Ipomoea batatas*, wound-induced stress elevates miR828 expression, which contributes to increased lignin synthesis and H_2O_2 accumulation. These physiological responses strengthen the plant's structural defences and activate protective mechanisms against environmental damage (Lin et al. 2012) (Table 7.1).

7.3.2 siRNAs for Abiotic Stress Resilience in Plants

Recent investigations have highlighted the critical role played by small interfering RNAs (siRNAs) in modulating plant reactions during plant adaptation, particularly abiotic stresses (Table 7.2) (Sunkar and Zhu 2004). In *Arabidopsis thaliana*, heritable stress responses are modulated via DNA methylation pathways, with heterochromatic siRNAs (hc-siRNAs) playing a central role in this epigenetic regulation (Boyko and Kovalchuk 2010). A notable example involves the natural antisense siRNA (nat-siRNA) derived from *SRO5*, which interacts with the *P5CDH* gene to regulate proline metabolism, thereby reducing ROS levels under high salinity (Borsani et al. 2005). In response to salt stress, several miRNAs—including miR159, miR168, miR169, miR172, miR393, miR395, miR396, miR399, and miR408—demonstrate dynamic expression patterns across various plant species (Xu et al. 2021). These expression changes are often species specific, contributing uniquely to salt tolerance mechanisms. For instance, in *Nicotiana tAbscisic Acid(ABA)cum*, elevated levels of miR396 enhance osmotic balance by promoting water retention and proline synthesis, thus mitigating ROS damage (Chen et al. 2015). However, in *Oryza sativa* and *Arabidopsis*, increased miR396 levels have the opposite effect, reducing tolerance to salinity (Gao et al. 2010). Moreover, specific siRNAs such as 002061_0636_3054.1 and 005047_0654_1904.1 are downregulated under salt exposure, as observed by Yao et al. (2010), underscoring the complex and varied functions performed by small RNAs during managing salt-induced stress in plants.

The contribution of small RNAs in regulating heat responses in plants has been progressively acknowledged. One example is the *ONSEN* retrotransposon, which accumulates in siRNA biogenesis-deficient mutants, underscoring the essential role of siRNAs in heat stress regulation (Ito et al. 2011). In *Arabidopsis thaliana*, a specific group of phased siRNAs called tasiRNAs—originating from TAS loci—are significantly reduced during heat exposure, pointing to their role in thermal stress adaptation. Investigations have since focused on their functions in heat tolerance

Table 7.1 Role of miRNAs in mediating abiotic stress resilience in plants

miRNA	Plant	Response to abiotic stress	References
Drought stress			
miRNA8	*Zea mays*	Upregulated	Liu et al. (2019a, b)
miRNA13	*Zea mays*	Downregulated	Liu et al. (2019a, b)
miR156	*Arabidopsis, Triticum dicoccoides, Hordeum vulgare, Populus euphratica, Prunus persica*	Upregulated	Eldem et al. (2012), Kantar et al. (2011), Ren et al. (2012), Wu and Poethig (2006) and Zhou et al. (2010)
miR156	*Oryza sativa, Populus tomentosa*	Downregulated	Eldem et al. (2012), Kantar et al. (2011), Ren et al. (2012), Wu and Poethig (2006) and Zhou et al. (2010)
csi-miR156, csi-miR159a, csi-miR6170, csi-miR165a-3p, csi-miR408b-5p, and csi-miR854	*Camellia sinensis*	Upregulated	Guo et al. (2017)
csi-miR7	*Camellia sinensis*	Upregulated	Guo et al. (2017)
miR156	*Arabidopsis thaliana*	Increased	Guo et al. (2017)
miR156	*Musa* sp.	Upregulated	Muthusamy et al. (2015)
miR157	*Prunus persica*	Upregulated and downregulated	Eldem et al. (2012)
miR159	*Arabidopsis*	Upregulated	Arenas-Huertero et al. (2009), Eldem et al. (2012), Jones-Rhoades and Bartel (2004) and Zhou et al. (2010)
miR159	*Oryza sativa*	Downregulated	Arenas-Huertero et al. (2009), Eldem et al. (2012), Jones-Rhoades and Bartel (2004), Reyes and Chua (2007) and Zhou et al. (2010)
miR159	*Prunus persica*	Downregulated	Arenas-Huertero et al. (2009), Eldem et al. (2012), Jones-Rhoades and Bartel (2004) and Zhou et al. (2010)

miRNA159	*Zea mays* and *Arabidopsis*	Downregulated	Liu et al. (2019a, b) and Akdogan et al. (2016)
miR164	*Musa trunculata*	Downregulated	Wang et al. (2011a, b)
miR164	*P. trichocarpa*	Downregulated	Shuai et al. (2013)
csi-miR166a	*Camellia sinensis*	Downregulated	Guo et al. (2017)
miR166	*Triticum dicoccoides*	Downregulated	Kantar et al. (2011)
miR169	*Musaceae truncatula*	Upregulated	Muthusamy et al. (2015)
csi-miR398	*Camellia sinensis*	Upregulated	Guo et al. (2017)
miRNA399	*Zea mays* and *Arabidopsis*	Downregulated	Liu et al. (2019a, b) and Akdogan et al. (2016)
miR403	*Populus tomentosa*	Upregulated	Allen et al. (2005) and Ren et al. (2012)
miR408	*Cicer arietinum*	Overexpression of Athpre-miR408	Hajyzadeh et al. (2015)
miR528	*Zea mays*	Downregulated	Wei et al. (2009)
miR827	*Zea mays*	Upregulated	Zhang et al. 2009a, b, c
miR1432	*Triticum dicoccoides*	Upregulated	Kantar et al. (2011) and Zhang et al. 2009a, b, c
cin-miR2118	*Caragana intermedia* and *tobacco*	Overexpression	Wu et al. (2015)
miR2118	*Musa* sp.	Upregulated	Muthusamy et al. (2016)
miR2118	*Medicago truncatula*	Upregulated	Jagadeeswaran et al. (2009) and Wang et al. (2011a, b)
miR1444	*Populus trichocarpa*	Downregulated	Khraiwesh et al. (2012), Shuai et al. (2013) and Thipyapong et al. (2004)
miR164c, miR319b, and miR1861d	*Oryza sativa*	Downregulated	Zhang et al. 2009a, b, c
miR166h, miR172d, and miR408	*Oryza sativa*	Upregulated	Zhang et al. 2009a, b, c

(continued)

Table 7.1 (continued)

miRNA	Plant	Response to abiotic stress	References
miR1861n, miR5072, miR810a, miR6253, miR6255, miR6250, miR164c, miR319b, miR2863b, miR7692-3p, miR1861d, miR4-44d.2, miR444a-3p.2, miR444e, miR1861h, and miR18-61j	*Oryza sativa*	Upregulated	Zhang et al. (2016)
miR166h-5p, miR172d-5p, miR167h-3p, miR160f-5p, miR5828, miR2864.2, miR5076, miR5805, miR2871a-5p, miR18-46e, miR818a, miR2871b, miR5505, miR2871a-3p, miR1861c, miR408-3p, and miR1846d-3p	*Oryza sativa*	Downregulated	Zhang et al. (2016)
hvu-miR156a, hvu-miR166, hvu-miR171, and hvu-miR408	Barley leaves	Upregulated	Kantar et al. (2010)
	Barley roots	Downregulated	Kantar et al. (2010)
miR5213 and miR5294	*Medicago sativa*	Upregulated	Li et al. (2017a, b)
miR3512, miR3630, miR5368, and miR6173	*Medicago sativa*	Downregulated	Li et al. (2017a, b)
Hv-miR827	*Hordeum vulgare*	Increased	Ferdous et al. (2017)
Ath-miR169b, Osa-miR1432, Hv-miRx5, and Hv-miR166b/c	*Hordeum vulgare*	Increased	Ferdous et al. (2016)
miR172, miR396a, miR396c, and miR4233	Potato	Regulate P5CS gene	Yang et al. (2013)
miR2673, and miR6461	Potato	Regulate P5CR and ProDH gene	Yang et al. (2013)
Drought + salt stress			
miR172	*Arabidopsis*	Overexpression of gma-miR172c	Li et al. (2016a, b)
miR319	*Agrostis stolonifera*	Constitutive overexpression of osa-miR319a	Zuo et al. (2021)
Dehydration stress			
miR393, miR397b, and miR402	*Arabidopsis thaliana*	Upregulated	Lu and Huang (2008)
miR319c and miR389a	*Arabidopsis thaliana*	Downregulated	Lu and Huang (2008)

Salt stress			
miR167a, miR399a, and miR535a/b	*Gossypium hirsutum*	Upregulated	Yin et al. (2011)
miR156a/d/e, miR169, miR397a/b, and miR827b	*Gossypium hirsutum*	Downregulated	Yin et al. (2011)
ghr-miR399 and ghr-156e	*Gossypium hirsutum*	lnc_973 and lnc_253 are regulated by these as a target mimic	Deng et al. (2015)
miR827b, miR535a/b, miR169, and miR156a/d/e	*Gossypium hirsutum*	Decreased	Yin et al. (2012)
cin-miR165a, cin-miR396a, cin-miR172b, cin-miR157a, cin-miR390a, cin-miR159a, and cin-miR167b	*Caragana intermedia*	Increased	Zhu et al. (2013)
miR414	*Gossypium hirsutum*	Downregulated	He et al. (2014)
cin-miR398a	*Caragana intermedia*	Decreased	Zhu et al. (2013)
miR156	*Arabidopsis thaliana*	Increased	Liu et al. (2008)
Heat stress			
miRNA 164	*Arabidopsis*	Overexpression Heat stress decreased heat-inhibition of hypocotyl lengths	Tsai et al. (2023)
ath-miR169g-3p	Radish	Downregulated	Yang et al. (2019a, b)
ath-miR169g-3p	*Populus tomentosa*	Downregulated	Lu et al. (2008)
ath-miR171b-3p	Radish	Downregulated	Yang et al. (2019a, b)
ath-miR171b-3p	*Populus tomentosa*	Downregulated	Chen et al. (2012)
ath-miR398a-3p	Radish	Downregulated	Yang et al. (2019a, b)
ath-miR159b-3p, ath-miR159c	Radish	Induced	Yang et al. (2019a, b)
miR167, miR396, and miR482	Tomato	Significantly expressed	Pan et al. (2017)
miR156	*Alfa alfa*	Overexpression	Arshad et al. (2020)

(continued)

Table 7.1 (continued)

miRNA	Plant	Response to abiotic stress		References
Cold stress				
miR156	*Oryza sativa*	Overexpression of OsmiR156k		Cui et al. (2015)
miR319	*Oryza sativa*	Overexpression		Wang et al. (2014a, b, c)
miR858	*Arabidopsis thaliana*	Overexpression		Tripathi et al. (2019)
miR319c	*Cucumis melo*	Increased		Bustamante et al. (2018)
Stu-mi164	Potato	Overexpression		Zhang et al. (2018)
Low nitrate stress				
miR160, miR164, miR172, and miR827	Maize	Upregulated		Xu et al. (2017)
miR169, miR397, miR398, miR399, miR408, and miR528	Maize	Downregulated		Xu et al. (2017)
Heavy metal stress				
miR164, miR319, miR393, miR394	Flax	Aluminium stress	Upregulated	Dmitriev et al. (2017)
miR159, miR167, and miR408	Flax	Aluminium stress	Downregulated	Dmitriev et al. (2017)
miR390	*Oryza sativa*	Cadmium stress	Downregulated	Ding et al. (2016)
Musa-miR397	*Musa* sp.	Copper deficiency	Upregulated	Patel et al. (2019)
miR399	*Solanum lycopersicum*	Phosphorus deficiency and low temperatures	Overexpression of ath-miR399d under control of rd29A promoter	Gao et al. (2015)

Table 7.2 Role of siRNAs in mediating abiotic stress responses in plants

siRNAs	Plant	Expression of siRNA to abiotic stress	References
Salt stress			
SRO5-P5CDH nat-siRNA	*Arabidopsis*	Decreased	Borsani et al. (2005)
TAS1, TAS2, and TAS3 ta-siRNA	*Arabidopsis*	Increased	Allen et al. (2005), Rajagopalan et al. (2006) and Lurin et al. (2004)
Heat stress			
HTT1 and HTT2-TAS1	*Arabidopsis*	Increased	Kume et al. (2010) and Li et al. (2014)
080621_1340_0098.1 siRNA	*Triticum aestivum*	Decreased	Yao et al. (2010)
hcsiRNA (ONSEN)	*Arabidopsis*	Increased	Matsunaga et al. (2012), Ito et al. (2011) and Stief et al. (2014)
nat-siRNAs	*Brassica oleracea*	Increased	Wang et al. (2011a, b)
nat-siRNAs	*Brassica rapa*	Increased & decreased	Yu et al. (2013)
Dehydration stress			
CDT1-siRNA	*Craterostigma*	Increased	Furini et al. (1997)
Cold stress			
005047_0654_1904.1 siRNA, 080621_1340_0098.1 siRNA	*Triticum aestivum*	Increased	Yao et al. (2010)
3 ta-siRNAs, 2 nat-siRNA	*Manihot esculenta*	Increased & decreased	Xia et al. (2014)
Other abiotic stress			
TAS4 ta-siRNAs	*Arabidopsis*	Increased	Hsieh et al. (2009), Luo et al. (2012)
TAS4-siR81(−)	*Arabidopsis*	Increased	Luo et al. (2012)

(Zhong et al. 2013; Li et al. 2014). For instance, under low oxygen (hypoxic) conditions, levels of tasiRNAs derived from TAS1, TAS2, and TAS3 increase markedly, indicating their potential role in stress-related signalling pathways (Allen et al. 2005; Rajagopalan et al. 2006). These changes are closely linked to the activity of miR173 and miR390, which initiate tasiRNA biogenesis by targeting TAS loci (Allen et al. 2004). TAS2-derived tasiRNAs were found to regulate mRNAs encoding pentatricopeptide repeat (PPR) proteins, reducing mitochondrial damage during hypoxia (Lurin et al. 2004). During heat stress, two TAS1-targeted genes—*HTT1* and *HTT2* (heat-induced TAS1 targets)—are activated in *Arabidopsis*. Suppressing TAS1 increases the expression of these targets and improves heat resilience, while overexpression of TAS1 reduces *HTT* levels, diminishing thermotolerance. Notably, *HTT* genes stimulate the activity of heat shock factors (HSFs), especially when the master regulator HSFA1a is overexpressed, which physically interacts with HTT promoters to enhance their transcription (Li et al. 2014). Additionally, HTT1 interacts with heat shock proteins HSP70-14 and HSP40, functioning within the HSP70-14 complex and contributing to thermal tolerance through HSFA1a-mediated pathways. In phosphate-starved *Arabidopsis*, miR828 regulates tasiRNAs from the TAS4 locus, which modulate anthocyanin biosynthesis. Under nitrogen deficiency, TAS4-siR81(−) levels rise, targeting MYB transcription factors that drive anthocyanin accumulation (Hsieh et al. 2009; Luo et al. 2012). Besides this, the activation of ONSEN transposons by heat in *Arabidopsis* seedlings alters tasiRNA expression, influencing heat memory and immune responses via HTT5 regulation. These changes can even affect progeny fitness (Ito et al. 2011; Liu et al. 2019). In *Camellia*, a recently evolved phasiRNA locus produces 21-nt phasiRNAs during heat stress, potentially targeting genes like *lipoxygenase*, *RAN GTPase*, *xyloglucan modification enzyme*, and *ATPase*, suggesting a role in thermal adaptation (Hu et al. 2020). Experimental validation is needed to confirm these interactions. In *Zea mays* (maize), elevated temperatures significantly reduce 24-nt siRNAs associated with transposons, and neighbouring genes exhibit reduced expression, implying that transposon-derived siRNAs help regulate heat-responsive genes, though the underlying mechanisms remain elusive (He et al. 2019). Stress-induced transposon insertions like those of ONSEN have been detected in mutants lacking siRNA biogenesis, and these insertions are stably passed on to the next generation, enabling adjacent genes to become heat responsive. This emphasizes the importance of siRNA pathways in regulating stress-activated transposition across generations (Matsunaga et al. 2012; Steif et al. 2014). In *Brassica rapa* (Chinese cabbage), elevated high temperatures alter the transcriptional patterns of chloroplast-derived small RNAs (csRNAs) and natural antisense siRNAs (nat-siRNAs). Specifically, nat-siRNAs from 12 *cis*-NAT pairs were elevated, while their complementary transcripts were suppressed (Wang et al. 2011a; Yu et al. 2013). Similarly, in the resurrection plant *Craterostigma plantagineum*, an Abscisic Acid(ABA)-responsive gene facilitates the generation of endogenous siRNAs crucial for surviving dehydration stress (Furini et al. 1997). In wheat (*Triticum aestivum*), siRNA expression patterns vary significantly under cold, heat, salt, and dehydration stress. Specific siRNAs—such as siRNA 080621_1340_0098.1, 002061_0636_3054.1, and 005047_0654_1904.1—are suppressed during heat and salt exposure, suggesting their potential roles in stress signalling pathways (Yao et al. 2010). Interestingly, heat-induced miRNA expression differs between species. For example, miR397 is upregulated in *Musa* species but downregulated in both rice (*Oryza sativa*) and tomato (*Solanum lycopersicum*) under heat stress (Pan et al. 2017; Zuo et al. 2021). Comparative studies have revealed novel, species-specific small RNAs with roles in heat response. In *Saccharina japonica*, 25 previously unidentified heat-responsive miRNAs—including sja-novel-mir-5 and sja-novel-mir-59—were discovered and are believed to contribute to thermotolerance (Liu et al. 2014). Continued exploration of these small RNAs will provide deeper insights into the molecular frameworks that govern plant responses to high-temperature stress across diverse taxa.

During drought stress, histone deacetylases (HDACs) such as HD2C and HDA6 are crucial for orchestrating plant responses to both environmental stressors and abscisic acid (Abscisic Acid(ABA)) signalling pathways (Chen et al. 2010; Sridha and Wu 2006). Inactivation of HDA6 leads to the re-expression of genes typically suppressed by RNA-directed DNA methylation (RdDM), indicating a regulatory overlap (Aufsatz et al. 2002). Notably, HDA6 and HD2C interact with key DNA methyltransferases, MET1 and AtDNMT2, respectively, suggesting a synergistic relationship between histone modifications and DNA methylation in controlling gene expression under stress (To et al. 2011; Song et al. 2010). This interplay plays a foundational role in small non-coding RNA (sncRNA)-guided networks that enable plants to adapt to drought and hormonal stimuli. In *Arabidopsis thaliana*, disruption of miR159 function improves drought resistance by decreasing stomatal aperture, thereby limiting transpirational water loss (Jiang et al. 2022). On the contrary, plants overexpressing miR169a exhibit greater dehydration and heightened vulnerability to water-deficit stress when compared to those maintained under optimal watering conditions (Li et al. 2008). In *Triticum aestivum* (wheat), suppression of miR1119 expression in roots enhances MYC2 transcription, which subsequently boosts the activity of major antioxidant enzymes such as superoxide dismutase (SOD), catalase (CAT), and peroxidase (POD), thereby strengthening the plant's water-deficit tolerance (Shamloo-Dashtpagerdi et al. 2023). Collectively, these findings underscore the sophisticated regulatory functions of specific small RNAs and their targets in reinforcing plant defence mechanisms during water scarcity.

Trans-acting small interfering RNAs (ta-siRNAs) play a crucial part in facilitating plant adaptation to cold stress and is closely linked with auxin-mediated signalling pathways. One such ta-siRNA, TAS3a-5D6(+), facilitates cold adaptation in wheat by promoting the cleavage of *auxin response factor* (ARF) transcripts, thereby modulating auxin signalling under low-temperature conditions (Tang et al. 2012). Similarly, in cassava (*Manihot esculenta*), three ta-siRNAs derived from TAS3 loci were found to target ARF gene transcripts in response to cold exposure, indicating a conserved regulatory mechanism in chilling stress adaptation (Xia et al. 2014). Additionally, two natural antisense siRNAs (nat-siRNAs) exhibited differential expression in cassava during cold stress, with one linked to a transcription factor containing the NAC domain—an important regulator in cold-responsive gene expression (Xia et al. 2014). Advanced sequencing technologies efforts by Zeng et al. (2018) identified miR166e, miR319, and a novel small RNA, Bra-novel-miR3936-5p, as key regulators potentially involved in cold stress signalling networks. In *Triticum aestivum*, the small RNA siRNA 080621_1340_0098.1 showed marked upregulation under cold conditions, suggesting its involvement in orchestrating stress-responsive pathways during temperature decline (Yao et al. 2010). Together, these discoveries reveal the complex and dynamic participation of small RNAs—ranging from ta-siRNAs and nat-siRNAs to miRNAs—in fine-tuning gene expression and signalling processes crucial for plant resilience under cold stress.

Matthews et al. (2019) demonstrated that miR156 functions critically in enhancing thermotolerance in *Medicago sativa*. The overexpression of this microRNA in alfalfa resulted in a significant enhancement of its antioxidant capacity, thereby minimizing oxidative stress by efficiently neutralizing reactive oxygen species. In a related study, Shi et al. (2019) developed tomato plants to overexpress sha-miR319d, which enhanced tolerance to thermal extremes, including both heat and cold. This evidence emphasizes the importance of distinct miRNAs in strengthening host protective pathways during thermally induced adverse environments.

7.3.3 Functional Contributions of lncRNAs to Environmental Stress Adaptation in Plants

Numerous investigations have underscored the contribution of plant lncRNAs in orchestrating adaptations to various abiotic stresses (Table 7.3) (Helliwell et al. 2011; Xuan et al. 2015). These lncRNAs contribute to plant adaptation through multiple functional mechanisms. A prominent strategy is target mimicry, wherein lncRNAs functions as molecular decoys that interact with and trap microRNAs (miRNAs), thereby preventing their activity. This sequestration prevents miRNAs from binding to their typical mRNA counterparts, thereby influencing the plant-adaptive signalling mechanisms under stress (Wang et al. 2017). Furthermore, under abiotic stress conditions, antisense lncRNAs can hybridize with complementary sense mRNAs to form double-stranded RNA complexes. This duplex formation affects the expression patterns of the corresponding genes on the opposite strand (Wang et al. 2014a, b, c).

Multiple regulatory pathways involving long non-coding RNAs (lncRNAs) have been associated with water-deficit tolerance in plants, underscoring their functional significance and interaction as competitive endogenous RNA (ceRNA) pairs under drought stress conditions (Shuai et al. 2014). In *Oryza sativa*, directional RNA profiling and short RNA analysis have permitted the discovery of various lncRNAs, miRNAs, and mRNAs associated with dessication tolerance, supporting the development of a comprehensive ceRNA interaction framework (Yang et al. 2022). This study found 191 lncRNAs and 32 miRNAs exhibiting differential expression in response to drought, with 16 lncRNAs uniquely implicated in drought stress. Notably, these lncRNAs are predicted to regulate the drought-responsive gene *Os05g0586700*, which may play a role in membrane repair during water deficit (Yang et al. 2022). Another investigation identified a drought-responsive ceRNA network in rice comprising 40 lncRNAs, 23 miRNAs, and 103 mRNAs (Chen et al. 2021). Among these, the lncRNA *TCONS_00021861* acts as a molecular sponge for miR528-3p, thereby regulating *YUCCA7* and activating the auxin biosynthesis pathway, which contributes to enhanced drought tolerance. In *Arachis hypogaea*, comparative transcriptomic assessments across genotypes exhibiting contrasting drought responses have contributed to the establishment of integrated ceRNA networks (Ren et al. 2022). The drought-tolerant variety exhibited 673 ceRNA interactions, with six lncRNAs (MSTRG.70535.2, MSTRG.86570.2, MSTRG.86570.1, MSTRG.100618.1, MSTRG.81214.2, and MSTRG.30931.1) identified as critical regulatory hubs within this network, highlighting their potential roles in drought resilience (Ren et al. 2022). Likewise, RNA sequencing of native Tibetan barley

Table 7.3 Role of lncRNAs in mediating abiotic stress responses in plants

lncRNAs	Plant	Expression response to abiotic stress	References
Salt stress			
lncRNAs	*Arabidopsis thaliana*	Decreased	Xu et al. (2015)
npc536 and npc60	*Arabidopsis thaliana*	Induced	Ben Amor et al. 2009
TCONS_00046739, TCONS_00100258, and TCONS_00118328	*M. truncatula*	Induced	Wang et al. (2015)
AK370814	*H. vulgare*	Induced	Karlik and Gozukirmizi (2018)
Drought + salt stress			
DRIR	*Arabidopsis thaliana*	Induced	Qin et al. (2017)
Heat stress			
lncRNAs	*Arabidopsis thaliana*	Decreased	Popova et al. (2013)
asHSFB2a	*Arabidopsis thaliana*	Increased	Wunderlich et al. (2014)
TahlnRNA52, TahlnRNA47, TahlnRNA42, TahlnRNA41, TahlnRNA36, TahlnRNA19, TahlnRNA21, TahlnRNA27, and TahlnRNA5	*Triticum aestivum*	Increased	Yao et al. (2010) and Xin et al. (2011)
MSTRG4636, MSTRG.38321	*Z. mays*	Repressed	Lv et al. (2019)
MSTRG.11125, MSTRG.1555, MSTRG.31362, and MSTRG.63799	*Z. mays*	Induced	Lv et al. (2019)
Drought stress			
LincRNA 2962, LincRNA 2762, LincRNA 2623, LincRNA 1795, LincRNA 3018, LincRNA 1393, and LincRNA 1128	*Populus trichocarpa*	Decreased	Shuai et al. (2014)
LincRNA 1039, LincRNA 1534, LincRNA 2962, LincRNA 2085, LincRNA 2131, LincRNA 2198, LincRNA 179, LincRNA 20, and LincRNA 2752	*Populus trichocarpa*	Increased	Shuai et al. (2014)
Si NAT 80	*S. italica*	Induced	Qi et al. (2013)
c70772_g2_i1	*T. turgidum*	Induced	Cagirici et al. (2017)
c90557_g1_i1	*T. turgidum*	Induced	Cagirici et al. (2017)
LncRNA-tomato_535, LncRNA-tomato_146, LncRNA_tomato_467, and LncRNA-tomato_178	*S. lycopersicum*	Induced	Eom et al. (2019)

(continued)

Table 7.3 (continued)

lncRNAs	Plant	Expression response to abiotic stress	References
MSTRG.6838.1	*Z. mays*	Repressed	Pang et al. (2019)
MSTRG65848, MSTRG27834, MSTRG46301	*Oryza sativa*	Downregulated	Weidong et al. (2020)
MSTRG41712, MSTRG69391, and MSTRG68635	*Oryza sativa*	Upregulated	Weidong et al. (2020)
MSTRG.43964.1	*Cleistogenes songorica*	Upregulated	Yan et al. (2019)
Cold stress			
COOLAIR	*Arabidopsis thaliana*	Increased	Csorba et al. (2014)
COLDAIR	*Arabidopsis thaliana*	Increased	Heo and Sung (2011)
COLDWRAP	*Arabidopsis*	Reduced	Kim and Sung (2017)
TAS1a	*Arabidopsis*	Induced	Calixto et al. (2019)
MAS	*Arabidopsis*	Induced and activation of *MAF4*	Zhao et al. (2018)
BdCOOLAIR1 and BdCOOLAIR2	*Brachypodium*	Induced	Jiao et al. (2019)
CIL1	*Arabidopsis*	Induced	Liu et al. (2022)
SVALKA	*Arabidopsis*	Suppresses CBF1 expression	Kindgren et al. (2018); Kiger and Schroeder (2024)
lincRNA 159	Cassava	Decreased expression of miR164-targeted *NAC*	Li et al. (2017a, b)
VIT_203s0017n00360, VIT_207s0031n00070, VIT_201s0011n00530	*Grapevine*	Upregulated	Wang et al. (2019b)
VIT_209s0002n00340, VIT_213s0158n00020, and VIT_213s067n00110	‘*Cubernet Sauvignon*’	Downregulated	Wang et al. (2019b)
AT1G34844 and AT3G266121	*Arabidopsis thaliana*	Induced	Calixto et al. (2019)

Other abiotic stress			
IPS1	*Arabidopsis thaliana*	Increased	Aung et al. (2006), Franco-Zorrilla et al. (2007) and Wu et al. (2013)
Cis-NAT PHO1;2	*Oryza sativa*	Induced	Jabnoune et al. (2013)
Pti-MIR5471 and Pti-MIR5472	*Phaeodactylum tricornutum*	Increased	Cruz de Carvalho et al. (2016)
AtR8	*Arabidopsis thaliana*	Repressed	Wu et al. (2012)
XLOC_011965, XLOC_054416	*Oryza sativa*	Induced	He et al. (2015)
XLOC_048220, XLOC_001126	*Oryza sativa*	Repressed	He et al. (2015)
TCONS_00052316, TCONS_00069233, TCONS_00052315, TCONS_00064021, TCONS_00049805, TCONS_00017288, TCONS_0002186, and TCONS_00021860	*P. tomentosa*	Repressed	Chen et al. (2016)
TAS3	*Arabidopsis thaliana*	Repressed	Fukuda et al. (2019)
TCONS_00043651	*H. vulgare*	Induced	Karakulah and Unver (2017)
lncRNAs	*Arabidopsis thaliana*	Increased	Wang et al. 2014a, b, c

exposed to water-deficit conditions identified 535 lncRNAs exhibiting variable abundance and unique regulatory trends. The analysis predicted 503 *cis*-acting and 776 *trans*-acting lncRNA-mRNA regulatory pairs enriched in pathways including hormone signalling, kinase cascades, and ascorbate/aldarate metabolism. A serine/threonine-protein kinase, SMG1, was pinpointed as a major target regulated by several lncRNAs, potentially contributing to drought tolerance in specific cultivars (Qiu et al. 2019). In *Brassica napus*, RNA expression profiling across water resilient and water-susceptible varieties unveiled notable distinctions in lncRNA abundance patterns under limited moisture conditions (Tan et al. 2020). Network analyses indicated 126 drought-responsive lncRNAs interacting with 34 regulatory proteins in the resilient variety, while the susceptible variety exhibited 359 lncRNAs associated with 45 transcription factors. These findings enrich our understanding of the involvement of lncRNAs in drought stress adaptation. In *Arabidopsis thaliana*, an lncRNA designated as drought-induced lncRNA (DRIR) shows strong induction by drought, salinity, and abscisic acid (Abscisic Acid(ABA)), with minimal expression under normal conditions (Qin et al. 2017). Functional characterization using T-DNA insertion mutants (drirD) and overexpression lines demonstrated that DRIR enhances drought tolerance by reducing water loss through transpiration. Both mutant and overexpressing plants heightened responsiveness to Abscisic Acid(ABA) relative to wild-type counterparts, and gene expression profiling highlighted DRIR's involvement in modulating genes associated with Abscisic Acid(ABA)-mediated pathways, water transport, and protective stress responses (Qin et al. 2017). Collectively, these studies emphasize the intricate and multifaceted roles of lncRNAs in modulating drought responses via both Abscisic Acid(ABA)-mediated and Abscisic Acid(ABA)-independent signalling routes (Xiong et al. 2002; Zhu 2016).

In *Populus trichocarpa*, the drought-responsive long intergenic non-coding RNA lincRNA2752 has been demonstrated to inhibit the activity of ptc-miR169 by acting as a molecular decoy (Shuai et al. 2014). Since miR169 functions as a key regulator in drought stress regulation by modulating the transcription factor NF-YA (Ni et al. 2013), lincRNA2752 likely enhances drought tolerance through modulating both miR169 and NF-YA-related pathways. In maize, studies examining the spatiotemporal expression of drought-responsive lncRNAs across various tissues and developmental stages have revealed significant changes, particularly during the transition from vegetative to reproductive growth (Pang et al. 2019). A key discovery from this work was the lncRNA-mRNA pair MSTRG.6838.1 and *vpp4*, the latter encoding a vacuolar (H^+)-pumping ATPase subunit homolog, with their strongly correlated expression suggesting a role in coordinating drought response in maize (Pang et al. 2019). Moreover, *Populus trichocarpa* lncRNAs were also found to target specific miRNAs including ptc-miR482a.1, ptc-miR476a, and ptc-miR156c (Shuai et al. 2014). In maize, five antisense lncRNAs were identified as drought-responsive (Zhang et al. 2014). Certain lncRNAs contribute to environmental stress adaptation by means of RNA-mediated DNA methylation (RdDM) pathway. For example, in tomato, silencing of the key RdDM gene *SlAGO4A* enhanced resistance to drought and salinity relative to wild type plants and *SlAGO4A* overexpressing lines (Huang et al. 2016). Extensive profiling in maize across ear, tassel, and leaf tissues at various growth stages (V12, V14, V18, and R1) under both control and drought stress conditions identified 3488 high-confidence lncRNAs (Thatcher et al. 2016; Pang et al. 2019). Among these, MSTRG.2834.1 showed a pronounced drought-specific upregulation in the ear tissue at the R1 stage, while MSTRG.43642.1 exhibited significant downregulation at the same stage. Additionally, 1177 lncRNA-mRNA pairs were discovered, including the notable *cis*-acting pair of MSTRG.6838.1 and *vpp4* (GRMZM2G028432), both showing coordinated downregulation under drought stress, implying potential co-regulatory roles. Furthermore, multiple maize lncRNAs were recognized as originating templates for small RNAs, including miRNAs, siRNAs, and ShRNAs-under water-deficit conditions (Zhang et al. 2014). The lncRNA DRIR functions as a pivotal element in enhancing tolerance to both drought and salt stresses (Qin et al. 2017; Dong et al. 2022). Circular RNAs (circRNAs) have also been profiled in maize and *Arabidopsis thaliana*, with 2174 and 1354 high-confidence circRNAs identified, respectively (Zhang et al. 2019a, b). Among these, circRNAs derived from the genes *GRMZM2G032852* and *GRMZM5G817173*, encoding calcium-dependent protein kinase (CPK) and cytokinin oxidase/dehydrogenase (CKX), respectively, were highlighted. These circRNAs, named circCPK and circCKX, arise via exon backsplicing and exhibit allele-specific insertions or deletions in maize populations. Plants harbouring insertion alleles exhibited higher circRNA expression levels and improved survival rates across optimal and dry conditions, pointing to a likely correlation between circRNA expression and drought resilience. Another circRNA, circGORK, generated from the *GORK* gene encoding a potassium efflux channel essential for Abscisic Acid(ABA)-mediated signalling and drought responses (Ache et al. 2000; Becker et al. 2003), was found to negatively regulate Abscisic Acid(ABA) signalling. Transgenic *Arabidopsis* plants overexpressing circGORK showed reduced seed germination rates on Abscisic Acid(ABA)-containing media, suggesting its role in modulating water stress responses (Yadav et al. 2024). Regarding hormone-mediated drought responses, the lncRNA DRIR modulates drought and salt tolerance by regulating Abscisic Acid(ABA)-associated genes, promoting stomatal closure and proline accumulation (Qin et al. 2017). In soybean, lncRNA77580 controls multiple transcription factors under combined drought and salt stresses (Chen et al. 2023). MicroRNAs are also vital in drought adaptation; for example, miRP11, miRP14, miR858, and miR2118 enhance drought tolerance by targeting transcription factors such as MYB and oxidation-related zinc finger 1 (OZF1) (Gao et al. 2016). In *Camellia sinensis*, miR156, miR166, and miR398 are upregulated during drought stress and contribute to tolerance mechanisms (Guo et al. 2017). In apple, differential expression of miR35 and miR61 was observed between drought-tolerant and sensitive genotypes, highlighting their potential role in drought resilience (Niu et al. 2019).

lncRNAs function as pivotal regulators in plant adaptation to heat-induced stress. Co-expression networks involving lncRNAs and mRNAs under heat stress conditions have revealed that lncRNAs often regulate transcription factor genes through *cis*-acting mechanisms (Bhatia et al. 2020; He et al. 2020; Hu et al. 2022). In *Arabidopsis*, five antisense lncRNAs have been discovered that respond to both light and heat stress (Wang et al. 2014a, b, c; Wunderlich et al. 2014). Mutants deficient in the RNA-dependent RNA polymerase 2 (rdr2) and Dicer-like 3 (*dcl3*) genes exhibit significantly lower survival rates under heat stress, highlighting the importance of small RNA pathways in heat tolerance (Popova et al. 2013). Elevated temperature conditions likewise induce the transcription of the vacuolar protein sorting 4 (*Vps4*) gene in *Arabidopsis*, with notably elevated expression levels observed in the nrpd2 mutant background (Popova et al. 2013).

In *Triticum aestivum*, numerous lncRNAs have been linked to the regulation of miRNA, siRNA, and shRNA pathways during heat stress responses (Yao et al. 2010; Xin et al. 2011). Similarly, in poplar, 204 heat-responsive lncRNA transcripts have been predicted to target various genes, emphasizing their potential regulatory roles (Song et al. 2020). Among wheat lncRNAs identified under heat stress, 77 candidates were found, including TahlnRNA27, which is proposed to act as a precursor for certain miRNAs (Xin et al. 2011).

lncRNAs are classified into polyadenylated (poly(A)+) and non-polyadenylated (poly(A)-) types, with the latter generally shorter, expressed at lower levels, and exhibiting distinct stress-specific expression profiles compared to poly(A) + transcripts (Di et al. 2014). In *Arabidopsis*, 245 poly(A) + and 58 poly(A)-lncRNAs exhibited variable expression patterns under diverse stress conditions, although only 15 were specifically responsive to heat (Di et al. 2014). In *Brassica rapa*, 34 differentially expressed (DE) lncRNAs detected under heat stress were associated with 192 predicted target genes, many of which exhibited responsiveness to elevated temperatures (Song et al. 2016a). Transcriptomic comparisons of a heat-tolerant maize inbred line exposed to heat stress revealed 993 DE lncRNAs co-expressed with nearly 1000 predicted *cis*-target genes. These lncRNAs are enriched in functions related to stress response, hormone signalling, metabolism, photosynthesis, and spliceosome activity (Hu et al. 2022). DNA methylation is another critical regulatory mechanism in plant stress adaptation. In *Populus simonii*, integrated methylome and transcriptome analyses under cold, osmotic, heat, and salt stresses uncovered stress-specific differentially methylated regions (SDMRs) overlapping with 17 lncRNA genes. For example, SDMR162 harbours miR396e and the lncRNA PsiLNCRNA00268512, which exhibit stable binding energies, suggesting that PsiLNCRNA00268512 may function as a molecular decoy modulating miR396e activity (Song et al. 2016b). In Chinese cabbage, the heat-responsive lncRNA TCONS_00048391 acts as an endogenous target mimic (eTM) to regulate bra-miR164a. This lncRNA is also predicted to serve as a competing endogenous RNA (ceRNA) or miRNA sponge, impacting the expression of NAC1, a key transcription factor (Wang et al. 2019a). Heat stress transcription factors (HSFs) constitute the central regulators of heat response pathways in plants (Scharf et al. 2012; Zhu 2016; Chen et al. 2018). In *Arabidopsis*, HSFB2a is specifically expressed in the female gametophyte and is vital for vegetative and reproductive development (Wunderlich et al. 2014). Functional studies with transgenic plants demonstrated that the natural antisense transcript (NAT) lncRNA asHSFB2a negatively regulates HSFB2a expression. Overexpression of asHSFB2a leads to increased biomass, whereas HSFB2a overexpression alone reduces growth. Upon heat exposure, asHSFB2a mitigates HSFB2a's suppressive effects, restoring normal development

(Wunderlich et al. 2014). Another key player is HSP90, which modulates brassinosteroid signalling and controls BR-responsive gene expression via interaction with transcription factors BES1 (BRASSINOSTEROID INSENSITIVE 1-EMS-SUPPRESSOR 1) and BZR1 (BRASSINAZOLE RESISTANT 1) (Shigeta et al. 2015). Correspondingly, in Chinese cabbage, the lncRNA TCONS_00016454 is implicated in heat tolerance by influencing brassinosteroid signalling, displaying an inverse expression pattern with its target gene, a BES1/BZR1 homolog, under elevated temperature conditions (Wang et al. 2019a).

In *Arabidopsis* exposed to cold stress, the lncRNAs COOLAIR and COLDAIR play crucial roles in suppressing the expression of the FLOWERING LOCUS C (*FLC*) gene through lncRNA-mediated epigenetic modifications (Csorba et al. 2014; Heo and Sung 2011). Specifically, COOLAIR contributes to the reduction of active histone marks H3K36me3 and H3K4me2 at the FLC locus (Csorba et al. 2014; Heo and Sung 2011), whereas COLDAIR facilitates the recruitment of the Polycomb Repressive Complex 2 (PRC2), promoting the enrichment of the repressive histone mark H3K27me3 at FLC (Csorba et al. 2014; Heo and Sung 2011). More recently, Liu et al. (2022) identified a novel cold-responsive lncRNA named Cold-Induced lncRNA 1 (CIL1), which functions as a positive regulator enhancing *Arabidopsis* tolerance to low temperatures. The expression of CIL1 was notably elevated under cold conditions. Similarly, Zhao et al. (2023) reported the discovery of MtCIR2, a cold-responsive long intergenic non-coding RNA in *Medicago truncatula*, a legume species. MtCIR2 is located near *CBF/DREB1* genes within a key quantitative trait locus (QTL) for freezing tolerance. The transcriptional response of MtCIR2 occurred earlier than that of the adjacent *MtCBF/DREB1* genes during cold exposure. Functional studies revealed that overexpression of MtCIR2 increased freezing tolerance, resulting in higher survival rates and less cellular membrane damage compared to wild-type plants. Conversely, loss-of-function mutants of MtCIR2 showed increased sensitivity to cold stress. Additionally, MtCBF/DREB1 gene expression was significantly upregulated in MtCIR2 overexpressing lines, while it was suppressed in the mutant plants, indicating that MtCIR2 positively regulates these cold-responsive transcription factors.

Under saline stress, extensive transcriptomic profiling in *Populus trichocarpa* revealed 1183 lncRNAs exhibiting altered expression across various plant tissues, including leaves, stems, and roots (Ye et al. 2022). Among these, Ptlinc-NAC72 was found to suppress salt tolerance by modulating salt-responsive genes via both *cis*- and *trans*-regulatory mechanisms. It directly interacted with tandem GAAAAA motifs in the 5′-UTR of *PtNAC72.A/B*, enhancing their transcription and contributing to salt adaptation. In *Arabidopsis thaliana*, lncRNA npc536 was markedly induced in leaves and roots under salt conditions. Its overexpression improved root development under salinity, potentially by modulating the translation of the *AtCOG5* mRNA, a component of the conserved oligomeric Golgi complex (Ben Amor et al. 2009). In upland cotton, lncRNA354 was identified as an endogenous target mimic (eTM) of miR160b. Overexpressing lncRNA354 in *Arabidopsis* led to salt hypersensitivity by downregulating miR160b and consequently affecting *GhARF17/18*. In contrast, silencing lncRNA354 or its ARF targets resulted in increased plant height and improved salt resilience (Zhang et al. 2021). Another salt- and drought-inducible lncRNA, DRIR, was found to enhance salinity resistance in *Arabidopsis*. Transgenic seedlings expressing DRIR showed elevated levels of stress-related genes such as *P5CS1*, *RD29A*, *RD29B*, and *NAC3*, though Abscisic Acid(ABA)-associated markers like *ABI5* were not induced, suggesting a distinct signalling route (Qin et al. 2017). In cotton, salt-responsive lncRNA973 contributes to stress tolerance by regulating the miR399-*GhPHO2* module. Loss-of-function studies showed that knocking out lncRNA973 elevated miR399 levels and compromised salt resilience (Zhang et al. 2019a, b). Additionally, the RNA-directed DNA methylation (RdDM) pathway in *Arabidopsis* modulates the expression of the *MYB74* transcription factor under salinity (Xu et al. 2015). Cotton lncRNAs such as lnc_388, lnc_883, and lnc_973 are also linked to miRNAs activated under salt stress (Chen et al. 2023), and transcriptome studies identified 44 lincRNAs with salt-responsive expression profiles (Zhang et al. 2019a, b). In *Glycine max* (soybean), lncRNA77580 was implicated in salinity adaptation (Chen et al. 2023). Among 76 npcRNAs identified in *Arabidopsis*, npc536, npc60, and npc82 were notably responsive after 3 h of salt exposure, with npc72 showing transient expression peaks (Ben Amor et al. 2009). lnc_388 and its target gene *LRR8* were significantly induced under stress, suggesting *cis*-regulation of protein-coding genes. Similar patterns were observed with lnc_388 and *msD3*, indicating a role in salt-responsive transcriptional regulation (Zhang et al. 2019a, b). In barley, the expression of lncRNAs AK363462 and AK370506 varied between roots and shoots in response to saline environments, suggesting genotype- and tissue-specific responses (Karlik and Gozukirmizi 2018). Moreover, temporal expression patterns of salt-induced lncRNAs differ across species. For instance, duckweed lncRNAs responded rapidly to early salt exposure, while quinoa and wheat lncRNAs showed sustained differential expression over longer durations. Diverse plant species—including *Beta vulgaris*, *Medicago truncatula*, *Camellia sinensis*, *Vitis vinifera*, and *Melilotus albus*—displayed hundreds of differentially regulated lncRNAs under saline stress, reflecting the widespread involvement of these molecules in abiotic stress adaptation (Li et al. 2020). In grapevine, lncRNAs like TCONS_00024229 and TCONS_00045028 were tied to Abscisic Acid(ABA) signalling and auxin response factors (ARFs), suggesting their role in transcription factor and hormone regulation. TCONS_00018576 and TCONS_00037548 were associated with redox balance and chloroplast lipocalin regulation, underlining the role of lncRNAs in oxidative stress mitigation (Fu et al. 2020). In *Vitis vinifera* roots, several lncRNAs, including *VIT_201s0011n00350*, *VvLnc14.2231*, and *VIT_200s0282n00010*, were significantly upregulated under saline conditions. Other lncRNAs—such as *Vvlnc_14.529*, *Vvlnc_16.1373.1*, and *VIT_202s0025n00280*—were closely correlated with their target gene expression, indicating their functional involvement in salt stress modulation (Jin et al. 2020). This study reported 455 salt-induced lncRNAs acting as *cis*-regulators of 546 mRNAs, and 258 as *trans*-regulators, emphasizing the comprehensive regulatory landscape orchestrated by lncRNAs during salinity stress.

miRNAs are pivotal regulators in plant defence mechanisms against heavy metal stress, playing essential roles in minimizing metal-induced toxicity. One prominent example is the suppression of miR398, which results in enhanced expression of Cu/Zn superoxide dismutase genes (*CSD1* and *CSD2*)—key components of the antioxidant defence system—thus supporting plant survival in copper-rich environments (Noman and Aqeel 2017). Additionally, miRNAs like miR396, miR393, and miR164 are instrumental in modulating defence-related gene expression, thereby strengthening plant tolerance under metal stress conditions (Song et al. 2019). In rice, high-throughput RNA sequencing of roots exposed to cadmium (Cd) stress revealed dynamic changes in long non-coding RNA (lncRNA) expression, with 75 lncRNAs being downregulated and 69 upregulated following Cd treatment. Functional enrichment analyses indicated that these lncRNAs likely regulate Cd response by controlling genes involved in cysteine-rich peptide biosynthesis through *cis*-acting mechanisms, and genes associated with photosynthesis and secondary metabolite production via *trans*-acting mechanisms (Chen et al. 2018). Similarly, in barley, researchers uncovered 195 newly identified lncRNAs with altered expression in roots under Cd exposure. Among these, lncRNAs LTCONS_00177709 and LTCONS_00185175 were predicted to act as endogenous target mimics (eTMs) for miR319a_1 and miR319b_3p, respectively. These lncRNA-miRNA interactions influence the expression of the *HvGAMYB* gene, a known target of miR319. *HvGAMYB* appears to play a significant role in cadmium tolerance by promoting metal ion uptake and redistribution, enhancing photosynthetic efficiency, and boosting antioxidant capacity, thereby enabling better adaptation to toxic metal environments (Zhou et al. 2023).

Furthermore, lncRNA also have been increasingly recognized for their role in epigenetic regulation, particularly in histone modifications and acetylation, contributing to plant responses under light stress conditions. In *Arabidopsis thaliana*, approximately 40–50% of light-regulated sense and antisense transcripts associated with histone marks H3K9ac and/or H3K27ac were found to be differentially expressed under intense light exposure, indicating a strong link between lncRNA activity and chromatin remodelling during high-light stress (Wang et al. 2014a, b). In addition to light stress, lncRNAs also play significant roles in nutrient deprivation responses. Under nitrogen-deficient conditions in *Populus*, researchers identified five antisense lncRNAs associated with nitrogen stress adaptation (Song et al. 2016b). Further analysis uncovered nine long intergenic non-coding RNAs (lincRNAs) that serve as precursors to 11 known miRNAs, along with five lincRNAs that generate 14 novel miRNAs, highlighting their contribution to nitrogen-responsive regulatory networks (Chen et al. 2016). In the context of phosphorus scarcity, *Arabidopsis* utilizes a well-characterized lncRNA known as IPS1 (Induced by Phosphate Starvation 1). IPS1 functions as a target mimic for miR399, imitating its binding site without being cleaved. This mimicry interferes with miR399's ability to bind to its actual target, PHOSPHATE2 (PHO2), thereby mitigating miRNA's repressive effect and maintaining PHO2 expression under low-phosphate conditions (Wu et al. 2013; Aung et al. 2006; Franco-Zorrilla et al. 2007). This regulatory mechanism showcases how lncRNAs orchestrate finely tuned gene expression adjustments in response to nutrient limitations.

7.4 High-Throughput Sequencing Approaches for Identifying Stress-Responsive ncRNAs

The rise of advanced sequencing platforms has greatly accelerated the identification of ncRNAs, enabling comprehensive and detailed analysis of the transcriptome. Breakthroughs in genomic tools, such as microarrays and NGS, have expanded our ability to identify novel RNA molecules, many of which are not linked to annotated protein-coding genes. In *Arabidopsis thaliana*, for example, high-throughput sequencing techniques enabled the discovery of 746 lncRNAs expressed throughout leaf formation. A subset of 28 lncRNAs was implicated in a sophisticated regulatory network involving competing endogenous RNAs (ceRNAs) and circRNAs, underscoring their functional relevance in leaf morphogenesis (Meng et al. 2018). This technology has been pivotal in identifying ncRNAs responsive to environmental stress in other crops. In tomato, researchers discovered 467 circRNAs expressed under drought, heat, and their combined stress conditions (Zhou et al. 2020). Notably, 156 such circRNAs consistently appeared across all stress treatments, with circRNA197 and circRNA29 exhibiting markedly elevated transcript abundance. These molecules are believed to contribute to key physiological functions, including photosynthesis and stress adaptation.

Sunkar et al. (2008) documented approximately 714,202 sequence entries obtained from rice plantlets subjected to multiple conditions—normal, water-deficit and salinity stress—using high-throughput sequencing. In *Brachypodium*, this advanced technique led to the discovery of 28 cold-responsive miRNAs (Zhang et al. 2009a, b, c; Xin et al. 2010), while 12 heat-responsive miRNAs were identified in wheat under elevated temperature conditions (Xin et al. 2010). Additionally, Wang et al. (2011a, b) constructed two small RNA libraries from *Medicago truncatula*, a model legume, representing control and drought-treated samples. By integrating high-throughput sequencing with RT-qPCR analysis, they uncovered both established and novel miRNAs. Among these, 29 were either newly discovered or categorized into existing miRNA families, with eight showing distinct expression changes under drought conditions. Collectively, various high-throughput sequencing approaches have proven instrumental in uncovering stress-responsive ncRNAs, highlighting their critical functions during plant stress signalling and adaptive mechanisms (Table 7.4).

RNA-seq stands out as a transformative technology for gene expression analysis and the discovery of novel lncRNAs (Wang et al. 2009). This approach also enables the detection of alternative splicing events, splice junctions, and transcript isoforms (Trapnell et al. 2009). The RNA-seq process involves converting RNA into complementary DNA (cDNA), which is subsequently used to create a sequencing-ready library. Initially, RNA-seq relied on oligo(dT) primers, enriching for transcripts with poly-A tails while underrepresenting non-polyadenylated RNAs like ribosomal RNAs (rRNAs). To address this limitation, modern protocols utilize random primers and rRNA-depleted RNA to capture a broader range of transcripts, including non-polyadenylated RNAs. RNA-seq typically produces short reads, ranging from

Table 7.4 High-throughput approaches used for identification of non-coding RNAs in plants under abiotic stress

ncRNA	Plant species	Methods	References
Salt stress			
miR165, miR166	*Arabidopsis thaliana*	qRT-PCR	Scintu et al. (2022)
miR156a	*Solanum lycopersicum*	qRT-PCR	Çakır et al. (2021)
miR12477	*Oryza sativa*	Library sequencing	Parmar et al. (2021)
miR156, miR165, miR319, miR393, miR396, miR167, miR168, miR171, miR152, and miR394	*Arabidopsis thaliana*	Microarray	Liu et al. (2008)
miRNAs, miR156, miR396, and miR319	Chickpea	qRT-PCR	Kohli et al. (2014)
miR393, miR855, miR3, and miR34	*Broccoli*	Solexa sequencing	Tian et al. (2014)
Cold stress			
miR828	*Eucalyptus grandis*	Sequencing	Farias et al. (2023)
miR5809a/miRN2230	*Oryza sativa*	Sequencing	Kesawat et al. (2024)
161 conserved and 236 novel miRNAs	*S. habrochaites*	qRT-PCR	Cao et al. (2014)
Drought stress			
miR156, miR395a, miR164c, miR16c, miR712a, and miR172c, miR1711	*Malus domestica*	qRT-PCR, protein purification	Bao et al. (2022)
105 conserved and 179 novel microRNAs	Tomato	Illumina sequencing technology	Liu et al. (2017)
sly-newmiR26-3p	*Solanum lycopersicum*	RT-qPCR	Lopez-Galiano et al. (2019)
miR160, miR164, miR167, miR394, and miR398	*Vigna radiate*	Deep RNA sequencing	Kumar et al. (2022)
Other abiotic stress			
miR390a-5p, miR390b-5p, and miR24249-5p	*Zea mays*	RNA sequencing	Teng et al. (2023)
miR156, miR168, and miR396	*Saccharum officinarum*	qRT-PCR	Gao et al. (2022)
miR319, miR398, miR408, miR9670-3p, and miRNA9773	*Triticum aestivum*	RNA sequencing	Zhao et al. (2020)

30 to 400 base pairs depending on the sequencing platform employed. Common platforms include Illumina, ABI SOLiD, Ion Torrent, Oxford Nanopore, and Pacific Biosciences. Notably, Pacific Biosciences and Oxford Nanopore leverage single-molecule sequencing technologies that allow long-read generation—up to 60 kb—though they often come with elevated error rates, which can compromise data fidelity. Two primary methods are employed for RNA-seq library construction: one

targets polyadenylated RNA using oligo(dT) beads, capturing mRNAs and around 60% of polyadenylated lncRNAs; the other depletes rRNAs from total RNA, allowing enrichment of non-polyadenylated transcripts, including circular RNAs (circRNAs), which lack poly-A tails and are not captured using oligo(dT)-based methods (Jeck and Sharpless 2014; Boeckel et al. 2015; Militello et al. 2016). RNA-seq provides an unbiased, genome-wide overview of transcription, making it more versatile than microarrays, which are limited to known sequences. However, RNA-seq data analysis is more complex. While microarrays benefit from established and user-friendly analytical pipelines, RNA-seq requires powerful computational infrastructure and bioinformatics expertise. Several normalization algorithms—such as RPKM, FPKM, and TPM—are in use, yet no single method is universally accepted as optimal (Ergstrom et al. 2013; Eder et al. 2016; Evans 2018). Moreover, sequencing depth, typically 10–20 million reads or more, is essential for obtaining reliable results comparable to those from microarrays (Wang et al. 2011a, b, c). Despite its strengths, RNA-seq has limitations, especially when it comes to comprehensive lncRNA profiling. It captures only a subset (~60%) of lncRNAs with poly-A tails, potentially missing others including circRNAs. Furthermore, many newly discovered lncRNAs contain repetitive sequences, complicating downstream analyses (Weirick et al. 2016). The accuracy of RNA-seq also depends heavily on the quality of the reference genome; newer versions like GRCh38 have shown improved mapping efficiency. PCR amplification bias, particularly in GC-rich regions, can further affect data reliability. Although RNA-seq is invaluable for exploring alternative splicing and post-transcriptional modifications such as A-to-I RNA editing (John et al. 2017; Stellos et al. 2016), microarrays may still be more appropriate for studies targeting known transcripts and pathways due to their established workflows. In applied research, such as in rice, RNA-seq has been instrumental in uncovering stress-responsive lncRNAs. For example, analysis revealed 98 natural antisense transcript (NAT) lncRNAs responsive to drought conditions. Among them, NAT Os02g0250700-01 and NAT Os02g0180800-01 were shown to regulate genes involved in late embryogenesis abundant proteins and cinnamoyl-CoA reductase, respectively (Chung et al. 2016), emphasizing their functional relevance in abiotic stress responses.

Microarray technology serves as a high-throughput method for detecting ncRNAs, including miRNAs and lncRNAs. miRNA-specific microarrays allow researchers to simultaneously assess the expression of thousands of genes, enabling detailed profiling of miRNA activity in plant tissues under stress conditions. These expression changes can be further confirmed using methods like quantitative real-time PCR or northern blotting. Although conventional techniques such as cloning, northern hybridization, and microarrays may occasionally yield differing expression levels, microarrays continue to be a valuable tool for comprehensive miRNA screening at the genomic scale. Nonetheless, qRT-PCR is widely regarded as the most accurate and efficient technique for validating mature miRNAs due to its speed, specificity, and sensitivity, particularly when studying the functional roles of novel stress-responsive miRNAs in crops (Kumar 2014). For example, microarray-based analysis of *Zea mays* roots under submergence stress identified several miRNAs, including Zm-miR166, Zm-miR167, Zm-miR171, Os-miR396, Zm-miR399, Zm-miR159, At-miR395, Pt-miR474, and Os-miR528 (Zhang et al. 2008). Similarly, next-generation sequencing in *Arabidopsis* roots uncovered 19 miRNA families responsive to hypoxic conditions, further expanding the landscape of regulatory RNAs involved in environmental adaptation (Moldovan et al. 2009). Beyond miRNAs, microarrays have also been adapted for the detection of lncRNAs, particularly as many arrays originally designed for protein-coding transcripts inadvertently contained probes aligning with lncRNA sequences. This is due to re-annotation efforts, revealing that several genes previously classified as protein coding were actually non-coding. Notably, platforms such as the Affymetrix GeneChip, although initially tailored to protein-coding genes, have been found to contain a significant number of probes aligning with non-coding RNA regions (Gellert et al. 2013). In one such study, Xin et al. (2011) used a wheat Affymetrix microarray in combination with Solexa sequencing to identify 54 potential lncRNAs responsive to heat stress. Among these, two transcripts—TahlnRNA27 and TalnRNA5—were notably upregulated following heat exposure and showed expression patterns that were consistent across both microarray and qRT-PCR analyses. These findings demonstrate the versatility of microarray platforms in investigating not just coding RNAs but also the diverse repertoire of non-coding transcripts involved in plant stress responses.

7.5 Clustered Regularly Interspaced Short Palindromic Repeats (CRISPR)-Based Exploration of Regulatory Non-coding Sequences Involved in Stress Resilience

The Clustered Regularly Interspaced Short Palindromic Repeats (CRISPR) (Clustered Regularly Interspaced Short Palindromic Repeats) genome engineering platforms, including Clustered Regularly Interspaced Short Palindromic Repeats/ Clustered Regularly Interspaced Short Palindromic Repeats (CRISPR)-associated proteins(Clustered Regularly Interspaced Short Palindromic Repeats (CRISPR)/ Cas) and Clustered Regularly Interspaced Short Palindromic Repeats (CRISPR)/ Cpf1 (also known as Cas12a), have been effectively tailored for precise genetic alterations in plants, guided by short RNA molecules known as guide RNAs (gRNAs) (Osakabe and Oskabe 2015; Osakabe et al. 2016; Wang et al. 2018). Clustered Regularly Interspaced Short Palindromic Repeats/Clustered Regularly Interspaced Short Palindromic Repeats (CRISPR)-associated proteins(Clustered Regularly Interspaced Short Palindromic Repeats (CRISPR)/Cas)9 primarily introduces double-strand breaks, which are typically repaired through the error-prone non-homologous end joining (NHEJ) pathway. This repair mechanism often results in insertions or deletions (indels), particularly at regions such as pre-miRNA loci, miRNA maturation sites, or miRNA binding domains on target mRNAs—thereby affecting miRNA biosynthesis or function (Chang et al. 2016; Zhou et al. 2017).

Alternatively, the homology-directed repair (HDR) mechanism may be utilized to accurately delete or integrate miRNA loci or their associated regulatory upstream elements (Zhao et al. 2016a). In standard plant genome editing protocols, a codon-optimized Cas9 enzyme is usually driven by strong promoters like CaMV 35S, while gRNAs are transcribed from U3 or U6 small RNA promoters. These genetic elements, often accompanied by a selectable marker, are introduced into plant tissues using transformation methods like *Agrobacterium tumefaciens*–mediated gene transfer or biolistic delivery (Liang et al. 2017). A notable application of this system is in rice, where site-specific mutations were introduced into the miR156 target site of the *IPA1* (Ideal Plant Architecture 1) gene. These modifications led to beneficial changes in plant architecture, improving characteristics like tiller number and overall stature (Li et al. 2016b).

Clustered Regularly Interspaced Short Palindromic Repeats/Clustered Regularly Interspaced Short Palindromic Repeats (CRISPR)-associated proteins(Clustered Regularly Interspaced Short Palindromic Repeats (CRISPR)/Cas)9 technology has become a transformative tool in plant biotechnology, particularly for generating knockout mutants of microRNAs (miRNAs). This system has enabled reverse genetics studies with greater precision, offering a promising route for developing superior alleles and accelerating crop improvement initiatives (Song et al. 2019). Both single guide RNA (sgRNA) and dual guide RNA (dgRNA) configurations have proven effective for editing miRNA genes (Zhao et al. 2016a; Damodharan et al. 2018; Bi et al. 2020). Modifying miRNAs via Clustered Regularly Interspaced Short Palindromic Repeats/Clustered Regularly Interspaced Short Palindromic Repeats (CRISPR)-associated proteins(Clustered Regularly Interspaced Short Palindromic Repeats (CRISPR)/Cas)9 can significantly affect plant responses to various abiotic stresses by altering downstream biochemical and regulatory circuits. Targeted deletions in miRNA loci have yielded valuable insights into their biological roles. For example, editing *OsMIR528* led to deletions of 4 and 54 base pairs within the mature miRNA sequence, effectively nullifying its function (Zhou et al. 2017). Similarly, large deletions—such as 934 bp in *AtMIR169a* and 607–618 bp in *AtMIR827a*—resulted in the complete loss of miRNA activity (Zhao et al. 2016a). In another case, a 1-bp insertion in *miR160* caused moderate leaf serration, while a 5-bp deletion entirely disrupted the function of *DCL1*, producing severe leaf morphology defects (Bi et al. 2020). Clustered Regularly Interspaced Short Palindromic Repeats/Clustered Regularly Interspaced Short Palindromic Repeats (CRISPR)-associated proteins(Clustered Regularly Interspaced Short Palindromic Repeats (CRISPR)/Cas)9 generates double-stranded DNA breaks, which are usually resolved through non-homologous end joining (NHEJ) or homologous recombination (HR), frequently resulting in small sequence modifications that impair gene activity. Such precise edits enable the production of true null mutants, facilitating detailed phenotypic analysis (Jinek et al. 2012; Reyon et al. 2012). The dual gRNA strategy for targeting *miR160* has successfully created deletions ranging from 5 to 48 bp, resulting in phenotypes similar to those obtained using single-guide approaches (Bi et al. 2020). Importantly, selecting gRNA sites within the mature miRNA sequence is critical to achieving functional knockouts. For instance, attempts to edit *miR1514* and *miR1509* in soybean yielded no functional change when mismatches were tolerated outside the mature region (Jacobs et al. 2015). This indicates that precise targeting within mature miRNA sequences is essential for effective gene silencing (Damodharan et al. 2018). In tomato, deletion of *SlmiR172d* via Clustered Regularly Interspaced Short Palindromic Repeats/ Clustered Regularly Interspaced Short Palindromic Repeats (CRISPR)-associated proteins(Clustered Regularly Interspaced Short Palindromic Repeats (CRISPR)/ Cas)9 caused abnormal floral organ development in a dose-dependent manner, likely due to deregulation of AP2-like target genes (Lin et al. 2021). Such studies have extended across several plant species, including *Arabidopsis thaliana* (Bi et al. 2020), *Marchantia polymorpha* (Tsuzuki et al. 2019), rice (Bi et al. 2020), wheat (Wang et al. 2020), soybean (Jacobs et al. 2015), and tomato (Lin et al. 2021). The majority of these efforts aim to enhance agronomic characteristics like stress tolerance, nutrient use efficiency, and yield potential (Deng et al. 2022). As *MIR* genes follow transcriptional pathways similar to protein-coding genes, their expression can also be fine-tuned using Clustered Regularly Interspaced Short Palindromic Repeats (CRISPR) activation (Clustered Regularly Interspaced Short Palindromic Repeats (CRISPR)a) or interference (Clustered Regularly Interspaced Short Palindromic Repeats (CRISPR)i) systems (Manghwar et al. 2019; Moradpour and Abdulah 2020; Pan et al. 2021). Additionally, newer genome editing strategies like base editing and prime editing are being explored to precisely dissect the functions of miRNA families with high sequence similarity (Deng et al. 2022). A case study by Chung et al. (2020) used Clustered Regularly Interspaced Short Palindromic Repeats/Clustered Regularly Interspaced Short Palindromic Repeats (CRISPR)-associated proteins(Clustered Regularly Interspaced Short Palindromic Repeats (CRISPR)/Cas)9 to analyse *OsmiR818b* in rice under drought conditions. A 24-bp deletion in the *OsMIR818b* gene, induced by recombinant Cas9 and sgRNA constructs, reduced the accumulation of both primary and mature forms of *OsmiR818b*. Edited plants displayed early leaf rolling and wilting during drought and recovered more slowly post-rehydration, establishing the role of *OsmiR818b* in drought resilience. In the same study, editing *OsMIR171f* resulted in small deletions (2–16 bp) that significantly lowered mature miRNA levels without impacting primary transcript levels, confirming the precision of the system in post-transcriptional regulation. Another miRNA, *OsmiR535*, was edited to confer increased tolerance to salt stress, Abscisic Acid(ABA), dehydration, and PEG-induced osmotic stress (Yue et al. 2020). Since *OsmiR818b* operates independently of Abscisic Acid(ABA) signalling, its disruption further validates its unique role in stress adaptation. Interestingly, deletion of individual miRNAs, such as *OsmiR408* or *OsmiR528*, modulated the activity of other miRNAs and genes associated with stress responses. For instance, knocking out *OsmiR528* led to reduced salt tolerance (Zhou et al. 2017), and disruption of *miR1320* elevated *OsERF096* expression, promoting cold tolerance via inhibition of the jasmonic acid pathway (Sun et al. 2022). Moreover, editing *miR393a* and *miR396c* improved salt tolerance in rice, demonstrating the broad utility of Clustered Regularly Interspaced Short Palindromic Repeats (CRISPR)-mediated miRNA editing in abiotic stress mitigation. Notably, homozygous T3 generation rice lines lacking *MIR169a* and *MIR827a*, developed through a dual-sgRNA/Cas9 approach, displayed enhanced drought tolerance, reinforcing the functional relevance of miRNA gene knockouts in crop stress physiology (Zhao et al. 2016a).

Enhancing our knowledge of miRNA functionality, along with improving Clustered Regularly Interspaced Short Palindromic Repeats (CRISPR)-based genome editing strategies, is critical for achieving accurate and efficient modifications of *MIR* genes in agricultural crops (Song et al. 2019). One of the major challenges lies in the high sequence similarity among miRNA family members, which complicates the design of precise guide RNAs. Selecting an appropriate protospacer adjacent motif (PAM) sequence becomes especially important, as it determines whether the editing goal is to target the entire miRNA family or a specific miRNA isoform. Moreover, miRNA genes are typically short in length, posing limitations for traditional Clustered Regularly Interspaced Short Palindromic Repeats/Clustered Regularly Interspaced Short Palindromic Repeats (CRISPR)-associated proteins(Clustered Regularly Interspaced Short Palindromic Repeats (CRISPR)/Cas)9 systems, which rely on the presence of 5′-NGG PAM motifs (Adli 2018). As a result, not all miRNA targets are compatible with the conventional Clustered Regularly Interspaced Short Palindromic Repeats/Clustered Regularly Interspaced Short Palindromic Repeats (CRISPR)-associated proteins(Clustered Regularly Interspaced Short Palindromic Repeats (CRISPR)/Cas)9 platform. However, the discovery and development of novel Cas protein variants have significantly expanded the editing capabilities, allowing researchers to overcome these constraints (Manghwar et al. 2019). For instance, Niu et al. (2021) employed a dual-sgRNA Clustered Regularly Interspaced Short Palindromic Repeats/Clustered Regularly Interspaced Short Palindromic Repeats (CRISPR)-associated proteins(Clustered Regularly Interspaced Short Palindromic Repeats (CRISPR)/Cas)9 approach to introduce substantial excisions within *lncRNA77580*, lncRNA linked to saline stress adaptation in soybean. This targeted excision led to modified expression of adjacent gene-coding regions implicated in stress signalling, demonstrating the regulatory significance of *lncRNA77580* in regulating saline stress tolerance mechanism in soybean root systems (Table 7.5).

7.6 Conclusion

ncRNAs, including miRNAs, lncRNAs, and circRNAs, have been recognized as vital regulators of plant responses to various abiotic stresses such as heat, cold, drought, and salinity. These ncRNAs control complex regulatory networks by influencing gene expression at both transcriptional and post-transcriptional levels, thereby affecting plant growth, development, and adaptation to adverse environmental conditions. Additionally, ncRNAs play important roles in stress adaptation by modulating gene expression, signal transduction pathways, and physiological responses. They also contribute to epigenetic regulation by impacting chromatin

Table 7.5 Edited miRNA and lncRNA using Clustered Regularly Interspaced Short Palindromic Repeats (CRISPR)/ Cas technology for functional characterization and plant improvement under abiotic stress

Plants	Edited miRNA & lncRNA	Key findings	References
Arabidopsis	miR535	miR535 knockout improves the tolerance to NaCl, drought, Abscisic Acid(ABA), and PEG stresses	Yue et al. (2020)
Arabidopsis	miR171f and miR818b	miR818b is implicated in drought response	Chung et al. (2020)
Arabidopsis	miR171f	miR171f promotes drought tolerance via modulating flavonoid biosynthesis genes	Um et al. (2022)
Soyabean	miR398c	miR398c is negatively implicated in drought response	Zhou et al. (2020b)
Soyabean	lncRNA77580	Deletions in lncRNA77580 expression affect the occurrence of a number of genes implicated in salt stress response	Niu et al. (2021)
Apple	miR17li	miR171i suppression improves drought stress tolerance via modulating ascorbic acid metabolism	Wang et al. (2020)
Switchgrass (*Panicum virgatum*)	miR169	Targeted and deleted regions of the miRNA genes for enhanced drought resistance	Zhao et al. (2016)

remodelling, DNA methylation, and histone modification patterns. Interactions between ncRNAs and plant hormonal signalling pathways—including auxin, cytokinin, abscisic acid, gibberellins, ethylene, and JA—further fine-tune growth and stress response mechanisms. Recent advances in high-throughput sequencing and computational biology have led to the discovery of numerous novel miRNAs and their targets, revealing their significant functions in enhancing plant tolerance under diverse abiotic stresses. miRNAs mainly function by post-transcriptionally regulating gene expression through mRNA cleavage or translational repression. Under stress conditions, certain miRNAs are upregulated to suppress their target transcripts, acting as negative regulators, while others are downregulated, resulting in the accumulation of target mRNAs that serve as positive regulators. Despite these findings, the full complexity of ncRNA-mediated regulatory networks in plants remains to be elucidated, necessitating further detailed research.

Looking ahead, future studies should aim to develop genetically engineered crops that incorporate stress-responsive miRNAs or their corresponding targets to improve tolerance against abiotic stresses such as drought, salinity, and extreme temperatures. Employing miRNA-mediated gene silencing approaches has the potential to enhance crop resilience, conserve water resources in agriculture, and increase yields, especially in drought-prone regions. In the broader scope, a deeper understanding of ncRNA functions will be critical for advancing sustainable agricultural practices and securing global food supply amid climate challenges. Integrating knowledge about ncRNAs into breeding programs—augmented by marker-assisted selection and genome editing techniques—can accelerate the creation of high-yielding, stress-resilient crop varieties. Precise manipulation of ncRNAs thus represents a promising avenue for improving crop productivity, quality, and tolerance to unfavourable environmental factors. Further investigations should focus on unravelling the molecular mechanisms underlying ncRNA functions and translating this knowledge into practical agricultural applications to promote sustainable crop production. miRNA interference (miRNAi) technology offers a valuable tool for functional genomics, enabling researchers to dissect plant developmental pathways and stress response mechanisms while facilitating agricultural innovations. By targeting stress-responsive genes through genetic engineering or selective breeding, it is possible to generate crop varieties with superior agronomic traits. To fully decipher the roles of ncRNAs in plant growth and stress adaptation, comprehensive approaches encompassing genetics, genomics, and biochemistry are required. Combining multi-omics technologies—including transcriptomics, proteomics, and metabolomics—will provide a holistic understanding of ncRNA regulatory networks throughout plant development and under stress conditions. Continued progress in sequencing technologies, bioinformatics, and genome editing will further empower the identification and functional manipulation of ncRNAs for crop improvement. Harnessing these advances will unlock the potential of ncRNAs to enhance food production and sustainability in the face of increasingly unpredictable environmental stresses.

References

Ache P, Becker D, Ivashikina N, Dietrich P, Roelfsema MRG, Hedrich R (2000) GORK, a delayed outward rectifier expressed in guard cells of *Arabidopsis thaliana*, is a K+-selective, K+-sensing ion channel. FEBS Lett 486:93–98

Adli M (2018) The Clustered Regularly Interspaced Short Palindromic Repeats (CRISPR) tool kit for genome editing and beyond. Nat Commun 9:1911

Ahmed W, Xia Y, Zhang H, Li R, Bai G, Siddique KHM, Guo P (2019) Identification of conserved and novel miRNAs responsive to heat stress in flowering Chinese cabbage using high-throughput sequencing. Sci Rep 9:14922

Akdogan G, Tufekci ED, Uranbey S, Unver T (2016) miRNA-based drought regulation in wheat. Funct Integr Genomics 16:221–233

Allen E, Xie Z, Gustafson AM, Sung GH, Spatafora JW, Carrington JC (2004) Evolution of microRNA genes by inverted duplication of target gene sequences in *Arabidopsis thaliana*. Nat Genet 36:1282–1290. https://doi.org/10.1038/ng1478

Allen E, Xie ZX, Gustafson AM, Carrington JC (2005) microRNA directed phasing during trans-acting siRNA biogenesis in plants. Cell 121:207–221

Altieri MA, Nicholls CI, Dinelli G, Negri L (2024) Towards an agroecological approach to crop health: reducing pest incidence through synergies between plant diversity and soil microbial ecology. NPJ Sustain Agric 2:6. https://doi.org/10.1038/s44264-024-00016-2

Alzahrani SM, Alaraidh IA, Migdadi H, Alghamdi SS (2019) Physiological, biochemical, and antioxidant properties of two genotypes of *Vicia fAbscisic Acid(ABA)* grown under salinity stress. Pak J Bot 51(3):786–798. https://doi.org/10.30848/PJB2019-3(3)

Arenas-Huertero C, Pérez B, RAbscisic Acid(ABA)nal F, Blanco-Melo D, Rosa CDL, Estrada-Navarrete G, Sanchez F, Covarrubias AA, Reyes JL (2009) Conserved and novel miRNAs in

the legume *Phaseolus vulgaris* in response to stress. Plant Mol Biol 70:385–401. https://doi.org/10.1007/s11103-009-9480-3
Arshad M, Puri A, Simkovich AJ, Renaud J, Gruber MY, Marsolais F, Hannoufa A (2020) Label-free quantitative proteomic analysis of alfalfa in response to microRNA156 under high temperature. BMC Genomics 21:758. https://doi.org/10.1186/s12864-020-07161-1
Aufsatz W, Mette M, van der Winden J, Matzke M, Matzke AJ (2002) HDA6, a putative histone deacetylase needed to enhance DNA methylation induced by double-stranded RNA. EMBO J 21:6832–6841
Aung K, Lin SI, Wu CC, Huang YT, Su CL, Chiou TJ (2006) pho2, a phosphate over accumulator, is caused by a nonsense mutation in a microRNA399 target gene. Plant Physiol 141:1000–1011
Bailey-Serres J, Voesenek LACJ (2008) Flooding stress: acclimations and genetic diversity. Ann Rev Plant Biol 59:313–339
Bao C, Qin G, Cao F, He J, Shen X, Chen P, Niu C, Zhang D, Ren T, Zhi F, Ma L, Ma F, Guan Q (2022) MdZAT5 regulates drought tolerance via mediating accumulation of drought-responsive miRNAs and mRNAs in apple. New Phytol 236(6):2131–2150
Beauclair L, Yu A, Bouché N (2010) microRNA-directed cleavage and translational repression of the copper chaperone for superoxide dismutase mRNA in *Arabidopsis*. Plant J 62:454–462
Becker D, Hoth S, Ache P, Wenkel S, Roelfsema MRG, Meyerhoff O, Hartung W, Hedrich R (2003) Regulation of the Abscisic Acid(ABA)-sensitive *Arabidopsis* potassium channel gene GORK in response to water stress. FEBS Lett 554:119–126
Ben Amor B, Wirth S, Merchan F, Laporte P, D'Aubenton-Carafa Y, Hirsch J, Maizel A, Mallory A, Lucas A, Deragon JM, Vaucheret H, Thermes C, Crespi M (2009) Novel long non-proteins coding RNAs involved in *Arabidopsis* differentiation and stress responses. Genome Res 19(1):57–69. https://doi.org/10.1101/gr.080275.108
Bhatia G, Singh A, Verma D, Sharma S, Singh K (2020) Genome-wide investigation of regulatory roles of lncRNAs in response to heat and drought stress in *Brassica juncea* (Indian mustard). Environ Exp Bot 171:103922e. https://doi.org/10.1016/j.envexpbot.2019.103922
Bi H, Fei Q, Li R, Liu B, Xia R, Char SN, Meyers BC, Yang B (2020) Disruption of miRNA sequences by TALENs and Clustered Regularly Interspaced Short Palindromic Repeats/ Clustered Regularly Interspaced Short Palindromic Repeats (CRISPR)-associated proteins(Clustered Regularly Interspaced Short Palindromic Repeats (CRISPR)/Cas)9 induces varied lengths of miRNA production. Plant Biotechnol J 18:1526–1536
Boeckel JN, Jae N, Heumuller AW, Chen W, Boon RA, Stellos K, Zeiher AM, John D, Uchida S, Dimmeler S (2015) Identification and characterization of hypoxia-regulated endothelial circular RNA. Circ Res 117:884–890
Bonnet E, Wuyts J, Rouze P, Van de PY (2004) Detection of 91 potential conserved plant microRNAs in *Arabidopsis thaliana* and *Oryza sativa* identifies important target genes. Proc Nat Acad Sci U S A 101:11511–11516
Borsani O, Zhu J, Verslues PE, Sunkar R, Zhu JK (2005) Endogenous siRNAs derived from a pair of natural cis-antisense transcripts regulate salt tolerance in *Arabidopsis*. Cell 123:1279–1291
Boyko A, Kovalchuk I (2010) Transgenerational response to stress in *Arabidopsis thaliana*. Plant Signal Behav 5:995–998
Bustamante A, Marques MC, Sanz-Carbonell A, Mulet JM, Gomez G (2018) Alternative processing of its precursor is related to miR319 decreasing in melon plants exposed to cold. Sci Rep 8:15538
Cagirici HB, Alptekin B, Budak H (2017) RNA sequencing and co-expressed long non-coding RNA in modern and wild wheats. Sci Rep 7:10670. https://doi.org/10.1038/s41598-017-11170-8
Çakır Ö, Arıkan B, Karpuz B, Turgut-Kara N (2021) Expression analysis of miRNAs and their targets related to salt stress in *Solanum lycopersicum* H-2274. Biotechnol Biotechnol Equip 35(1):275–282
Calixto CPG, Tzioutziou NA, James AB, Hornyik C, Guo W, Zhang R, Nimmo HG (2019) Cold-dependent expression and alternative splicing of *Arabidopsis* long non-coding RNAs. Front Plant Sci 10:235. https://doi.org/10.3389/fpls.2019.00235

Cao X, Wu Z, Jiang F, Zhou R, Yang Z (2014) Identification of chilling stress-responsive tomato microRNAs and their target genes by high-throughput sequencing and degradome analysis. BMC Genomics 15:1130. https://doi.org/10.1186/1471-2164-15-1130

Chang H, Yi B, Ma R, Zhang X, Zhao H, Xi Y (2016) Clustered Regularly Interspaced Short Palindromic Repeats/Clustered Regularly Interspaced Short Palindromic Repeats (CRISPR)-associated proteins(Clustered Regularly Interspaced Short Palindromic Repeats (CRISPR)/Cas)9 a novel genomic tool to knock down microRNA in vitro and in vivo. Sci Rep 6:22312

Cheah BH, Nadarajah K, Divate MD, Wickneswari R (2015) Identification of four functionally important micro-RNA families with contrasting differential expression profiles between drought-tolerant and susceptible rice leaf at vegetative stage. BMC Genomics 16:692

Chen M, Lv S, Meng Y (2010) Epigenetic performers in plants. Dev Growth Differ 52:555–566

Chen L, Luan Y, Zhai J (2015) Sp-miR396a-5p acts as a stress-responsive genes regulator by conferring tolerance to abiotic stresses and susceptibility to *Phytophthora nicotianae* infection in transgenic tobacco. Plant Cell Rep 34:2013–2025. https://doi.org/10.1007/s00299-015-1847-0

Chen M, Wang C, Bao H, Chen H, Wang Y (2016) Genome-wide identification and characterization of novel lncRNAs in Populus under nitrogen deficiency. Mol Gen Genomics 291:1663–1680. https://doi.org/10.1007/s00438-016-1210-3

Chen B, Feder ME, Kang L (2018) Evolution of heat-shock protein expression underlying adaptive responses to environmental stress. Mol Ecol 27(15):3040–3054. https://doi.org/10.1111/mec.14769

Chen L, Ren Y, Zhanf Y, Xu J, Sun J, Sun F, Zhanf Z, Wan Y 2012 Genome wide identification and expression analysis of heat responsive and novel microRNAs in Populus tomentosa. Gene 504(2):160–165

Chen J, Pan A, He S, Su P, Yuan X, Zhu S, Liu Z (2020) Different microRNA families involved in regulating high temperature stress response during cotton (*Gossypium hirsutum* L.) anther development. Int J Mol Sci 21:1280. https://doi.org/10.3390/ijms21041280

Chen J, Zhong Y, Qi X (2021) LncRNA TCONS_00021861 is functionally associated with drought tolerance in rice (*Oryza sativa* L.) via competing endogenous RNA regulation. BMC Plant Biol 21(410):1–12. https://doi.org/10.1186/s12870-021-03195-z

Chen P, Song Y, Liu X, Xiao L, Bu C, Liu P, Zhao L, Ingvarsson PK, Wu HX, El-Kassaby YA, Zhang D (2022) LncRNA PMATP to MYB46 module represses PtoMATE and PtoARF2 promoting Pb($^{2+}$) uptake and plant growth in poplar. J Hazard Mater 433:128769

Chen Z, Jiang X, Niu F, Sun X, Hu Z, Gao F (2023) Overexpression of lncRNA77580 regulates drought and salinity stress responses in soyabean. Plants 12:181

Chung PJ, Jung H, Jeong DH, Ha SH, Choi YD, Kim JK (2016) Transcriptome profiling of drought responsive noncoding RNAs and their target genes in rice. BMC Genomics 17:563. https://doi.org/10.1186/s12864-016-2997-3

Chung PJ, Chung H, Oh N, Choi J, Bang SW, Jung SE, Jung H, Shim JS, Kim JK (2020) Efficiency of recombinant Clustered Regularly Interspaced Short Palindromic Repeats (CRISPR)/rCas9-mediated miRNA gene editing in rice. Int J Mol Sci 21:9606. https://doi.org/10.3390/ijms21249606

Cordeiro D, Canhoto J, Correia S (2022) Regulatory non-coding RNAs: emerging roles during plant cell reprogramming and in vitro regeneration. Front Plant Sci 13:1049631

Correa JPO, Silva EM, Nogueira FTS (2018) Molecular control by non-coding RNAs during fruit development: from gynoecium patterning to fruit ripening. Front Plant Sci 9:1760

Covarrubias AA, Reyes JL (2010) Post-transcriptional gene regulation of salinity and drought responses by plant microRNAs. Plant Cell Environ 33:481–489

Cruz de Carvalho MH, Sun HX, Bowler C, Chua NH (2016) Noncoding and coding transcriptome responses of a marine diatom to phosphate fluctuations. New Phytol 210:497–510

Csorba T, Questa JI, Sun Q, Dean C (2014) Antisense COOLAIR mediates the coordinated switching of chromatin states of FLC during vernalization. Proc Natl Acad Sci USA 111:16160–16165

Cui N, Sun X, Sun M, Jia B, Duanmu H, Lv D, Duan X, Zhu Y (2015) Overexpression of *OsmiR156k* leads to reduced tolerance to cold stress in rice (*Oryza Sativa*). Mol Breed 35:214. https://doi.org/10.1007/s11032-015-0402-6

D'Ario M, Griffiths-Jones S, Kim M (2017) Small RNAs: big impact on plant development. Trends Plant Sci 22:1056–1068

Damodharan S, Corem S, Gupta SK, Arazi T (2018) Tuning of SlARF10A dosage by sly-miR160a is critical for auxin-mediated compound leaf and flower development. Plant J 96:855–868. https://doi.org/10.1111/tpj.14073

Deng P, Wang L, Cui L, Feng K, Liu F, Du X, Tong W, Nie X, Ji W, Weining S (2015) Global identification of MicroRNAs and their targets in barley under salinity stress. PLoS One 10:e0137990

Deng F, Zeng F, Shen Q, Abbas A, Cheng J, Jiang W, Chen G, Shah AN, Holford P, Tanveer M, Zhang D, Chen ZH (2022) Molecular evolution and functional modification of plant miRNAs with Clustered Regularly Interspaced Short Palindromic Repeats (CRISPR). Trends Plant Sci 27:890–907

Di C, Yuan J, Wu Y, Li J, Lin H, Hu L, Hu L, Zhang T, Qi Y, Gerstein MB, Guo Y, Lu ZJ (2014) Characterization of stress responsive lncRNAs in *Arabidopsis thaliana* by integrating expression, epigenetic and structural features. Plant J 80(5):848–861. https://doi.org/10.1111/tpj.12679

Ding D, Zhang L, Wang H, Liu Z, Zhang Z, Zheng Y (2009) Differential expression of miRNAs in response to salt stress in maize roots. Ann Bot 103:29–38

Ding Y, Chen Z, Zhu C (2011) Microarray-based analysis of cadmium-responsive microRNAs in rice (*Oryza sativa*). J Exp Bot 62(10):3563–3573

Ding Y, Ye Y, Jiang Z, Wang Y, Zhu C (2016) MicroRNA390 is involved in cadmium tolerance and accumulation in rice. Front Plant Sci 7:235. https://doi.org/10.3389/fpls.2016.00235

Ding Y, Huang L, Jiang Q, Zhu C (2020) MicroRNAs as important regulators of heat stress responses in plants. J Agric Food Chem 68(41):11320–11326

Dmitriev AA, Kudryavtseva AV, Bolsheva NL, Zyablitsin AV, Rozhmina TA, Kishlyan NV, Krasnov GS, Speranskaya AS, Krinitsina AA, Sadritdinova AF, Snezkhina AV, Fedorova MS, Yurkevich OY, Muravenko OV, Belenikin MS, Melnikova NV (2017) miR319, miR390, and miR393 are involved in aluminium response in flax (*Linum usitatissimum* L.). Biomed Res Int 2017:1–6

Dong S-M, Xiao L, Li ZB, Shen J, Yan HB, Li SX, Liao WB, Peng M (2022) A novel long non-coding RNA, DIR, increases drought tolerance in cassava by modifying stress-related gene expression. J Integr Agric 21:2588–2602

Eldem V, Akcay UC, Ozhuner E, Bakir Y, Uranbey S, Unver T (2012) Genome-wide identification of miRNAs responsive to drought in peach (*Prunus persica*) by high-throughput deep sequencing. PLoS One 7:e50298. https://doi.org/10.1371/journal.pone.0050298

Ergstrom PG, Steijger T, Sipos B, Grant GR, Kahles A, Consortium THE RGASP, Ratsch G, Goldman N, Hubbard TJ, Harrow J, Guigo R, Bertone P 2013 Systematic evaluation of spliced alignment programs for RNA seq data. Nature Methods 10, 1185–1191

Evans C, Hardin J, Stoebel DM 2018 Selecting between sample RNA seq normalization methods from the perspective of their normal assumptions. Briefings in Bioinformatics 19(5):776–792

Eder T, Grebien F, Rattei T (2016) NVT: a fast and simple tool for the assessment of RNA-seq normalization strategies. Bioinformatics 32(23):3682–2684

Eom SH, Lee HJ, Lee JH, Wi SH, Kim SK, Hyun TK (2019) Identification and functional prediction of drought-responsive long non-coding RNA in tomato. Agronomy 9:629. https://doi.org/10.3390/agronomy9100629

FAO (2023) Ukraine: impact of the war on agricultural enterprises– findings of a nationwide survey of agricultural enterprises with land up to 250 hectares

Farias J, Basile P, Giorello F (2023) Identification of cold stress-responsive microRNAs and mRNA in *Eucalyptus grandis*, vol 27. Agrociencia Uruguay

Ferdous J, Hussain SS, Shi BJ (2015) Role of microRNAs in plant drought tolerance. Plant Biotechnol J 13(3):293–305. https://doi.org/10.1111/pbi.12318

Ferdous J, Sanchez-Ferrero JC, Langridge P, Milne L, Chowdhury J, Brien C, Tricker PJ (2016) Differential expression of microRNAs and potential targets under drought stress in barley. Plant Cell Environ 40:11–24

Ferdous J, Whitford R, Nguyen M, Brien C, Langridge P, Tricker PJ (2017) Drought-inducible expression of Hv-miR827 enhances drought tolerance in transgenic barley. Funct Integr Genomics 17(2–3):279–292. https://doi.org/10.1007/s10142-016-0526-8

Franco-Zorrilla JM, Valli A, Todesco M, Mateos I, Puga MI, Rubio-Somoza I, Leyva A, Weigel D, García JA, Paz-Ares J (2007) Target mimicry provides a new mechanism for regulation of microRNA activity. Nat Genet 39:1033–1037

Fu L, Ding Z, Tan D, Han B, Sun X, Zhang J (2020) Genome wide discovery and functional prediction of salt responsive in *Camellia sinensis*. Front Plant Sci 11:218

Fukuda M, Nishida S, Kakei Y, Shimada Y, Fujiwara T (2019) Genome-wide analysis of long intergenic noncoding RNAs responding to low-nutrient conditions in *Arabidopsis thaliana*: possible involvement of trans-acting siRNA3 in response to low nitrogen. Plant Cell Physiol 60:1961–1973

Furini A, Koncz C, Salamini F, Bartels D (1997) High level transcription of a member of a repeated gene family confers dehydration tolerance to callus tissue of *Craterostigma plantagineum*. EMBO J 16:3599–3608

Gao P, Bai X, Yang L, Lv D, Li Y, Cai H, Ji W, Guo D, Zhu Y (2010) Over-expression of osa-MIR396c decreases salt and alkali stress tolerance. Planta 231:991–1001. https://doi.org/10.1007/s00425-010-1104-2

Gao P, Bai X, Yang L. Lv D, pan X, Li Y, Cai H, Ji W, Chen Q, Zhu Y 2011 Osa-MIR393: a salinity and alkaline stress-related microRNA gene. Mol Biol Rep 38(1):237–42

Gao N, Qiang XM, Zhai BN, Min J, Shi WM (2015) Transgenic tomato overexpressing *ath-miR399d* improves growth under abiotic stress conditions. Russ J Plant Physiol 62:360–366. https://doi.org/10.1134/S1021443715030061

Gao F, Wang N, Li H, Liu J, Fu C, Xiao Z, Wei C, Lu X, Feng J, Zhou Y (2016) Identification of drought-responsive microRNAs and their targets in *Ammopiptanthus mongolicus* by using high-throughput sequencing. Sci Rep 6:34601. https://doi.org/10.1038/srep34601

Gao S, Yang Y, Yang Y, Zhang X, Su Y, Guo J, Que Y, Xu L (2022) Identification of low-nitrogen-related miRNAs and their target genes in sugarcane and the role of miR156 in nitrogen assimilation. Int J Mol Sci 23:13187

Gellert P, Ponomareva Y, Braun T, Uchida S (2013) Noncoder: a web interface for exon array-based detection of long non-coding RNAs. Nucleic Acids Res 41:e20

Guo Y, Zhao S, Zhu C, Chang X, Yue C, Wang Z, Lin Y, Lai Z (2017) Identification of drought-responsive miRNAs and physiological characterization of tea plant (*Camellia sinensis* L.) under drought stress. BMC Plant Biol 17:211. https://doi.org/10.1186/s12870-017-1172-6

Hajyzadeh M, Turktas M, Khawar KM, Unver T (2015) miR408 overexpression causes increased drought tolerance in chickpea. Gene 555:186–193. https://doi.org/10.1016/j.gene.2014.11.002

He F, Liu Q, Zheng L, Cui Y, Shen Z, Zheng L (2015) Rna-seq analysis of rice roots reveals the involvement of post-transcriptional regulation in response to cadmium stress. Front Plant Sci 6:1136. https://doi.org/10.3389/fpls.2015.01136

He J, Jiang Z, Gao L, You C, Ma X, Wang X, Xu X, Mo B, Chen X, Liu L (2019) Genome-wide transcript and small RNA profiling reveals transcriptomic responses to heat stress. Plant Physiol 181:609–629. https://doi.org/10.1104/pp.19.00403

He X, Guo S, Wang Y, Wang L, Shu S, Sun J (2020) Systematic identification and analysis of heat-stress-responsive lncRNAs, circRNAs and miRNAs with associated co-expression and ceRNA networks in cucumber (*Cucumis sativus* L.). Physiol Plant 168(3):736–754. https://doi.org/10.1111/ppl.12997

He Q, Zhu S, Zhang B (2014) Micro-RNA target gene responses to lead induced stress in cotton (Gossypium hirsutum L.) Functional & Integrative Genomics 14:507–515

Helliwell CA, Robertson M, Finnegan EJ, Buzas DM, Dennis ES (2011) Vernalization-repression of Arabidopsis FLC requires promoter sequences but not antisense transcripts. PLoS One 6:e2151

Heo JB, Sung S (2011) Vernalization-mediated epigenetic silencing by a long intronic noncoding RNA. Science 331:76–79

Higashi Y, Takechi K, Takano H, Takio S (2013) Involvement of microRNA in copper deficiency-induced repression of chloroplastic Cu Zn-superoxide dismutase genes in the moss *Physcomitrella patens*. Plant Cell Physiol 54:1345–1355

Hou J, Lu D, Mason AS, Li B, Xiao M, An S (2019) Non-coding RNAs and transposable elements in plant genomes: emergence, regulatory mechanisms and roles in plant development and stress responses. Planta 250(1):23–40

Hsieh LC, Lin SI, Shih ACC, Chen JW, Lin WY, Tseng CY, Li WH, Chiou TJ (2009) Uncovering small RNA-mediated responses to phosphate deficiency in Arabidopsis by deep sequencing. Plant Physiol 51:2120–2132

Hu B, Wang W, Deng K, Li H, Zhang Z, Zhang L, Chu C (2015) MicroRNA399 is involved in multiple nutrient starvation responses in rice. Front Plant Sci 6. https://doi.org/10.3389/fpls.2015.00188

Hu Z, Lyu T, Yan C, Wang Y, Ye N, Fan Z, Li X, Li J, Yin H (2020) Identification of alternatively spliced gene isoforms and novel noncoding RNAs by single-molecule long-read sequencing in *Camellia*. RNA Biol 17:966–976. https://doi.org/10.1080/15476286.2020.1738703

Hu X, Wei Q, Wu H, Huang Y, Peng X, Han G, Ma Q, Zhao Y (2022) Identification and characterization of heat-responsive lncRNAs in maize inbred line CM1. BMC Genomics 23(1):208. https://doi.org/10.1186/s12864-022-08448-1

Huang W, Xian Z, Hu G, Li Z (2016) SLAGO4A, a core factor of RNA-directed DNA methylation (RdDM) pathway, plays an important role under salt and drought stress in tomato. Mol Breed 36:28

Huo K, Rua Y., Fan H, Guo C, and Cai H (2022) Spatial-temporal variation characteristics of cultivated land and controlling factors in the Yangtze River Delta region of China. Front. Env. Sci. 10, 2140. https://doi.org/10.3389/fenvs.2022.871482

Imaduwage I, Hewadikaram M (2024) Predicted roles of long non-coding RNAs in abiotic stress tolerance responses of plants. Mol Hortic 4(1):20

Ingram J, Bartels D (1996) The molecular basis of dehydration tolerance in plants. Annu Rev Plant Physiol Plant Mol Biol 47:377–403

Islam W, Idrees A, Waheed A, Zeng F (2022) Plant responses to drought stress: microRNAs in action. Environ Res 215(2):114282

Ito H, Gaubert H, Bucher E, Mirouze M, Vaillant I, Paszkowski J (2011) An siRNA pathway prevents transgenerational retrotransposition in plants subjected to stress. Nature 472:115–119. https://doi.org/10.1038/nature09861

Jabnoune M, Secco D, Lecapmion C, Robaglia C, Shu Q, Poirier Y (2013) A rice cis-natural antisense RNA acts as a translational enhancer for its cognate mRNA and contributes to phosphate homeostasis and plant fitness. Plant Cell 25:4166–4182

Jacobs TB, LaFayette PR, Schmitz RJ, Parrott WA (2015) Targeted genome modifications in soybean with Clustered Regularly Interspaced Short Palindromic Repeats (CRISPR)/ Cas9. BMC Biotechnol 15:16. https://doi.org/10.1186/s12896-015-0131-2

Jagadeeswaran G, Zheng Y, Li YF, Shukla LI, Matts J, Hoyt P, Macmil SL, Wiley GB, Roe B, Zhang W, Sunkar R (2009) Cloning and characterization of small RNAs from *Medicago truncatula* reveals four novel legume-specific microRNA families. New Phytol 184:85–98

Jeck WR, Sharpless NE (2014) Detecting and characterizing circular RNAs. Nat Biotechnol 32:453–461

Jinek M, Chylinski K, Fonfara I, Hauer M, Doudna JA, Charpentier E (2012) A programmable dual RNA guided DNA endonuclease in adaptive bacterial immunity. Science 337(6096):816–21

Jha UC, Nayyar H, Jha R, Khurshid M, Zhou M, Mantri N, Siddique KH (2020) Long non-coding RNAs: emerging players regulating plant abiotic stress response and adaptation. BMC Plant Biol 20:1–20

Jia XY, Wang WX, Ren LG, Chen QJ, Mendu V, Willcut B, Dinkins R, Tang X, Tang G (2009) Differential and dynamic regulation of miR398 in response to Abscisic Acid(ABA) and salt stress in *Populus tremula* and *Arabidopsis thaliana*. Plant Mol Biol 71:51–59

Jiang Y, Wu X, Shi M, Yu J, Guo C (2022) The miR159-MYB33-ABI5 module regulates seed germination in *Arabidopsis*. Physiol Plant 174:e13659. https://doi.org/10.1111/ppl.13659

Jiao F, Pahwa K, Manning M, Dichy N, Geuten K (2019) Cold induced antisense transcription of *FLOWERING LOCUS C* in distant grasses. Front Plant Sci 10:72. https://doi.org/10.3389/fpls.2019.00072

Jin Z, Gao S, Ma W, Lyu X, Cao X, Yao Y (2020) Identification and functional prediction of salt stress-related long noncoding RNAs in grapevine roots. Environ Exp Bot 179:104215

Jin X, Wang Z, Li X, Ai Q, Wong DCJ, Zhang F, Yang J, Zhang N, Si H (2024) Current perspectives of lncRNAs in abiotic and biotic stress tolerance in plants. Front Plant Sci 14:1334620

John D, Weirick T, Dimmeler S, Uchida S (2017) RNA editor: easy detection of RNA editing events and the introduction of editing islands. Brief Bioinform 18(6):993–1001

Jones-Rhoades MW, Bartel DP (2004) Computational identification of plant microRNAs and their targets, including a stress-induced miRNA. Mol Cell 14:787–799

Kang SH, Aknadibossian V, Kharel L, Mudiyanselage SDD, Wang Y, Folimonova SY (2021) The intriguing conundrum of a non-conserved multifunctional protein of citrus tristeza virus that interacts with a viral long non-coding RNA. Viruses 13:2129

Kantar M, Unver T, Budak H (2010) Regulation of barley miRNAs upon dehydration stress correlated with target gene expression. Funct Integr Genomics 10:493–507

Kantar M, Lucas S, Budak H (2011) miRNA expression patterns of *Triticum dicoccoides* in response to shock drought stress. Planta 233:471–484

Karakulah G, Unver T (2017) Barley long non-coding RNAs and their tissue-specific co-expression pattern with coding-transcripts. bioRxiv[preprint] https://doi.org/10.1101/229559

Karlik E, Gozukirmizi N (2018) Expression analysis of lncRNA AK370814 involved in the barley vitamin B6 salvage pathway under salinity. Mol Biol Rep 45:1597–1609. https://doi.org/10.1007/s11033-018-4289-

Kesawat MS, Manohar S, Kherawat BS, Kumar S, Lenka SK, Parameswaran C, Katara JL, Tripathy S, Kim HU, Chung SM, Siddiqui MH, Alamri S, Kumar M (2024) Genome-wide survey of peptides containing tyrosine sulfation (PSY) gene family and potential PSY specific miRNA revealed their role in plant development and diverse stress conditions in rice (*Oryza sativa*). Plant Stress 11:100412

Khraiwesh B, Zhu JK, Zhu J (2012) Role of miRNAs and siRNAs in biotic and abiotic stress responses of plants. Biochim Biophys Acta 1819:137–148

Kidner CA, Martienssen RA (2004) Spatially restricted microRNA directed leaf polarity through AGRONUATE1. Nature 428:81–84

Kiger NM, Schroeder SJ (2024) SVALKA: a long noncoding cis-natural antisense RNA that plays a role in the regulation of the cold response of *Arabidopsis thaliana*. Non-Coding RNA 10(6):59. https://doi.org/10.3390/ncrna10060059

Kim DH, Sung S (2017) Vernalization-triggered intragenic chromatin-loop formation by long noncoding RNAs. Dev Cell 40:302–12.e4

Kindgren P, Ard R, Ivanov M, Marquardt S (2018) Transcriptional read-through of the long noncoding RNA SVALKA governs plant cold acclimation. Nat Commun 9:4561

Kinoshita N, Wang H, Kasahara H, Liu J, Macpherson C, Machida Y, Kamiya Y, Hannah MA, Chua NH (2012) IAA-Ala Resistant3 an evolutionary conserved target of miR167, mediates Arabidopsis root architecture changes during high osmotic stress. Plant Cell 24:3590–3602

Knauer S, Holt AL, Rubio-Somoza I, Tucker EJ, Hinze A, Pisch M, Javelle M, Timmermans MC, Tucker MR, Laux T (2013) A protodermal miR394 signal defines a region of stem cell competence in the Arabidopsis shoot meristem. Dev Cell 24:125–132

Kohli D, Joshi G, Deokar AA, Bhardwaj AR, Agarwal M, Agarwal SK, Srinivasan R, Jain PK (2014) Identification and characterization of wilt and slat stress responsive MicroRNAs in chickpea through high-throughput sequencing. PLoS One 9(10):e108851. https://doi.org/10.1371/journal.pone.0108851

Kumar R (2014) Role of microRNAs in biotic and abiotic stress responses in crop plants. Appl Biochem Biotechnol 174:93

Kumar S, Das M, Sadhukhan A, Sahoo L (2022) Identification of differentially expressed mungbean miRNAs and their targets in response to drought stress by small RNA deep sequencing. Curr Plant Biol 30:100246

Kume K, Tsutsumi KI, Saitoh Y (2010) TAS1 trans-acting siRNA targets are differentially regulated at low temperature, and TAS1 trans-acting siRNA mediates temperature At1g51670 expression. Biosci Biotechnol Biochem 74:1435–1440

Kuska B (1998) Should scientists scrap the notion of junk DNA? J Natl Cancer Inst 90:1032–1033

Lakhotia N, Joshi G, Bhardwaj AR, Katiyar-Agarwal S, Agarwal M, Jagannath A, Goel S, Kumar A (2014) Identification and characterization of miRNAome in root, stem, leaf and tuber developmental stages of potato (*Solanum tuberosum* L.) by high-throughput sequencing. BMC Plant Biol 14:6

Li WX, Oono Y, Zhu J, He XJ, Wu JM, Iida K, Lu XY, Cui X, Jin H, Zhu JK (2008) The *Arabidopsis* NFYA5 transcription factor is regulated transcriptionally and post transcriptionally to promote drought resistance. Plant Cell 20:2238–2251. https://doi.org/10.1105/tpc.108.059444

Li JS, Fu FL, Ming AN, Zhou SF, She YH, Li WC (2013) Differential expression of microRNAs in response to drought stress in maize. J Integr Agric 12:1414–1422

Li S, Liu J, Liu Z, Li X, Wu F, He Y (2014) Heat-induced tas1 target1 mediates thermotolerance via heat stress transcription factor a1a-directed pathways in Arabidopsis. Plant Cell 26:1764–1780

Li W, Wang T, Zhang Y, Li Y (2016a) Overexpression of soyabean miR172c confers tolerance to water deficit and salt stress, but increases Abscisic Acid(ABA) sensitivity in transgenic *Arabidopsis thaliana*. J Exp Bot 67(1):175–194

Li M, Li X, Zhou Z, Wu P, Fang M, Pan X, Lin Q, Luo W, Wu G, Li H (2016b) Reassessment of the four yield related genes Gn1a, DEP1, GS3 and IPA1 in rice using a Clustered Regularly Interspaced Short Palindromic Repeats/Clustered Regularly Interspaced Short Palindromic Repeats (CRISPR)-associated proteins(Clustered Regularly Interspaced Short Palindromic Repeats (CRISPR)/Cas)9 system. Front Plant Sci 7:377

Li S, Yu X, Cheng Z, Zhao P, He Y, Wang W, Peng M (2017a) Genome-wide identification and functional prediction of cold and/or drought-responsive lncRNAs in cassava. Sci Rep 7:45981

Li Y, Wan L, Bi S, Wan X, Li Z, Cao J, Tong Z, Xu H, He F, Li X (2017b) Identification of drought-responsive micrornas from roots and leaves of alfalfa by high-throughput sequencing. Genes 8:119

Li X, Yang L, Chen LL (2018) The biogenesis functions and challenges of circular RNAs. Mol Cell 71:428–442

Li J, Cui J, Dai C, Luo C (2020) Whole transcriptome RNA sequencing reveals the global molecular responses and CeRNA regulatory network of mRNAs, lncRNAs, miRNAs and circRNAs in response to salt stress in sugar beet (*Beta vulgaris*). Int J Mol Sci 22:289

Liang Z, Chen K, Li T, Zhang Y, Zhao Q, Zhang H, Liu C, Ran Y, Gao C (2017) Efficient DNA-free genome editing of bread wheat using Clustered Regularly Interspaced Short Palindromic Repeats/Clustered Regularly Interspaced Short Palindromic Repeats (CRISPR)-associated proteins(Clustered Regularly Interspaced Short Palindromic Repeats (CRISPR)/Cas) 9 nucleoprotein complexes. Nat Commun 8:14261

Lin JS, Lin CC, Lin HH, Jeng ST (2012) MicroR828 regulates lignin and H2O2 accumulation in sweet potato on wounding. New Phytol 196(2):427–440

Lin JS, Kuo CC, Yang IC, Tsai WA, Shen YH, Lin CC, Liang YC, Li YC, Kuo YW, King YC, Lai HM, Jeng ST (2018) MicroRNA160 modulates plant development and heat shock protein gene expression to mediate heat tolerance in *Arabidopsis*. Front Plant Sci 9. https://doi.org/10.3389/fpls.2018.00068

Lin W, Gupta SK, Arazi T, Spitzer-Rimon B (2021) MIR172d is required for floral organ identity and number in tomato. Int J Mol Sci 22:4659

Liu HH, Tian X, Li YJ, Wu CA, Zheng CC (2008) Microarray-based analysis of stress-regulated microRNAs in *Arabidopsis thaliana*. RNA 14(5):836–843

Liu M, Yu H, Zhao G, Huang Q, Lu Y, Ouyang B (2017) Profiling of drought-responsive micro RNA and mRNA in tomato using high-throughput sequencing. BMC Genomics 18:481. https://doi.org/10.1186/s12864-017-3869-1

Liu X, Zhang X, Sun B, Hao L, Liu C, Zhang D, Tang H, Li C, Li Y, Shi Y, Xie X, Song Y, Wang T, Li Y (2019a) Genome-wide identification and comparative analysis of drought related microRNAs in two maize inbred lines with contrasting drought tolerance by deep sequencing. PLoS One 14(7):e0219176. https://doi.org/10.1371/journal.pone.0219176

Liu J, Feng L, Gu X, Deng X, Qiu Q, Li Q, Zhang Y, Wang M, Deng Y, Wang E, He Y, Baurle I, Li J, Cao X, He Z (2019b) An H3K27me3 demethylase-HSFA2 regulatory loop orchestrates transgenerational thermomemory in *Arabidopsis*. Cell Res 29:379–390. https://doi.org/10.1038/s41422-019-0145-8

Liu G, Liu F, Wang Y, Liu X (2022) A novel long noncoding RNA CIL1 enhances cold stress tolerance in Arabidopsis. Plant Sci 323:11370

Lopez-Galiano MJ, Sentandreu V, Martinez-Ramirez AC, Rausell C, Real MD, Camanes G, Ruiz-Rivero O, Crespo-Salvador O, Garcia-Robles I (2019) Identification of stress associated microRNAs in *Solanum lycopersicum* by high-throughput sequencing. Genes 10(6):475. https://doi.org/10.3390/genes10060475

Lu XY, Huang XL (2008) Plant miRNAs and abiotic responses. Biochem Biophys Commun 368(3):458–462

Lu SF, Sun YH, Shi R, Clark C, Li LG, Chiang VL (2005) Novel and mechanical stress-responsive micro-RNAs in *Populus trichocarpa* that are absent from Arabidopsis. Plant Cell 17:2186–2203

Lu YZ, Zhen F, Bian LY, Xie H, Liang JS (2010) MiR398 regulation in rice of the responses to abiotic and biotic stresses depends on CSD1 and CSD2 expression. Funct Plant Biol 38:44–53

Luo QJ, Mittal A, Jia F, Rock CD (2012) An autoregulatory feedback loop involving PAP1 and TAS4 in response to sugars in *Arabidopsis*. Plant Mol Biol 80:117–129

Liu F, Wang W, Sun X, Liang Z, Wang F 2014 Conserved and novel heat stress responsive microRNAs were identified by deep sequencing in Saccharina japonica (Laminariales, Phaeophyta). Plant Cell & Environment 38(7):1357–1367

Lu S, Sun YH, Chiang VL (2008) Stress responsive microRNAs in Populus. The plant Journal 55(1):131-151 Chen L, Ren Y, Zhanf Y, Xu J, Sun J, Sun F, Zhanf Z, Wan Y 2012. Genome wide identificarui and expression analysis of heat responsive and novel microRNAs in Populus tomentosa. Gene 504(2):160–165

Lurin C, Andrés C, Aubourg S, Bellaoui M, Bitton F, Bruyère C, Caboche M, Debast C, Gualberto J, Hoffmann B, Lecharny A, Le Ret M, Martin-Magniette ML, Mireau H, Peeters N, Renou JP, Szurek B, Taconnat L, Small I (2004) Genome-wide analysis of *Arabidopsis* pentatricopeptide repeat proteins reveals their essential role in organelle biogenesis. Plant Cell 16:2089–2103

Lv Y, Hu F, Zhou Y, Wu F, Gaut BS (2019) Maize transposable elements contribute to long noncoding RNAs that are regulatory hubs for abiotic stress response. BMC Genomics 20:864. https://doi.org/10.1186/s12864-019-6245-5

Manghwar H, Lindsey K, Zhang X, Jin S (2019) Clustered Regularly Interspaced Short Palindromic Repeats/Clustered Regularly Interspaced Short Palindromic Repeats (CRISPR)-associated proteins(Clustered Regularly Interspaced Short Palindromic Repeats (CRISPR)/Cas) system: recent advances and future prospects for genome editing. Trends Plant Sci 24:1102–1125

Matsunaga W, Kobayashi A, Kato A, Ito H (2012) The effects of heat induction and the siRNA biogenesis pathway on the transgenerational transposition of ONSEN, a copia-like retrotransposon in *Arabidopsis thaliana*. Plant Cell Physiol 53:824–833

Matthews C, Arshad M, Hannoufa A (2019) Alfalfa response to heat stress is modulated by microRNA156. Physiol Plant 165:830–842. https://doi.org/10.1111/ppl.12787

Mattrick JS, Rinn JL (2015) Discovery and annotation of long noncoding RNAs. Nat Struct Mol Biol 22(1):5–7

Megha S, Basu U, Kav NNV (2018) Regulation of low temperature stress in plants by microRNAs. Plant Cell Environ 41:1–15

Meng X, Zhang P, Chen Q, Wang J, Chen M (2018) Identification and characterization of ncRNA-associated ceRNA networks in *Arabidopsis* leaf development. BMC Genomics 19:1–10. https://doi.org/10.1186/s12864-018-4993-2

Militello G, Weirick T, John D, Doring C, Dimmeler S, Uchida S (2016) Screening and validation of lncRNAs and circRNAs as miRNA sponges. Brief Bioinform 18:bbw053

Moldovan D, Spriggs A, Yang J, Pogson BJ, Dennis ES, Wilson IW (2009) Hypoxia-responsive microRNAs and trans-acting small interfering RNAs in *Arabidopsis*. J Exp Bot 61:165–177

Mondal TK, Panda AK, Rawal HC, Sharma TR (2018) Discovery of microRNA-target modules of African rice (*Oryza glaberrima*) under salinity stress. Sci Rep 8:570. https://doi.org/10.1038/s41598-017-18206-z

Moradpour M, Abdulah SNA (2020) Clustered Regularly Interspaced Short Palindromic Repeats (CRISPR)/dCas9 platforms in plants: strategies and applications beyond genome editing. Plant Biotech J 18:32–44

Muhammad A, Margaret YG, Ken W (2017) An insight into microRNA-156 role in salinity stress responses of alfalfa. Front Plant Sci 8:356–359

Muhammad M, Waheed A, Wahab A, Majeed M, Nazim M, Liu YH, Li L, Li WJ (2024) Soil salinity and drought tolerance: an evaluation of plant growth, productivity, microbial diversity, and amelioration strategies. Plant Stress 11:100319. https://doi.org/10.1016/j.stress.2023.100319

Muthusamy M, Uma S, Backiyarani S, Saraswathi MS (2015) Genome-wide screening for novel, drought stress-responsive long non-coding RNAs in drought-stressed leaf transcriptome of drought-tolerant and -susceptible banana (*Musa* spp) cultivars using Illumina high-throughput sequencing. Plant Biotechnol Rep 9:279–286

Muthusamy M, Uma S, Backiyarani S, Saraswati MS (2016) Computational prediction, identification and expression profiling of microRNAs in banana (*Musa spp.*) during soil moisture deficit stress. The journal of Horticultural Science and Biotechnology 89(2):208–214

Nai J, Ma T, Liu Y, Zhou Y (2023) Identification of small RNAs associated with salt stress in Chrysanthemums through high-throughput sequencing and bioinformatics analysis. Genes 14(3):561. https://doi.org/10.3390/genes14030561

Naya L, Paul S, Valdés-López O, Mendoza-Soto AB, Nova-Franco B, Sosa-Valencia G, Reyes JL, Hernández G (2014) Regulation of copper homeostasis and biotic interactions by microRNA 398b in common bean. PLoS One 9:e84416

Nezhadahmadi A, Prodhan ZH, Faruq G (2013) Drought tolerance in wheat. Sci World J 2013:610721. https://doi.org/10.1155/2013/610721

Nguyen G, Rothstein S, Spangenberg G, Kant S (2015) Role of microRNAs involved in plant response to nitrogen and phosphorous limiting conditions. Front Plant Sci 6:629

Ni Z, Hu Z, Jiang Q, Zhang H (2013) GmNFYA3, a target gene of miR169, is a positive regulator of plant tolerance to drought stress. Plant Mol Biol 82:113–129

Niu C, Li H, Jiang L, Yan M, Li C, Geng D, Xie Y, Yan Y, Shen X, Chen P, Dong J, Ma F, Guan Q (2019) Genome-wide identification of drought responsive microRNAs in two sets of *Malus* from interspecifc hybrid progenies. Hortic Res 6:75. https://doi.org/10.1038/s41438-019-0157-z

Niu F, Jiang Q, Sun X, Hu Z, Wang L, Zhang H (2021) Large DNA fragment deletion in lncRNA77580 regulates neighboring gene expression in soybean (Glycine max). Funct Plant Biol 48:1139–1147. https://doi.org/10.1071/FP20400

Nodine MD, Bartel DP (2010) MicroRNAs prevent precocious gene expression and enable pattern formation during plant embryogenesis. Genes Dev 24:2678–2692

Noman A, Aqeel M (2017) miRNA-based heavy metal homeostasis and plant growth. Environ Sci Pollut Res 24:10068–10082. https://doi.org/10.1007/s11356-017-8593-5

Osakabe Y, Oskabe K (2015) Genome editing with engineered nucleases in plants. Plant Cell Physiol 56:389–400

Osakabe Y, Watanabe T, Sugano SS, Ueta R, Ishihara R, Shinozaki K, Osakabe K (2016) Optimization of Clustered Regularly Interspaced Short Palindromic Repeats/Clustered Regularly Interspaced Short Palindromic Repeats (CRISPR)-associated proteins(Clustered Regularly Interspaced Short Palindromic Repeats (CRISPR)/Cas)9 genome editing to modify abiotic stress responses in plants. Sci Rep 6:26685

Pan C, Ye L, Zheng Y, Wang Y, Yang D, Liu X, Chen L, Zhang Y, Fei Z, Lu G (2017) Identification and expression profiling of microRNAs involved in the stigma exsertion under high temperature stress in tomato. BMC Genomics 18:843. https://doi.org/10.1186/s12864-017-4238-9

Pan C, Wu X, Markel K, Malzahn AA, Kundagrami N, Sretenovic S, Zhang Y, Cheng Y, Shih PM, Qi Y (2021) Clustered Regularly Interspaced Short Palindromic Repeats (CRISPR)-Act3.0 for highly efficient multiplexed gene activation in plants. Nat Plants 7:942–953

Pandita D (2019) Plant MIRnome: miRNA biogenesis and abiotic stress response. In: Plant abiotic stress response. Springer, Berlin/Heidelberg, pp 449–474

Pang J, Zhang X, Ma X, Zhao J (2019) Spatio-temporal transcriptional dynamics of maize long non-coding RNAs responsive to drought stress. Genes (Basel) 10:138. https://doi.org/10.3390/genes10020138

Panigrahy M, Panigrahi KCS, Poli Y, Ranga A, Majeed N (2022) Integrated expression analysis of small RNA, degradome and microarray reveals complex regulatory action of miRNA during prolonged shade in swarnaprabha rice. Biology 11:798

Parmar S, Gharat SA, Tagirasa R, Chandra T, Behera L, Dash SK, Shaw BP (2020) Identification and expression analysis of miRNAs and elucidation of their role in salt tolerance in rice varieties susceptible and tolerant to salinity. PLoS One 15(4):e0230958. https://doi.org/10.1371/journal.pone.0230958

Parmar S, Gharat SA, Tagirasa R, Chandra T, Behera L, Dash SK, Shaw BP (2021) Identification and expression analysis of miRNAs and elucidation of their role in salt tolerance in rice varieties susceptible and tolerant to salinity. PLoS One 15(4):e0230958

Patel P, Yadav K, Srivastava AK, Suprasanna P, Ganapathi TR (2019) Overexpression of native Musa-miR397 enhances plant biomass without compromising abiotic stress tolerance in banana. Sci Rep 9:16434. https://doi.org/10.1038/s41598-019-52858-3

Patop IL, Wust S, Kadener S (2019) Past present and future of circRNAs. EMBO J 38:e100836

Paul S, Kundu A, Pal A (2011) Identification and validation of conserved microRNAs along with their differential expression in roots of *Vigna unguiculata* grown under salt stress. Plant Cell Tissue Organ Cult 105:233–242

Popova OV, Dinh HQ, Aufsatz W, Jonak C (2013) The RdDM pathway is required for basal heat tolerance in *Arabidopsis*. Mol Plant 6:396–410

Qi X, Xie S, Liu Y, Yi F, Yu J (2013) Genome-wide annotation of genes and noncoding RNAs of foxtail millet in response to simulated drought stress by deep sequencing. Plant Mol Biol 83:459–473. https://doi.org/10.1007/s11103-013-0104-6

Qin T, Zhao H, Cui P, Albesher N, Xiong L (2017) A nucleus-localized long non-coding RNA enhances drought and salt stress tolerance. Plant Physiol 175(3):1321–1336. https://doi.org/10.1104/pp.17.00574

Qiu CW, Zhao J, Chen Q, Wu F (2019) Genome-wide characterization of drought stress responsive long non-coding RNAs in Tibetan wild barley. Environ Exp Bot 164:124–134. https://doi.org/10.1016/j.envexpbot.2019.05.002

Ragupathy R, Ravichandran S, Mahdi MSR, Huang D, Reimer E, Domaratzki M, Cloutier S (2016) Deep sequencing of wheat sRNA transcriptome reveals distinct temporal expression pattern of miRNAs in response to heat, light and UV. Sci Rep 6:39373

Rahman MA, Afridi S, Hossain MB, Rana M, Masum AA, Rahman MM, Al Maruf A (2024) Nexus between heat wave, food security and human health (HFH): developing a framework for livelihood resilience in Bangladesh. Environ Challenges 14:100802. https://doi.org/10.1016/j.envc.2023.10080

Rajagopalan R, Vaucheret H, Trejo J, Bartel DP (2006) A diverse and evolutionarily fluid set of microRNAs in *Arabidopsis thaliana*. Genes Dev 20:3407–3425

Ravichandran S, Ragupathy R, Edwards T, Domaratzki M, Cloutier S (2019) MicroRNA-guided regulation of heat stress response in wheat. BMC Genomics 20:488. https://doi.org/10.1186/s12864-019-5799-6

Ren Y, Chen L, Zhang Y, Kang X, Zhang Z, Wang Y (2012) Identification of novel and conserved *Populus tomentosa* microRNA as components of a response to water stress. Funct Integr Genomics 12:327–339

Ren J, Jiang C, Zhang H, Shi X, Ai X, Li R, Dong J (2022) LncRNA-mediated ceRNA networks provide novel potential biomarkers for peanut drought tolerance. Physiol Plant 174(1):e13610. https://doi.org/10.1111/ppl.13610

Rey-Burusco MF, Daleo GR, Feldman ML (2019) Identification of potassium phosphite responsive miRNAs and their targets in potato. PLoS One 14:e0222346

Reyon D, Tsai SQ, Khayter C, Foden JA, Sander JD, Joung JK (2012) FALSH assembly of TALENs for high throughput genome editing. Nat Biotechnol 30(5):460–5

Reyes JL and Chua NH (2007) Abscisic Acid(ABA) induction of miR159 controls transcript levels of two MYB factors during Arabidopsis seed germination. Plant J 49(4):592–606

Sachin AG, Birendra Prasad S (2015) Novel and conserved miRNAs in the halophyte *Suaeda maritima* identified by deep sequencing and computational predictions using the ESTs of two mangrove plants. BMC Plant Biol 15:1–18

Scharf KD, Berberich T, Ebersberger I, Nover L (2012) The plant heat stress transcription factor (Hsf) family: structure, function and evolution. Biochim Biophys Acta (BBA) 1819:104–119. https://doi.org/10.1016/j.bbagrm.2011.10.002

Scintu D, Scacchi E, Cazzaniga F, Vinciarelli F, De Vivo M, Shtin R, Svolacchia N, Bertolotti G, Unterholzener S, Del Bianco M, Timmermans MCP, Di Mambro R, SAbscisic Acid(ABA)tini S, Costantino P, Dello Ioio R (2022) microRNA165 and 166 modulate salt stress response of the *Arabidopsis* root. BioRxiv, In press https://doi.org/10.1101/2022.11.25.517945

Shamloo-Dashtpagerdi R, Shahriari AG, Tahmasebi A, Vetukuri RR (2023) Potential role of the regulatory miR1119-MYC2 module in wheat (*Triticum aestivum* L.) drought tolerance. Front Plant Sci 4:1161245. https://doi.org/10.3389/fpls.2023.1161245

Shi X, Jiang F, Wen J, Wu Z (2019) Overexpression of *Solanum habrochaites* microRNA319d (sha-miR319d) confers chilling and heat stress tolerance in tomato (*S. lycopersicum*). BMC Plant Biol 19:214. https://doi.org/10.1186/s12870-019-1823-x

Shigeta T, Zaizen Y, Sugimoto Y, Nakamura Y, Matsuo T, Okamoto S (2015) Heat shock protein 90 acts in brassinosteroid signaling through interaction with BES1/BZR1 transcription factor. J Plant Physiol 178:69–73. https://doi.org/10.1016/j.jplph.2015.02.003

Shuai P, Liang D, Zhang Z, Yin W, Xia X (2013) Identification of drought-responsive and novel *Populus trichocarpa* microRNAs by high throughput sequencing and their targets using degradome analysis. BMC Genomics 14:233. https://doi.org/10.1186/1471-2164-14-233

Shuai P, Liang D, Tang S, Zhang Z, Ye CY, Su Y, Xia X (2014) Genome-wide identification and functional prediction of novel and drought-responsive lincRNAs in *Populus trichocarpa*. J Exp Bot 65(17):4975–4983. https://doi.org/10.1093/jxb/eru256

Shukla LI, Chinnusamy V, Sunkar R (2008) The role of microRNAs and other endogenous small RNAs in plant stress responses. Biochim Biophys Acta 1779:743–748

Sihag P, Sagawal V, Kumar A, Balyan P, Mir RR, Dhankher OP, Kumar U (2021) Discovery of miRNAs and development of heat-responsive miRNA-SSR markers for characterization of wheat germplasm for terminal heat tolerance breeding. Front Genet 12. https://doi.org/10.3389/fgene.2021.699420

Song Y, Wu K, Dhaubhadel S, An L, Tian L (2010) *Arabidopsis* DNA methyltransferase AtDNMT2 associates with histone deacetylase AtHD2s activity. Biochem Biophys Res Commun 396:187–192

Song X, Liu G, Huang Z, Duan W, Tan H, Li Y, Hou X (2016a) Temperature expression patterns of genes and their coexpression with LncRNAs revealed by RNASeq in non-heading Chinese cabbage. BMC Genomics 17:297. https://doi.org/10.1186/s12864-016-2625-2

Song Y, Ci D, Tia M, Zhang D (2016b) Stable methylation of a non-coding RNA gene regulates gene expression in response to abiotic stress in *Populus simonii*. J Exp Bot 67(5):1477–1492. https://doi.org/10.1093/jxb/erv543

Song X, Li Y, Cao X, Qi Y (2019) microRNAs and their regulatory roles in plant environment interactions. Annu Rev Plant Biol 70:489–525. https://doi.org/10.1146/annurev-arplant-050718-100334

Song Y, Chen P, Liu P, Bu C, Zhang D (2020) High-temperature responsive poplar lncrnas modulate target gene expression via RNA interference and act as RNA scaffolds to enhance heat tolerance. Int J Mol Sci 21(18). https://doi.org/10.3390/ijms21186808

Sridha S, Wu K (2006) Identification of AtHD2C as a novel regulator of abscisic responses in Arabidopsis. Plant J 46:124–133

Srivastava S, Srivastava AK, Suprasanna P, D'Souza SF (2013) Identification and profiling of arsenic stress-induced microRNAs in *Brassica juncea*. J Exp Bot 64(1):303–315

Steif A, Altmann S, Hoffmann K, Pant BD, Scheible WR, Baurle I (2014) Arabidopsis miR156 regulates tolerance to recurring environmental stress through SPL transcription factors. The Plant Cell 26: 1792–1807

Stellos K, Gatsiou A, Stamatelopoulos K, Perisic Matic L, John D, Lunella FF, Jae N, Rossbach O, Amrhein C, Sigala F, Boon RA, Furtig B, Manavski Y, You X, Uchida S, Keller T, Boeckel JN, Franco-Cereceda A, Maegdefessel L, Chen W, Schwalbe H, Bindereif A, Eriksson P, Hedin U, Zeiher AM, Dimmeler S (2016) Adenosine-to-inosine RNA editing controls cathepsin S expression in atherosclerosis by enabling HuR-mediated post-transcriptional regulation. Nat Med 22:1140–1150

Stief A, Brzezinka K, Lämke J, Bäurle I (2014) Epigenetic responses to heat stress at different time scales and the involvement of small RNAs. Plant Signal Behav 9:e970430

Sun M, Shen Y, Chen Y, Wang Y, Cai X, Yang J, Jia B, Dong W, Chen X, Sun X (2022) Osa-miR1320 targets the ERF transcription factor OsERF096 to regulate cold tolerance via JA-mediated signaling. Plant Physiol 189:2500–2516

Sunkar R, Zhu JK (2004) Novel and stress-regulated microRNAs and other RNAs from *Arabidopsis*. Plant Cell 16(8):2001–2019

Sunkar R, Kapoor A, Zhu J (2006) Posttranscriptional induction of two cu/Zn superoxide dismutase genes in Arabidopsis is mediated by down-regulation of miR398 and important for oxidative stress tolerance. Plant Cell 18:2051–2065

Sunkar R, Chinnusamy V, Zhu J, Zhu JK (2007) Small RNAs as big players in plant abiotic stress responses and nutrient deprivation. Trends Plant Sci 12:301–309

Sunkar R, Zhou X, Zheng Y, Zhang W, Zhu JK (2008) Identification of novel and candidate miRNAs in rice by high throughput sequencing. BMC Plant Biol 8:25. https://doi.org/10.1186/1471-2229-8-25

Sunkar R, Li YF, Jagadeeswaran G (2012) Functions of microRNAs in plant stress responses. Trends Plant Sci 17:196–203

Tan X, Li S, Hu L, Zhang C (2020) Genome-wide analysis of long noncoding RNAs (lncRNAs) in two contrasting rapeseed (*Brassica napus* L.) genotypes subjected to drought stress and re-watering. BMC Plant Biol 20(1):81. https://doi.org/10.1186/s12870-020-2286-9

Tang J, Chu C (2017) MicroRNAs in crop improvement: fine-tuners for complex traits. Nat Plants 3:17077

Tang Z, Zhang L, Xu C, Yuan S, Zhang F, Zheng Y, Zhao C (2012) Uncovering small RNA-mediated responses to cold stress in a wheat thermosensitive genic male-sterile line by deep sequencing. Plant Physiol 159:721–738

Tarolli P, Luo J, Park E, Barcaccia G, Masin R (2024) Soil salinization in agriculture: mitigation and adaptation strategies combining nature-based solutions and bioengineering. iScience 27(2):108830. https://doi.org/10.1016/j.isci.2024.108830

Teng L, Zhang X, Wang R, Lin K, Zeng M, Chen H, Cao F (2023) miRNA transcriptome reveals key miRNAs and their targets contributing to the difference in Cd tolerance of two contrasting maize genotypes. Ecotoxicol Environ Saf 256:114881

Thatcher SR, Danilevskaya ON, Meng X, Beatty M, Zastrow-Hayes G, Harris C, Van Allen B, Habben J, Li B (2016) Genome-wide analysis of alternative splicing during development and drought stress in Maize1[OPEN]. Plant Physiol 170:586–599

Thipyapong P, Melkonian J, Wolfe DW, Steffens JC (2004) Suppression of polyphenol oxidases increases stress tolerance in tomato. Plant Sci 167:693–703

Tian Y, Tian Y, Luo X, Zhou T, Huang Z, Liu Y, Qiu Y, Hou B, Sun D, Deng H, Qian S, Yao K (2014) Identification and characterization of microRNAs related to salt stress in broccoli, using high-throughput sequencing and bioinformatics analysis. BMC Plant Biol 14:226. https://doi.org/10.1186/s12870-014-0226-2

To TK, Kim JM, Matsui A, Kurihara Y, Morosawa T, Ishida J, Tanaka M, Endo T, Kakutani T, Toyoda T, Kimura H, Yokoyama S, Shinozaki K, Seki M (2011) *Arabidopsis* HDA6 regulates locus-directed heterochromatin silencing in cooperation with MET1. PLoS Genet 7:e1002055

Trapnell C, Pachter L, Salzberg SL (2009) TopHat: discovering splice junctions with RNA-Seq. Bioinformatics 25:1105–1111

Traubenik S, Charon C, Blein T (2024) From environmental responses to adaptation: the roles of plant lncRNAs. Plant Physiol 195(1):232–244. https://doi.org/10.1093/plphys/kiae034

Trindade I, Capitao C, Dalmay T, Fevereiro MP, Santos DM (2010) miR398 and miR408 are up-regulated in response to water deficit in *Medicago truncatula*. Planta 231:705–716

Tripathi AM, Singh A, Singh R, Verma AK, Roy S (2019) Modulation of miRNA expression in natural populations of *A. thaliana* along a wide altitudinal gradient of Indian Himalayas. Sci Rep 9:441. https://doi.org/10.1038/s41598-018-37465-y

Tsai WA, Sung PH, Kuo YW, Chen MC, Jeng ST, Lin JS (2023) Involvement of microRNA164 in responses to heat stress in *Arabidopsis*. Plant Sci 329:111598

Tsuzuki M, Futagami K, Shimamura M, Inoue C, Kunimoto K, Oogami T, Tomita Y, Inoue K, Kohchi T, Yamaoka S, Araki T, Hamada T, Watanabe Y (2019) Early arising role of the microRNA156/529-SPL module in reproductive development revealed by the liverwort *Marchantia polymorpha*. Curr Biol 29:3307–3314

Um T, Choi J, Park T, Chung PJ, Jung SE, Shim JS, Kim YS, Choi IY, Park SC, Oh SJ, Seo JS, Kim JK (2022) Rice microRNA171f/SCL6 module enhances drought tolerance by regulation of flavonoid biosynthesis genes. Plant Direct 6:e374. https://doi.org/10.1002/pld3.374

Vakilian KA (2020) Machine learning improves our knowledge about miRNA functions towards plant abiotic stresses. Sci Rep 10:1–10

Vincze EB, Becze A, Laslo E, Mara G 2024 Beneficial soil microbiomes and their potential role in plant growth and soil fertility. Agriculture 2024, 14(1), 152. https://doi.org/10.3390/agriculture14010152

Wang Z, Gerstein M, Snyder M (2009) RNA-Seq: a revolutionary tool for transcriptomics. Nat Rev Genet 10:57–63

Wang L, Yu X, Wang H, Lu YZ, de Ruiter M, Prins M, He YK (2011a) A novel class of heat-responsive small RNAs derived from the chloroplast genome of Chinese cabbage (*Brassica rapa*). BMC Genomics 12:289

Wang T, Chen L, Zhao M, Tian Q, Zhang WH (2011b) Identification of drought-responsive microRNAs in *Medicago truncatula* by genome-wide high-throughput sequencing. BMC Genomics 12:367. https://doi.org/10.1186/1471-2164-12-367

Wang Y, Ghaffari N, Johnson CD, Braga-Neto UM, Wang H, Chen R, Zhou H (2011c) Evaluation of the coverage and depth of transcriptome by RNA-seq in chickens. BMC Bioinform 12(Suppl. 10):S5

Wang AT, Sun XL, Hoshino Y, Yu Y, Jia B, Sun ZW, Sun MZ, Duan XB, Zhu YM (2014a) MicroRNA319 positively regulates cold tolerance by targeting OsOCF6 and OsTCP21 in rice (*Oryza sativa* L.). PLoS One 9(3):e91357

Wang H, Chung PJ, Liu J, Jang IC, Kean MJ, Xu J, Chua NH (2014b) Genome-wide identification of long noncoding natural antisense transcripts and their responses to light in *Arabidopsis*. Genome Res 24:444–453

Wang YG, An M, Zhou SF, She YH, Li WC, Fu FL (2014c) Expression profile of maize microRNAs corresponding to their target genes under drought stress. Biochem Genet 52:474–493

Wang TZ, Liu M, Zhao MG, Chen R, Zhang WH (2015) Identification and characterization of long non-coding RNAs involved in osmotic and salt stress in *Medicago truncatula* using genome-wide high-throughput sequencing. BMC Plant Biol 15:131. https://doi.org/10.1186/s12870-015-0530-5

Wang J, Meng X, Dobrovolskaya OB, Orlov YL, Chen M (2017) Non-coding RNAs and their roles in stress response in plants. Genom Proteom Bioinform 15:301–312

Wang P, Zhang J, Sun L, Na Y, Xu J, Liang S, Deng J, Tan J, Zhang Q, Tu L, Daniell H, Jin S, Zhang X (2018) High efficient multisites genome editing in allotetraploid cotton (*Gossypium hirsutum*) using Clustered Regularly Interspaced Short Palindromic Repeats/Clustered Regularly Interspaced Short Palindromic Repeats (CRISPR)-associated proteins(Clustered Regularly Interspaced Short Palindromic Repeats (CRISPR)/Cas)9 system. Plant Biotechnol J 16(1):137–150

Wang A, Hu J, Gao C, Chen G, Wang B, Lin C, Song L, Ding Y, Zhou G (2019a) Genome-wide analysis of long non-coding RNAs unveils the regulatory roles in the heat tolerance of Chinese cabbage (*Brassica rapa* ssp. *chinensis*). Sci Rep 9(1):5002. https://doi.org/10.1038/s41598-019-41428-2

Wang P, Dai L, Ai J, Wang Y, Ren F (2019b) Identification and functional prediction of cold-related long noncoding RNA (lncRNA) in grapevine. Sci Rep 9:6638

Wang Y, Feng C, Zhai Z, Peng X, Wang Y, Sun Y, Li J, Shen X, Xiao Y, Zhu S, Huang X, Li T (2020) The apple microR171i-SCARE CROW-LIKE PROTEINS26.1 module enhances drought stress tolerance by integrating ascorbic acid metabolism. Plant Physiol 184:194–211. https://doi.org/10.1104/pp.20.00476

Wang ST, Sun XL, Hoshino Y, Yu Y, Jia B, Sun ZW, Sun MZ, Duan XB, Zhu YM (2014) MicroRNA319 positively regulates cold tolerance by targeting OsPCF6 and OsTCP21 in Rice (Oryza sativa L.). PLoS ONE 9(3): e91357. https://doi.org/10.1371/journal.pone.0091357

Wei L, Zhang D, Xiang F, Zhang Z (2009) Differentially expressed miRNAs potentially involved in the regulation of defense mechanism to drought stress in maize seedlings. Int J Plant Sci 170:979–989

Weidong Q, Hongping C, Zouzhen Y, Hu B (2020) Systematic characterization of long non-coding RNAs and their responses to drought stress in Dongxiang wild rice. Ric Sci 27:21–31

Weirick T, Militello G, Muller R, John D, Dimmeler S, Uchida S (2016) The identification and characterization of novel transcripts from RNA-seq data. Brief Bioinform 17:678–685

Wirsiy EL, Casir O (2024) Distinctive features of plant microRNAs in response to abiotic stresses. In: Marakli S, Yilmazer M (eds) The broad spectrum of biological sciences, pp 521–545

Wu G, Poethig RS (2006) Temporal regulation of shoot development in *Arabidopsis thaliana* by miR156 and its target SPL3. Development 133:3539–3547

Wu J, Okada T, Fukushima T, Tsudzuki T, Sugiura M, Yukawa Y (2012) A novel hypoxic stress-responsive long non-coding RNA transcribed by RNA polymerase III in *Arabidopsis*. RNA Biol 9:302–313. https://doi.org/10.4161/rna.19101

Wu HJ, Wang ZM, Wang M, Wang XJ (2013) Widespread long noncoding RNAs as endogenous target mimics for microRNAs in plants. Plant Physiol 161:1875–1884

Wu BF, Li WF, Xu HY, Qi LW, Han SY (2015) Role of cin-miR2118 in drought stress responses in *Caragana intermedia* and tobacco. Gene 574(1):34–40

Wunderlich M, Gross-Hardt R, Schoffl F (2014) Heat shock factor HSFB2a involved in gametophyte development of *Arabidopsis thaliana* and its expression is controlled by a heat-inducible long non-coding antisense RNA. Plant Mol Biol 85(6):541–550. https://doi.org/10.1007/s11103-014-0202-0

Xia J, Zeng C, Chen Z, Zhang K, Chen X, Zhou Y, Song S, Lu C, Yang R, Yang Z, Zhou J, Peng H, Wang W, Peng M, Zhang W (2014) Endogenous small-noncoding RNAs and their roles in chilling response and stress acclimation in Cassava. BMC Genomics 15:634

Xie F, Wang Q, Sun R, Zhang B (2015) Deep sequencing reveals important roles of microRNAs in response to drought and salinity stress in cotton. J Exp Bot 66(3):789–804

Xie YR, Liu Y, Wang H, Ma XJ, Wang BB, Wu GX, Wang HY (2017) Phytochrome-interacting factors directly suppress MIR156 expression to enhance shade-avoidance syndrome in *Arabidopsis*. Nat Commun 8:348

Xin M, Wang Y, Yao Y, Xie C, Peng H, Ni Z, Sun Q (2010) Diverse set of microRNAs are responsive to powdery mildew infection and heat stress in wheat (*Triticum aestivum* L.). BMC Plant Biol 10:123. https://doi.org/10.1186/1471-2229-10-123

Xin M, Wang Y, Yao Y, Song N, Hu Z, Qin D, Xie C, Peng H, Ni Z, Sun Q (2011) Identification and characterization of wheat long non-protein coding RNAs responsive to powdery mildew infection and heat stress by using microarray analysis and SBS sequencing. BMC Plant Biol 11:61

Xiong L, Schumaker KS, Zhu JK (2002) Cell signaling during cold, drought, and salt stress. Plant Cell 14(Suppl):S165–S183. https://doi.org/10.1105/tpc.000596

Xu R, Wang Y, Zheng H, Lu W, Wu C, Huang J, Yan K, Yang G, Zheng C (2015) Salt induced transcription factor MYB74 is regulated by the RNA directed DNA methylation pathway in *Arabidopsis*. J Exp Bot 66:5997–6008

Xu J, Li Y, Wang Y, Liu X, Zhu XG (2017) Altered expression profiles of microRNA families during de-etiolation of maize and rice leaves. BMC Res Notes 10:108

Xu T, Zhang L, Yang Z, Wei Y, Dong T (2021) Identification and functional characterization of plant MiRNA under salt stress shed light on salinity resistance improvement through MiRNA manipulation in crops. Front Plant Sci 12:665439. https://doi.org/10.3389/fpls.2021.665439

Xuan H, Zhang L, Liu X, Han G, Li J, Li X, Zhang S (2015) PLNlncRbase: a resource for experimentally identified lncRNAs in plants. Gene 573:328–332

Xue T, Liu Z, Dai X, Xiang F (2017) Primary root growth in *Arabidopsis thaliana* is inhibited by the miR159 mediated repression of MYB33, MYB65 and MYB101. Plant Sci 262:182–189

Yadav A, Mathan J, Dibey AK, Singh A (2024) The emerging role of non-coding RNAs (ncRNAs) in plant growth, development, and stress response signaling. Non-Coding RNA 10(1):13. https://doi.org/10.3390/ncrna10010013

Yan Q, Wu F, Yan Z, Li J, Ma T, Zhang Y, Zhao Y, Wang Y, Zhang J (2019) Differential co-expression networks of long non-coding RNAs and mRNAs in *Cleistogenes songorica* under water stress and during recovery. BMC Plant Biol 19:23

Yang J, Zhang N, Ma C, Qu Y, Si H, Wang D (2013) Prediction and verification of microRNAs related to proline accumulation under drought stress in potato. Comput Biol Chem 46:48–54

Yang R, Zeng Y, Yi X, Zhao L, Zhang Y (2015) Small RNA deep sequencing reveals the important role of microRNAs in the halophyte *Halostachys caspica*. Plant Biotechnol J 13:395–408

Yang R, Li P, Mei H, Wang D, Sun J, Yang C, Hao L, Cao S, Chu A, Hu S, Song X, Cao X (2019a) Finetuning of MiR528 accumulation modulates flowering time in rice. Mol Plant 12:1103–1113

Yang Z, Li W, Su X, Ge P, Zhou Y, Hao Y, Shu H, Gao C, Cheng S, Zhu G, Wang Z (2019b) Early response of radish to heat stress by strand-specific transcriptome and miRNA analysis. Int J Mol Sci 20:3321

Yang X, Liu C, Niu X, Wang L, Li L, Yuan Q, Pei X (2022) Research on lncRNA related to drought resistance of Shanlan upland rice. BMC Genomics 23(1):336. https://doi.org/10.1186/s12864-022-08546-0

Yao Y et al (2010) Non-coding small RNAs responsive to abiotic stress in wheat (*Triticum aestivum* L.). Funct Integr Genomics 10:187–190. https://doi.org/10.1007/s10142-010-0163-6

Ye X, Wang S, Zhao X, Gao N, Wang Y, Yang Y, Wu E, Jiang C, Cheng Y, Wu W, Liu S (2022) Role of lncRNAs in cis- and trans-regulatory responses to salt in *Populus trichocarpa*. Plant J 110:978–993. https://doi.org/10.1111/tpj.15714

Yin Z, Li Y, Yu J, Liu Y, Li C, Han X, Shen F (2011) Difference in miRNA expression profiles between two cotton cultivars with distinct salt sensitivity. Mol Biol Rep 39:4961–4970

Yin Z, Li Y, Yu J, Liu Y, Li C, Han X, Shen F (2012) Difference in miRNA expression profiles between two cotton cultivars with distinct salt sensitivity. Mol Biol Rep 39:4961–4970

Yu X, Yang J, Li X, Liu X, Sun C, Wu F, He Y (2013) Global analysis of cis-natural antisense transcripts and their heat-responsive nat-siRNAs in *Brassica rapa*. BMC Plant Biol 13:208

Yuan LZ, Tang JH, Liu JY, Song H, Zhang MB, Li HP, Li ZH (2016) Differential miRNA expression in maize ear subjected to shading tolerance. Acta Physiol Plant 38:1–12

Yue E, Cao H, Liu B (2020) OsmiR535, a potential genetic editing target for drought and salinity stress tolerance in *Oryza sativa*. Plants 9:1337. https://doi.org/10.3390/plants9101337

Zeng X, Xu Y, Jiang J, Zhang F, Ma L, Wu D, Wang Y, Sun W (2018) Identification of cold stress responsive microRNAs in two winter turnip rape (*Brassica rapa* L.) by high throughput sequencing. BMC Plant Biol 18:52. https://doi.org/10.1186/s12870-018-1242-4

Zhang JF, Yuan LJ, Shao Y, Du W, Yan DW, Lu YT (2008) The disturbance of small RNA pathways enhanced abscisic acid response and multiple stress responses in Arabidopsis. Plant Cell Environ 31:562–574

Zhang J, Xu Y, Huan Q, Chong K (2009a) Deep sequencing of *Brachypodium* small RNAs at the global genome level identifies microRNAs involved in cold stress response. BMC Genomics 10:449. https://doi.org/10.1186/1471-2164-10-449

Zhang L, Chia JM, Kumari S, Stein JC, Liu Z, Narechania A, Maher CA, Guill K, McMullen MD, Ware D (2009b) A genome-wide characterization of microRNA genes in maize. PLoS Genet 5(11):e1000716. https://doi.org/10.1371/journal.pgen.1000716

Zhang WW, Luo YP, Gong X, Zeng WH, Li SG (2009c) Computational identification of 48 potato microRNAs and their targets. Comput Biol Chem 33:84–93

Zhang RX, Marshall D, Bryan GJ, Hornyik C (2013) Identification and characterization of miRNA transcriptome in potato by high-throughput sequencing. PLoS One 8:e57233

Zhang W, Han Z, Guo Q, Liu Y, Zheng Y, Wu F, Jin W (2014) Identification of maize long non-coding RNAs responsive to drought stress. PLoS One 9:e98958

Zhang F, Luo X, Zhou Y, Xie J (2016) Genome-wide identification of conserved microRNA and their response to drought stress in Dongxiang wild rice (*Oryza rufipogon* Gri.). Biotechnol Lett 38:711–721

Zhang L, Yao L, Zhang N, Yang J, Zhu X, Tang X, Calderón-Urrea A, Si H (2018) Lateral root development in potato is mediated by Stu-mi164 regulation of NAC transcription factor. Front Plant Sci 9:383

Zhang P, Fan Y, Sun X, Chen L, Terzaghi W, Bucher E, Li L, Dai M (2019a) A large-scale circular RNA profiling reveals universal molecular mechanisms responsive to drought stress in maize and *Arabidopsis*. Plant J 98:697–713

Zhang X, Dong J, Deng F, Wang W, Cheng Y, Song L, Hu M, Shen J, Xu Q, Shen F (2019b) The long non-coding RNA lncRNA973 is involved in cotton response to salt stress. BMC Plant Biol 19(1):459. https://doi.org/10.1186/s12870-019-2088-0

Zhang X, Shen J, Xu Q, Dong J, Song L, Wang W, Shen F (2021) Long noncoding RNA lncRNA354 functions as a competing endogenous RNA of miR160b to regulate ARF genes in response to salt stress in upland cotton. Plant Cell Environ 44(10):3302–3321. https://doi.org/10.1111/pce.14133

Zhao M, Ding H, Zhu JK, Zhang F, Li WX (2011) Involvement of miR169 in the nitrogen-starvation responses in Arabidopsis. New Phytol 190:906–915

Zhao Y, Zhang C, Liu W, Gao W, Liu C, Song G, Li WX, Mao L, Chen B, Xu Y, Li X, Xie C (2016a) An alternative strategy for targeted gene replacement in plants using a dual-sgRNA/cas9 design. Sci Rep 6:23890–23890. https://doi.org/10.1038/srep23890

Zhao JG, He QS, Chen G, Wang L, Jin B (2016b) Regulation of non-coding RNAs in heat stress responses of plants. Front Plant Sci 7:1213. https://doi.org/10.3389/fpls.2016.01213

Zhao X, Li J, Lian B, Gu H, Li Y, Qi Y (2018) Global identification of Arabidopsis lncRNAs reveals and the regulation of MAF4 by a natural antisense RNA. Nat Commun 9:5056

Zhao J, Yuan S, Zhou M, Yuan N, Li Z, Hu Q, Bethea FG, Liu H, Li S, Luo H (2019) Transgenic creeping bentgrass overexpressing Osa-miR393a exhibits altered plant development and improved multiple stress tolerance. Plant Biotechnol J 17:233–251

Zhao Y, Xu K, Liu G, Li S, Zhao S, Liu X, Yang X, Xiao K (2020) Global identification and characterization of miRNA family members responsive to potassium deprivation in wheat (*Triticum aestivum*). Sci Rep 10:15812

Zhao M, Tian R, Sun X, Zhang WH (2023) lncRNA MtCIR2 positively regulates plant freezing tolerance by modulating CBF/DREB1gene clusters. Plant Cell Environ 46(8):2450–2469

Zhong SH, Liu JZ, Jin H, Lin L, Li Q, Chen Y, Yuan YX, Wang ZY, Haung H, Qi YJ, Chen XY, Vaucheret H, Chory J, Li J, He ZH (2013) Warm temperatures induce transgenerational epigenetic release of RNA silencing by inhibiting siRNA biogenesis in *Arabidopsis*. Proc Natl Acad Sci USA 110:9171–9176. https://doi.org/10.1073/pnas.1219655110

Zhou ZS, Huang SJ, Yang ZM (2008) Bioinformatic identification and expression analysis of new microRNAs from *Medicago truncatula*. Biochem Biophys Res Commun 374:538–542

Zhou L, Liu Y, Liu Z, Kong D, Duan M, Luo L (2010) Genome-wide identification and analysis of drought-responsive microRNAs in *Oryza sativa*. J Exp Bot 61:4157–4168

Zhou M, Li DY, Li ZG, Hu Q, Yang CH, Zhu LH, Luo H (2013) Constitutive expression of a *miR319* gene alters plant development and enhances salt and drought tolerance in transgenic creeping bentgrass. Plant Physiol 161:1375–1391. https://doi.org/10.1104/pp.112.208702

Zhou R, Wang Q, Jiang F, Cao X, Sun M, Liu M, Wu Z (2016) Identification of miRNAs and their targets in wild tomato at moderately and acutely elevated temperatures by high throughput sequencing and degradome analysis. Sci Rep 6:33777

Zhou J, Deng K, Cheng Y, Zhong Z, Tian L, Tang X, Zheng X, Zhang T, Qi Y, Zhang Y (2017) Clustered Regularly Interspaced Short Palindromic Repeats (CRISPR)-Cas9 based genome editing reveals new insights into microRNA function and regulation in rice. Front in Plant Sci 8:1598. https://doi.org/10.3389/fpls.2017.01598

Zhou R, Yu X, Ottosen CO, Zhao T (2020a) High throughput sequencing of circRNAs in tomato leaves responding to multiple stresses of drought and heat. Hortic Plant J 6:34–38

Zhou Y, Liu W, Li X, Sun D, Xu K, Feng C, Foka ICK, Ketehouli T, Gao H, Wang N, Dong Y, Wang F, Li H (2020b) Integration of sRNA, degradome, transcriptome analysis and functional investigation reveals gma miR398c negatively regulates drought tolerance via GmCSDs and GmCCS in transgenic *Arabidopsis* and soybean. BMC Plant Biol 20:190. https://doi.org/10.1186/s12870-020-02370-y

Zhou XY, Wang NH, Qiu CW, Luo L, Zhang M, Zhang S, Gao ZF, Ahmed IM, Feibo W (2023) Transcriptome profiling uncovers the lncRNA-mediated regulatory networks associated with tolerance to cadmium stress in barley. Environ Exp Bot 206:105156

Zhu JK (2016) Abiotic stress signaling and responses in plants. Cell 167(2):313–324. https://doi.org/10.1016/j.cell.2016.08.029

Zhu QH, Wang MB (2012) Molecular functions of long non-coding RNAs in plants. Genes 3:176–190

Zhu J, Li W, Yang W, Qi L, Han S (2013) Identification of microRNAs in *Caragana intermedia* by high throughput sequencing and expression analysis of 12 microRNAs and their targets under salt stress. Plant Cell Rep 32:1339–1349

Zhu H, Zhang Y, Tang R, Qu H, Duan X, Jiang Y (2019) Banana sRNAome and degradome identify microRNAs functioning in differential responses to temperature stress. BMC Genomics 20:33. https://doi.org/10.1186/s12864-018-5395-1

Zuo ZF, He W, Li J, Mo B, Liu L (2021) Small RNAs: the essential regulators in plant thermotolerance. Front Plant Sci 12. https://doi.org/10.3389/fpls.2021.726762

Chapter 8
Non-coding RNA-Mediated Signaling Pathways in Abiotic Stress Tolerance

Parneeta Mishra, Pei Jia Ng, Pradeep Kumar, Shalini Tiwari, and Dhananjay Gotarkar

Abstract Abiotic stresses including drought, salinity, temperature, and nutritional deficiencies are some of the serious threats to crop yield all over the world. Plants have developed complex molecular cascades of signaling to combat these stresses at developmental and physiological levels. Non-coding RNAs (ncRNAs) such as microRNAs (miRNAs), small interfering RNAs (siRNAs), and long non-coding RNAs (lncRNAs) are some of the key regulators of such cascades. Although they do not code for proteins, ncRNAs play regulatory roles in gene expression at the transcriptional, post-transcriptional, and epigenetic levels to modulate plant stress perception, signaling, and adaptation. Here in this chapter, we discuss the critical roles of ncRNAs in the regulation of abiotic stress signaling in plants. It discusses how certain ncRNAs are targeting stress-responsive genes and modulating key pathways, notably those of abscisic acid (ABA), in the regulation of antioxidant responses, osmotic adjustment, and growth under stress. The chapter also highlights the functional coordination between various classes of ncRNAs and their integration within complex regulatory networks. Through the use of high-throughput sequencing and functional genomics, many stress-responsive ncRNAs have been identified and characterized in plant species. This widened knowledge database pro-

P. Mishra
Molecular Biology and Biotechnology Division, CSIR-National Botanical Research Institute, Lucknow, India

P. J. Ng
Department of Biochemistry and Molecular Biology, Oklahoma State University, Stillwater, OK, USA

P. Kumar · S. Tiwari
Agricultural and Biosystems Engineering, South Dakota State University, Brookings, SD, USA

D. Gotarkar (✉)
Department of Fisheries, Wildlife and Conservation Biology, University of Minnesota, Minneapolis, MN, USA
e-mail: gotarkar@umn.edu

S. Tiwari et al. (eds.), *Decoding Non-Coding RNA*, Concepts and Strategies in Plant Sciences, https://doi.org/10.1007/978-981-96-9859-2_8

vides tremendous potential for molecular breeding and genetic engineering for developing stress tolerance and resilience in crops in a climate change context.

Keywords Abiotic stress · Non-coding RNA · Signaling · Stress tolerance · Transcriptomics

8.1 Introduction

Plants are constantly exposed to a tremendous variety of abiotic stresses, such as drought, salinity, temperature, and nutrient deficiencies, that can potentially exert a profound effect on plant growth, productivity, and survival. Against the backdrop of global climate change and environmental heterogeneity, the complex molecular mechanisms of plant perception, response, and adaptation to abiotic stresses have become an urgent area of investigation. One of the most exciting recent findings in plant molecular biology is the critical role played by non-coding RNAs (ncRNAs) in the modulation of stress-responsive signaling pathways (Yadav et al. 2024; Zhang et al. 2022a, b).

In contrast to messenger RNAs (mRNAs), which are employed to be translated into protein, ncRNAs do not encode proteins but rather exist as potent regulatory molecules. These RNAs are a diverse family of subclasses comprising microRNAs (miRNAs), small interfering RNAs (siRNAs), and long non-coding RNAs (lncRNAs). With their capacity to regulate gene expression at the transcriptional, post-transcriptional, and epigenetic levels, they are at the hub of plant stress signaling networks (Cao et al. 2024). By regulating master transcription factors, hormone signaling constituents, and stress response genes, ncRNAs are also master regulators that modulate plant stress responses (Yadav et al. 2024). Tolerance to abiotic stress is not regulated by any gene or pathway but by an interactive network of signals and regulatory circuits. Here, ncRNAs play the role of sensors and modulators of stress signals and incorporate the external stimuli into gene expression responses in a targeted way (Zhang et al. 2022a, b). For example, certain miRNAs are quickly induced by various stress signals and suppress negative regulators or activate defense pathways (Zhang et al. 2022a, b). In the same way, siRNAs are significant players in chromatin remodeling and transcriptional gene silencing that help to bring about stable expression of stress-adaptive genes (Puchta-Jasińska et al. 2025). Increasing evidence also indicates that lncRNAs act together with DNA, RNA, and proteins to create multifaceted regulatory scaffolds that model transcriptional landscapes in response to stress (and Chekanova 2017).

One of the most important areas of ncRNA-mediated regulation is in the signaling of the phytohormone abscisic acid (ABA), a key hormone in abiotic stress adaptation (Qin et al. 2017). In addition, crosstalk among the various classes of ncRNAs adds strength and specificity to the stress responses, which enable plants to prioritize and modulate their physiological adaptations under multiplicative stress environments (Jannesar et al. 2024). Through next-generation sequencing (NGS) and bioinformatics, the knowledge of the diversity and function of ncRNA in plants has expanded significantly (Liu et al. 2017). Genome-wide profiling has identified a plethora of stress-responsive ncRNAs in different species and shed useful light on conserved and species-specific regulation (Sunkar and Zhu 2004; Mirdar Mansuri et al. 2022; Song et al. 2021). Functional characterization via genetic and molecular tools such as CRISPR/Cas systems has also pointed towards the possibility of targeting ncRNAs for crop improvement (Bravo-Vázquez et al. 2024; Yadav et al. 2024; Zhang et al. 2021).

This chapter deals with the intricate and dynamic area of ncRNA-mediated plant abiotic stress tolerance signaling pathways. It discusses the biogenesis, classification, and regulatory roles of key classes of ncRNAs, their integration with core stress signaling pathways, and their activities in transcriptional regulation, epigenetic modification, and hormonal crosstalk. Furthermore, the chapter discusses the recent research tools and strategies for ncRNA manipulation and offers a perspective on their prospects for developing climate-resilient crops. Elucidating the functions of ncRNAs in stress signaling brings us one step closer to an understanding of plant adaptive responses that not only are at the center of plant biology but also vital to future-proofing global agriculture.

8.2 Types of Non-coding RNAs Involved in Stress Signaling

8.2.1 MicroRNAs (miRNAs)

MicroRNAs (miRNAs) are endogenous small non-coding RNAs with a typical length of 20–24 nucleotides that are essential regulators of gene expression in plants. By mRNA cleavage or repression of translation, miRNAs regulate the expression of a large number of genes implicated in stress perception, signal transduction, hormone responses, and downstream adaptation mechanisms (Voinnet 2009). Their role in abiotic stress signaling has become more apparent with the discovery of additional stress-responsive miRNAs and their functional analysis in different plant species (Table 8.1) (Sunkar and Zhu 2004).

Under drought stress, various miRNAs have been found to target central genes of protection pathways for regulation. MiR398, for instance, is downregulated under drought stress, and its target genes CSD1 and CSD2, which encode copper/zinc superoxide dismutases, are upregulated. These enzymes have fundamental roles in relieving oxidative stress through detoxifying ROS and thus improving drought tolerance (Sunkar et al. 2006). Likewise, miR169 is downregulated under drought stress, which leads to higher expression of NFYA5, a positive transcription regulator that induces the drought-responsive genes (Li et al. 2008).

MiR775 was detected to be upregulated under hypoxia stress (Mishra et al. 2022).

Table 8.1 MicroRNAs (miRNAs) involved in abiotic stress responses

miRNA	Stress	Plant	Role in stress response	References
miR159	Drought	Arabidopsis	Targets MYB transcription factors, modulating ABA signaling and drought response pathways	Reyes and Chua (2007)
miR393	Salinity	Arabidopsis	Regulates auxin signaling by targeting AFB2, influencing root architecture under salt stress	Iglesias et al. (2014)
miR172	Drought	Rice, wheat	Targets *IDS1* transcription factors, influencing salt tolerance	Cheng et al. (2021)
miR319	Cold	Rice	Targets TCP transcription factors, influencing cold stress responses	Wang et al. (2014)
miR393	Cold	Arabidopsis	Modulates TIR1 expression, affecting cold stress responses	Sunkar et al. (2007)

In response to salinity stress, miR393 is induced quickly and inhibits the auxin receptor gene TIR1, thereby regulating auxin signaling pathways that determine root architecture and facilitate stress adaptation (Navarro et al. 2006).

In response to temperature stress, miRNAs like miR156 and miR319 have been implicated in heat and cold stress, respectively. miR156 is engaged in targeting SPL (SQUAMOSA PROMOTER BINDING PROTEIN-LIKE) TFs and plays a role in heat stress memory and tolerance. It has been shown that overexpression of miR156 enhances heat tolerance in Arabidopsis and rice (Stief et al. 2014). Conversely, miR319 targets TCP (TEOSINTE BRANCHED/CYCLOIDEA/PCF) transcription factors that are involved in cell division and leaf development. Constitutive overexpression of miR319 in rice has been linked to increased cold tolerance and development alterations that confer stress adaptation (Zhou et al. 2013).

miRNAs also have close interactions with hormonal signaling pathways, particularly the abscisic acid (ABA), a stress-related hormone in abiotic stress responses. miR159 was shown to suppress MYB TFs that take part in ABA signaling. Overexpression of this miRNA leads to downregulated MYB expression, regulating ABA responses in seed germination and early seedling growth (Reyes and Chua 2007). miR160 and miR167, which target auxin response factors (ARFs), also demonstrate how miRNAs orchestrate hormone-mediated stress responses by modulating elements of auxin signaling (Liu et al. 2024).

The role of these miRNAs in function has also been validated using transgenic experiments. For instance, overexpression of osa-miR319 in rice enhanced cold and drought tolerance, mainly by regulating stress-related TFs (Yang et al. 2013). Similarly, manipulation of miR396 targeting GRF genes has been employed to improve drought resistance in rice (Zhou et al. 2013).

Collectively, miRNAs are hub nodes in abiotic stress signaling pathways that allow plants to respond rapidly and dynamically to environmental stresses. Their potential to precisely regulate gene expression and interact with hormone and stress signaling cascades make them attractive targets for crop improvement programs for inducing stress tolerance and resilience in the era of climate change (Chaudhary et al. 2021).

8.2.2 Long Non-coding RNAs (lncRNAs)

Long non-coding RNAs (lncRNAs) are becoming important regulators of plant abiotic stress responses to drought, salinity, heat, and cold stresses (Table 8.2). Unlike messenger RNAs, lncRNAs are not translated into proteins but rather act to regulate gene expression through a myriad of mechanisms such as acting as molecular scaffolds, binding to transcription factors, or facilitating the recruitment of chromatin-modifying complexes. These roles render lncRNAs central to stress signaling pathways in plants (Wang and Chekanova 2017). For instance, DRIR, a drought-induced lncRNA in Arabidopsis thaliana, increases tolerance to drought and salt stress by increasing the expression of stress-related genes, including abscisic acid (ABA) signaling (Qin et al. 2017). lncRNAs also function importantly under heat stress.

For example, the lncRNA SVALKA in a cold-sensitive region of the Arabidopsis genome, where mutations in SVALKA impact CBF1 expression and freezing tolerance (Kindgren et al. 2018). In Brassica rapa, heat-responsive lncRNAs are enriched with protein folding and chaperone activity-related genes that play a critical role in maintaining proteostasis during heat stress (Wang et al. 2019). Additionally, lncRNAs are usually engaged in crosstalk with other regulatory RNAs, including microRNAs (miRNAs), and epigenetic mechanisms. For example, in Arabidopsis, the highly characterized lncRNA COLDAIR directs the Polycomb Repressive Complex 2 (PRC2) to the FLC locus, resulting in histone modification and gene repression in vernalization (Heo and Sung 2011), indicating how lncRNAs can be

Table 8.2 Long non-coding RNAs (lncRNAs) involved in abiotic stress responses

lncRNA	Stress	Plant	Role in stress response	References
lncRNA973	Salt	Cotton	Enhance salt tolerance by regulating antioxidant enzymes and stress-responsive genes.	Zhang et al. (2019)
COOLAIR	Cold	Arabidopsis	Involved in cold-induced flowering through regulation of FLC	Zhao et al. (2021)
COLDAIR	Cold	Arabidopsis	Regulates flowering time and vernalization pathways through chromatin remodeling of FLC	Hung et al. (2022) and Kim et al. (2017)
DRIR	Drought, salt	Arabidopsis	Modulates genes involved in ABA signaling, water transport, and some stress-associated processes enhancing stress tolerance	Qin et al. (2017)
lincRNA1128	Drought	*Populus trichocarpa*	Targets miR482a.1, influencing drought stress responses	Shuai et al. (2014)
lincRNA1393	Drought	*Populus trichocarpa*	Targets miR6459b, affecting drought stress responses	Shuai et al. (2014)
lincRNA3018	Drought	*Populus trichocarpa*	Targets miR399i, influencing drought stress responses	Shuai et al. (2014)
lincRNA2752	Drought	*Populus trichocarpa*	Targets miR169o, affecting drought stress responses	Shuai et al. (2014)

incorporated into broad chromatin regulation networks. Functional characterization of lncRNAs under abiotic stress continues to be an ongoing field of study. Results like these point to the promise of lncRNA use as biomarkers or even as genetic engineering targets to create stress-resistant plants and thus provide optimistic approaches to improve agricultural resilience to climate change.

8.2.3 *Small Interfering RNAs (siRNAs)*

Small interfering RNAs (siRNAs) are key regulators of plant abiotic stress responses to drought, salinity, heat, and cold. These 20–24 nucleotide-long short RNAs, processed from long double-stranded RNA molecules, play a central role in post-transcriptional gene silencing by targeting the RNA-induced silencing complex (RISC) to degrade or suppress target mRNAs. In order to respond to drought stress and salinity, siRNAs modulate the expression of well-known stress-regulated genes participating in signaling pathways (Khraiwesh et al. 2012; Puchta-Jasińska et al. 2025).

In Arabidopsis thaliana, siRNAs also control salt tolerance genes such as NAT1 that mediates the expression of salt-responsive genes (Xu et al. 2017). During heat stress, siRNAs control heat shock proteins (HSPs) that are accountable for protein folding and stabilization and thereby enable plants to survive high temperatures. In Arabidopsis, siRNAs that target HSP70 play a role in thermotolerance by preserving protein homeostasis during heat stress (Wang et al. 2019). Besides this, siRNAs also play a role in cold stress regulation, for example, in Populus, cold-regulated siRNAs target cell membrane integrity and osmotic balance genes and increase cold tolerance (Khraiwesh et al. 2012). In addition to this, siRNAs coordinate with other regulatory RNAs, like miRNAs and long non-coding RNAs (lncRNAs), for the fine-tuning of stress responses. For instance, in plants, drought-responsive genes are coordinately regulated by miRNAs and siRNAs, and siRNAs provide an added layer of regulation via RNA interference (Khraiwesh et al. 2012). The findings not only highlight the significance of siRNAs in the regulation of stress tolerance but also their possible uses in crop improvement via genetic engineering and molecular breeding. Through the design of siRNA pathways, plants can be made more resilient to climate change, opening up new avenues to enhance agricultural sustainability.

8.3 ncRNA-Mediated Signaling in Abiotic Stress Responses

8.3.1 *Signal Perception and Initiation*

Plant tolerance to abiotic stress starts with the perception of drought, salinity, high temperature, and heavy metal toxicity environmental stress signals. It is the first and important step that triggers downstream signaling cascades to reprogram gene expression and cellular metabolism to adapt to stress conditions (Li et al. 2024). ncRNAs, indeed miRNAs and lncRNAs are directly implicated in perception and early transduction of these signals (Li et al. 2024; Yang et al. 2023). Plants perceive abiotic stress via specific receptors and sensors present on the plasma membrane and cytoplasm. As soon as the stress is sensed, there is a cascade of signaling with secondary messengers like calcium ions (Ca^{2+}), reactive oxygen species (ROS), and inositol phosphates (Li et al. 2024).

miRNAs control amounts of signaling proteins and Tfs that are directly induced as soon as stress signals are perceived (Li et al. 2024). miR398 is downregulated under oxidative stress (e.g., drought, high salinity), and thus its targets Cu/Zn superoxide dismutases (CSD1 and CSD2) are accumulated. The enzymes scavenge ROS, which are both detrimental molecules and signaling molecules (Sunkar et al. 2006). miR393 is induced by salt stress, in which it suppresses auxin receptors TIR1 and AFBs, thereby regulating the auxin signaling pathway, which is strongly linked to the early responses to stress, particularly root architecture reprogramming (Iglesias et al. 2014). miR319 targets TCP TFs that are characterized to control cell proliferation and hormonal signaling in the early phase of the stress response (Fang et al. 2021). Likewise, LncRNAs are more diversified in their role during early signaling, frequently by molecular scaffolding, chromatin remodeling, or as miRNA decoys (Yang et al. 2023).

DRIR (Arabidopsis Drought-Induced lncRNA) interacts with ABA signaling components and enhances early stress responses like stomata closure (Qin et al. 2017). lncRNA973 from cotton is salinity stress-induced and is reported to interact with regulatory proteins involved in ROS scavenging thereby modulating early detoxification signaling (Zhang et al. 2019).

8.3.2 ncRNA and Transcriptional Regulation

Transcriptional regulation is a fulcrum of plant abiotic stress response. Upon stress perception, the plant genome quickly reorganizes its transcriptional profile to induce protection-related processes and repress growth-related processes (Shinozaki and Yamaguchi-Shinozaki 2022). ncRNAs, particularly miRNAs and lncRNAs, function as master regulators in this process by targeting TFs, chromatin modifiers, and gene promoters and thus fine-tune transcriptional programs towards survival (Bhogireddy et al. 2021; Yadav et al. 2024).

Most miRNAs function post-transcriptionally by targeting mRNAs of transcription factors that regulate stress response genes. Through cleavage or inhibiting these mRNAs, miRNAs regulate the amplitude and longevity of transcriptional responses (Jatan et al. 2019). miR393 suppresses auxin receptors (TIR1, AFB2, AFB3), inhibiting auxin signaling. It thereby regulates transcriptional regulators of lateral root formation under salt and drought stress (Chen et al. 2012). miR396 suppresses GRF (Growth-Regulating Factor) transcription factors, suppressing cell growth and diverting energy toward stress defense under conditions of high salinity and during drought (Omidbakhshfard et al. 2015).

Apart from that, lncRNAs also control transcription through various mechanisms. LncRNAs are capable of recruiting chromatin-modifying complexes like PRC2—Polycomb Repressive Complex 2 to target gene loci, modifying histone marks and DNA methylation, and resulting in repression or activation of transcription (Yadav et al. 2023; Tiwari and Lata 2021). In Arabidopsis, COLDAIR lncRNA is PRC2-bound and recruits it to the FLC locus, resulting in its repression in vernalization (cold-induced flowering). This regulation makes plants flower at the appropriate time after cold stress (Yang et al. 2022). LDMAR lncRNA is necessary for male fertility under long days in rice, silence leads to premature programmed cell death during another development, a photoperiod-related stress (Ding et al. 2012). Certain lncRNAs also function as sponges or decoys for miRNAs, thereby inhibiting the miRNAs from binding to their target mRNAs (Tiwari et al. 2023). This indirect control enables the transcription factors to evade repression by miRNA and continue to be transcriptionally active. LncRNA IPS1 (INDUCED BY PHOSPHATE STARVATION 1), for instance, functions as a decoy for miR399, safeguarding its target PHO2 (a ubiquitin-conjugating enzyme) and thus controlling phosphate homeostasis during nutrient stress conditions (Franco-Zorrilla et al. 2007). siRNAs, particularly 24-nt heterochromatic siRNAs, are a key element of the RdDM pathway, directing DNA methylation to target genomic sites and regulating transcription (Pontier et al. 2012). Transposon-derived siRNAs are triggered by environmental stress and subsequently direct methylation of nearby stress-responsive genes. This is an epigenetic "switch" with the ability to temporarily repress or induce transcription based on stress conditions (Vatov and Gechev 2025). Some miRNA-TF modules function as stress-specific transcriptional switches. These modules enable dynamic regulation of gene expression in accordance with changing environmental conditions. Within the miR156–SPL loop, miR156 levels decrease with plant age, which elevates SPL TFs, inducing flowering. High miR156 under drought stress represses SPLs, postponing flowering as a drought avoidance mechanism (Jatan et al. 2019).

8.3.3 Crosstalk with Phytohormone Signaling

Phytohormones including abscisic acid (ABA), auxins, gibberellins (GAs), ethylene, salicylic acid (SA), and jasmonic acid (JA) are the principal plant growth, development, and stress response regulators (EL Sabagh et al. 2022). When plants are exposed to abiotic stress conditions (e.g., drought, salinity, heat, and cold), they quickly adapt hormonal homeostasis in favor of survival at the expense of growth (EL Sabagh et al. 2022). ncRNAs, especially miRNAs and lncRNAs, have come up as central regulators in this hormonal signaling pathway thereby regulating hormone-mediated plant stress response (Table 8.3; Jha et al. 2023).

Table 8.3 ncRNA-phytohormone crosstalk

ncRNA	Hormone	Target(s)	Stress type	Function	References
miR393	Auxin, JA	TIR1, AFBs	Salt, cold	Suppresses auxin signaling	Sunkar et al. (2007)
miR159	ABA, GA	MYB33, MYB101	Drought	Enhances ABA signaling, represses GA signaling	Reyes and Chua (2007)
miR160	Auxin	ARF10, ARF16, ARF17	Drought	Regulates root growth	Hao et al. (2022)
miR167	Auxin	ARF6, ARF8	Salt and drought	Affects the development of roots, stems, leaves, and flowers	Liu et al. (2021)
DRIR	ABA	ABA-responsive genes	Drought	Increases ABA sensitivity and drought tolerance	Qin et al. (2017)
ELENA1	SA	PR1 via MED19a	Biotic/abiotic stress	Enhances SA signaling	Seo et al. (2017)

8.3.3.1 miRNAs Regulating Phytohormone Signaling

ABA is a master regulator of drought, salt, and cold stress. It regulates stomatal closure, seed dormancy, and gene expression under stress (Wei et al. 2025). miR159 is targeted by ABA signaling positive regulators MYB33 and MYB101. During drought, ABA accumulation downregulates miR159, resulting in more MYB TFs and enhanced ABA sensitivity (Reyes and Chua 2007). miR168 is directed to AGO1, the central member of the RNA-induced silencing complex (RISC), and indirectly influences ABA-associated gene expression by controlling other miRNAs (Li et al. 2012).

Auxin regulates root growth, cell enlargement, and stress responses. It becomes reprogrammed in signaling and distribution during abiotic stress (Liu et al. 2024). miR393 is quickly induced by salt stress. It targets auxin receptors TIR1, AFB2, and AFB3, inhibits auxin signaling, and reshapes root architecture to reduce water loss (Iglesias et al. 2014). miR160 regulates ARF10, ARF16, and ARF17, auxin response factors that regulate root development and stress-responsive gene expression (Dai et al. 2021). miR167 regulates ARF6 and ARF8, which collectively modulate auxin-responsive genes under stress (Liu et al. 2021). GAs are growth-stimulating hormones that are typically suppressed under stress for resource partitioning (Margay et al. 2024). JA and SA regulate abiotic and biotic stress responses (Fujita et al. 2006). The miR319-TCP module regulates JA biosynthesis and JA signaling pathways (Schommer et al. 2008). When under salt or heat stress, the overexpression of miR319 causes JA accumulation and stress defense gene induction (Fang et al. 2021).

8.3.3.2 lncRNAs in Hormonal Crosstalk

DRIR (Drought-Induced lncRNA in Arabidopsis), which enhances ABA sensitivity by modulating the gene expression of ABA-responsive proteins such as RD29A and P5CS, leads to stomatal closure and the accumulation of osmoprotectants in response to drought (Qin et al. 2017). LncRNA-ELENA1 interacts with Mediator Complex subunit MED19a and enhances the expression of the PR1 gene, which is responsive to SA. Although characterized largely within the framework of pathogen responses, ELENA1 illustrates the degree to which lncRNAs integrate hormonal signaling under stress (Seo et al. 2017).

8.4 Stress-Specific Case Studies

8.4.1 Drought Stress

miR159, miR398, and miR169 are differentially expressed under drought. miR159 targets MYB transcription factors associated with ABA signaling, and the downregulation of miR398 promotes the biosynthesis of antioxidant enzymes (Reyes and Chua 2007; Li et al. 2022a, b). lncRNAs like DRIR (Drought-Induced lncRNA) modulate stomatal closure and water conservation by targeting transcription factors associated with stress- and hormone-signaling pathways (Qin et al. 2017).

miR159 targets MYB transcription factors, which promotes drought tolerance (López-Galiano et al. 2019). miR398 controls antioxidant enzymes, minimizing oxidative stress (Li et al. 2022a, b). miR156, overexpression in alfalfa, improves drought tolerance through the regulation of SPL13 (Arshad et al. 2017).

8.4.2 Salt Stress

miRNAs like miR393 and miR395 target ion transporters and detoxification genes (Mendoza-Soto et al. 2012; Zhang et al. 2013). miR393 represses auxin signaling, influencing root growth to limit the uptake of salt. miR393 targets auxin signaling, influencing root growth under salt conditions (Iglesias et al. 2014). LncRNA973 promotes salt tolerance by regulating antioxidant enzymes (Zhang et al. 2019).

8.4.3 Heat and Cold Stress

miR156 and miR398 are involved in heat stress regulation by targeting SPL transcription factors and antioxidant enzymes, respectively (Zhang et al. 2022a, b; Li et al. 2022a, b). lncRNAs possess the potential to modulate heat shock protein

(HSP) expression by binding to transcriptional regulators (Babaei et al. 2024). Cold-induced ncRNAs, including lncRNAs like COLDAIR, control vernalization pathways and flowering time through remodeling FLC chromatin (Kim et al. 2017). miR156, represses SPL transcription factors, modulating heat stress responses (Stief et al. 2014). SiRNAs trigger DNA methylation alterations, influencing cold stress tolerance (Chung et al. 2022).

8.5 ncRNAs as Biomarkers and Tools for Crop Improvement

ncRNAs, including miRNAs, lncRNAs, and siRNAs, are now effective tools in plant biotechnology for improving abiotic stress tolerance (Summanwar et al. 2020). Their tissue- or development-stage-specific expression profiles and stress-specificity, as well as their regulatory adaptability, make ncRNAs worthy candidates for use as biomarkers and molecular tools for genetic modification, breeding, and diagnostics (Nogoy et al. 2018; Raza et al. 2023; Bej and Basak 2014).

8.5.1 ncRNAs as Biomarkers

There are a few ncRNAs that do not exhibit stress-specific expression patterns, such as being used as molecular markers in various abiotic stresses like heat, salt, and drought. For instance, the downregulation of miR398 during oxidative stress is coupled with heightened reactive oxygen species (ROS), whereas the up-regulation of miR393 during salt stress reflects suppressed auxin signaling (Li et al. 2022a, b; Iglesias et al. 2014).

These expression changes are indicative of a molecular signature of the plant's stress status. Through profiling such ncRNA expressions using technologies like small RNA sequencing, researchers can distinguish between tolerant and stress-sensitive genotypes (Li et al. 2022a, b).

This technique also enables the development of diagnostic assays, such as qPCR-based panels, for in-field monitoring of crop stress response.

8.5.2 ncRNAs as Tools for Crop Genetic Engineering

ncRNAs are accessible to direct manipulation to improve stress tolerance by genetic engineering (Yadav et al. 2024). Stress-inducible miRNAs such as miR319 or miR160 can be overexpressed to enhance resilience by fine regulation of downstream transcription factors and signaling cascades (Şanlı and Öztürk Gökçe 2021; Wang et al. 2014). Toxic miRNA activity is also suppressed by antisense technology or target mimicry techniques, e.g., IPS1 mimic of miR399 (Franco-Zorrilla et al.

2007). In addition, CRISPR/Cas9 technologies now facilitate the precise editing of non-coding RNA or binding sites of their target mRNAs. While lncRNA editing is more challenging due to its complex structures and mechanisms, it is increasing as their functions enhance our insights (Basak and Nithin 2015; Zhao et al. 2024). Transgenic lines overexpressing or knocking down specific ncRNAs have exhibited increased tolerance to stresses of different natures, for example, modulation of miR156 expression affects plant development, phase change, and stress response (Jerome Jeyakumar et al. 2020).

8.5.3 Potential for Marker-Assisted Selection (MAS)

Integration of ncRNA profiles into breeding programs opens up new prospects for marker-assisted selection (MAS) of stress tolerance traits (Jaiswal et al. 2020). Expression quantitative trait locus (eQTL) mapping has the potential to enable the discovery of genomic loci associated with beneficial ncRNA expression patterns during stress, accelerating the generation of stress-resilient cultivars through genomics-assisted breeding strategies (Druka et al. 2010).

8.6 Conclusion and Future Perspectives

Non-coding RNAs (ncRNAs) have changed the landscape of plant abiotic stress response. Through the regulation of gene expression, modulation of signal transduction networks, and enhancement of epigenetic memory, ncRNAs are central to plant stress tolerance processes. Their biotechnological and breeding opportunities offer a thrilling avenue for crop improvement in tolerance towards climate change. Despite all these advancements, several crucial challenges remain to be addressed. These include (i) functional annotation of recently identified ncRNAs, particularly long non-coding RNAs (lncRNAs), (ii) comprehension of tissue- and stress-dependent expression profiles, (iii) unscrambling ncRNA interaction networks in epigenetic as well as in protein interaction frameworks, and (iv) translating outcomes from model systems into complex crop genomes. Subsequent research needs to emphasize integrative comprehension of ncRNA functions in stress signaling, development of ncRNA-based tools for crop improvement, and integration of the tools into current breeding programs. The use of multi-omics approaches-including transcriptomics, proteomics, and metabolomics and systems biology will be vital in disentangling the complex regulatory networks regulated by ncRNAs. Bridging these knowledge gaps is necessary to accelerate the deployment of climate-resilient crops and environmentally friendly agriculture.

References

Arshad M, Feyissa BA, Amyot L, Aung B, Hannoufa A (2017) MicroRNA156 improves drought stress tolerance in alfalfa (Medicago sativa) by silencing SPL13. Plant Sci 258:122–136

Babaei S, Bhalla PL, Singh MB (2024) Identifying long non-coding RNAs involved in heat stress response during wheat pollen development. Front Plant Sci 15:1344928

Basak J, Nithin C (2015) Targeting non-coding RNAs in plants with the CRISPR-Cas technology is a challenge yet worth accepting. Front Plant Sci 6:1001

Bej S, Basak J (2014) MicroRNAs: the potential biomarkers in plant stress response. Am J Plant Sci 2014:748

Bhogireddy S, Mangrauthia SK, Kumar R, Pandey AK, Singh S, Jain A, Budak H, Varshney RK, Kudapa H (2021) Regulatory non-coding RNAs: a new frontier in regulation of plant biology. Funct Integr Genomics 21(3):313–330

Bravo-Vázquez LA, Méndez-García A, Chamu-García V, Rodríguez AL, Bandyopadhyay A, Paul S (2024) The applications of CRISPR/Cas-mediated microRNA and lncRNA editing in plant biology: shaping the future of plant non-coding RNA research. Planta 259(2):32

Cao W, Yang L, Zhuang M, Lv H, Wang Y, Zhang Y, Ji J (2024) Plant non-coding RNAs: the new frontier for the regulation of plant development and adaptation to stress. Plant Physiol Biochem 208:108435

Chaudhary S, Grover A, Sharma PC (2021) MicroRNAs: potential targets for developing stress-tolerant crops. Life 11(4):289

Chen H, Li Z, Xiong L (2012) A plant microRNA regulates the adaptation of roots to drought stress. FEBS Lett 586(12):1742–1747

Cheng X, He Q, Tang S, Wang H, Zhang X, Lv M, Liu H, Gao Q, Zhou Y, Wang Q, Man X (2021) The miR172/IDS1 signaling module confers salt tolerance through maintaining ROS homeostasis in cereal crops. New Phytol 230(3):1017–1033

Chung S, Kwon C, Lee JH (2022) Epigenetic control of abiotic stress signaling in plants. Genes Genom 44:267–278

Dai X, Lu Q, Wang J, Wang L, Xiang F, Liu Z (2021) MiR160 and its target genes ARF10, ARF16 and ARF17 modulate hypocotyl elongation in a light, BRZ, or PAC-dependent manner in Arabidopsis: miR160 promotes hypocotyl elongation. Plant Sci 303:110686

Ding J, Lu Q, Ouyang Y, Mao H, Zhang P, Yao J, Xu C, Li X, Xiao J, Zhang Q (2012) A long non-coding RNA regulates photoperiod-sensitive male sterility, an essential component of hybrid rice. Proc Natl Acad Sci 109(7):2654–2659

Druka A, Potokina E, Luo Z, Jiang N, Chen X, Kearsey M, Waugh R (2010) Expression quantitative trait loci analysis in plants. Plant Biotechnol J 8(1):10–27

EL Sabagh A, Islam MS, Hossain A, Iqbal MA, Mubeen M, Waleed M, Reginato M, Battaglia M, Ahmed S, Rehman A, Arif M (2022) Phytohormones as growth regulators during abiotic stress tolerance in plants. Front Agron 4:765068

Fang Y et al (2021) Roles of miR319-regulated TCPs in plant development and response to abiotic stress. Crop J 9(1):17–28

Franco-Zorrilla JM, Valli A, Todesco M, Mateos I, Puga MI, Rubio-Somoza I, Leyva A, Weigel D, García JA, Paz-Ares J (2007) Target mimicry provides a new mechanism for regulation of microRNA activity. Nat Genet 39(8):1033–1037

Fujita M, Fujita Y, Noutoshi Y, Takahashi F, Narusaka Y, Yamaguchi-Shinozaki K, Shinozaki K (2006) Crosstalk between abiotic and biotic stress responses: a current view from the points of convergence in the stress signaling networks. Curr Opin Plant Biol 9(4):436–442

Hao K, Wang Y, Zhu Z, Wu Y, Chen R, Zhang L (2022) miR160: an indispensable regulator in plant. Front Plant Sci 13:833322

Heo, J.B. and Sung S (2011) Vernalization-mediated epigenetic silencing by a long intronic non-coding RNA. Science, 331(6013), pp.76–79.

Hung FY, Shih YH, Lin PY, Feng YR, Li C, Wu K (2022) WRKY63 transcriptional activation of COOLAIR and COLDAIR regulates vernalization-induced flowering. Plant Physiol 190(1):532–547

Iglesias MJ, Terrile MC, Windels D, Lombardo MC, Bartoli CG, Vazquez F, Estelle M, Casalongué CA (2014) MiR393 regulation of auxin signaling and redox-related components during acclimation to salinity in Arabidopsis. PLoS One 9(9):e107678

Jaiswal V, Rawoof A, Dubey M, Chhapekar SS, Sharma V, Ramchiary N (2020) Development and characterization of non-coding RNA based simple sequence repeat markers in Capsicum species. Genomics 112(2):1554–1564

Jannesar M, Seyedi SM, Botanga C (2024) Crosstalk between long non-coding RNAs and miRNAs regulates transferase activities in response to salt stress in Pistachio (Pistacia vera L.). Plant Stress 12:100474

Jatan R, Tiwari S, Asif MH, Lata C (2019) Genome-wide profiling reveals extensive alterations in Pseudomonas putida-mediated miRNAs expression during drought stress in chickpea (Cicer arietinum L.). Environ Exp Bot 157:217–227

Jerome Jeyakumar JM, Ali A, Wang WM, Thiruvengadam M (2020) Characterizing the role of the miR156-SPL network in plant development and stress response. Plants 9(9):1206

Jha UC, Nayyar H, Roychowdhury R, Prasad PV, Parida SK, Siddique KH (2023) Non-coding RNAs (ncRNAs) in plant: master regulators for adapting to extreme temperature conditions. Plant Physiol Biochem 205:108164

Khraiwesh B, Zhu JK, Zhu J (2012) Role of miRNAs and siRNAs in biotic and abiotic stress responses of plants. Biochim Biophys Acta 1819(2):137–148

Kim DH, Xi Y, Sung S (2017) Modular function of long noncoding RNA, COLDAIR, in the vernalization response. PLoS Genet 13(7):e1006939

Kindgren P, Ard R, Ivanov M, and Marquardt S (2018) Transcriptional read-through of the long non-coding RNA SVALKA governs plant cold acclimation. Nature communications, 9(1), p.4561.

Li W, Cui X, Meng Z, Huang X, Xie Q, Wu H, Jin H, Zhang D, Liang W (2012) Transcriptional regulation of Arabidopsis MIR168a and argonaute1 homeostasis in abscisic acid and abiotic stress responses. Plant Physiol 158(3):1279–1292

Li N, Liu T, Guo F, Yang J, Shi Y, Wang S, Sun D (2022a) Identification of long non-coding RNA-microRNA-mRNA regulatory modules and their potential roles in drought stress response in wheat (Triticum aestivum L.). Plant Sci 13:1011064

Li W.X, Oono Y, Zhu J, He X.J, Wu J.M, Iida K, Lu X.Y, Cui X, Jin H, and Zhu J.K (2008) The Arabidopsis NFYA5 transcription factor is regulated transcriptionally and posttranscriptionally to promote drought resistance. The Plant Cell, 20(8), pp.2238–2251.

Li J, Song Q, Zuo ZF, Liu L (2022b) MicroRNA398: a master regulator of plant development and stress responses. Int J Mol Sci 23(18):10803

Li Z, Yang J, Zou JJ, Cai X, Zeng X, Xing W (2024) A systematic review on the role of miRNAs in plant response to stresses under the changing climatic conditions. Plant Stress 14:100674

Liu D, Mewalal R, Hu R, Tuskan GA, Yang X (2017) New technologies accelerate the exploration of non-coding RNAs in horticultural plants. Hortic Res 4:17031

Liu X, Huang S, Xie H (2021) Advances in the regulation of plant development and stress response by miR167. Front Bioscience (Landmark) 26(9):655–665

Liu, L, Yahaya B.S, Li J, and Wu F (2024) Enigmatic role of auxin response factors in plant growth and stress tolerance. Frontiers in Plant Science, 15, p.1398818.

López-Galiano MJ, García-Robles I, González-Hernández AI, Camañes G, Vicedo B, Real MD, Rausell C (2019) Expression of miR159 is altered in tomato plants undergoing drought stress. Plants 8(7):201

Margay AR, Mehmood A, Bashir L (2024) Review on hormonal regulation of drought stress response in plants. Int J Plant Soil Sci 36:902–916

Mendoza-Soto AB, Sánchez F, Hernández G (2012) MicroRNAs as regulators in plant metal toxicity response. Front Plant Sci 3:105

Mirdar Mansuri R, Azizi AH, Sadri AH, Shobbar ZS (2022) Long non-coding RNAs as the regulatory hubs in rice response to salt stress. Sci Rep 12(1):21696

Mishra P, Singh A, Verma AK, Mishra SK, Singh R, Roy S (2022) MicroRNA775 targets a probable β-(1, 3)- galactosyltransferase to regulate growth and development in Arabidopsis thaliana. J Plant Growth Regul 41:3271–3284

Nogoy FM, Niño MC, Song JY, Jung YJ, Kang KK, Nou I, Cho YG (2018) Plant microRNAs in molecular breeding. Plant Biotechnol Rep 12:15–25

Navarro, L, Dunoyer P, Jay F, Arnold, B, Dharmasiri N, Estelle M, Voinnet O, and Jones J.D (2006) A plant miRNA contributes to antibacterial resistance by repressing auxin signaling. Science, 312(5772), pp.436–439

Omidbakhshfard MA, Proost S, Fujikura U, Mueller-Roeber B (2015) Growth-regulating factors (GRFs): a small transcription factor family with important functions in plant biology. Mol Plant 8(7):998–1010

Pontier D, Picart C, Roudier F, Garcia D, Lahmy S, Azevedo J, Alart E, Laudié M, Karlowski WM, Cooke R, Colot V (2012) NERD, a plant-specific GW protein, defines an additional RNAi-dependent chromatin-based pathway in Arabidopsis. Mol Cell 48(1):121–132

Puchta-Jasińska M, Bolc P, Pietrusińska-Radzio A, Motor A, Boczkowska M (2025) Small interfering RNAs as critical regulators of plant life process: new perspectives on regulating the transcriptomic machinery. Int J Mol Sci 26(4):1624

Qin T, Zhao H, Cui P, Albesher N, Xiong L (2017) A nucleus-localized long non-coding RNA enhances drought and salt stress tolerance. Plant Physiol 175(3):1321–1336

Raza A, Charagh S, Karikari B, Sharif R, Yadav V, Mubarik MS, Habib M, Zhuang Y, Zhang C, Chen H, Varshney RK (2023) miRNAs for crop improvement. Plant Physiol Biochem 201:107857

Reyes JL, Chua NH (2007) ABA induction of miR159 controls transcript levels of two MYB factors during Arabidopsis seed germination. Plant J 49(4):592–606

Ramanjulu Sunkar, Jian-Kang Zhu Novel (2004) and Stress-Regulated MicroRNAs and Other Small RNAs from Arabidopsis, The Plant Cell, Volume 16, Issue 8, August Pages 2001–2019

Şanlı BA, Öztürk Gökçe ZN (2021) Investigating effect of miR160 through overexpression in potato cultivars under single or combination of heat and drought stresses. Plant Biotechnol Rep 15(3):335–348

Schommer C, Palatnik JF, Aggarwal P, Chételat A, Cubas P, Farmer EE, Nath U, Weigel D (2008) Control of jasmonate biosynthesis and senescence by miR319 targets. PLoS Biol 6(9):e230

Seo JS, Sun HX, Park BS, Huang CH, Yeh SD, Jung C, Chua NH (2017) ELF18-induced long-noncoding RNA associates with mediator to enhance expression of innate immune response genes in Arabidopsis. Plant Cell 29(5):1024–1038

Shinozaki K, Yamaguchi-Shinozaki K (2022) Functional genomics in plant abiotic stress responses and tolerance: from gene discovery to complex regulatory networks and their application in breeding. Proc Jpn Acad Ser B Phys Biol Sci 98(8):470–492

Shuai P, Liang D, Tang S, Zhang Z, Ye CY, Su Y, Xia X, Yin W (2014) Genome-wide identification and functional prediction of novel and drought-responsive lincRNAs in Populus trichocarpa. J Exp Bot 65(17):4975–4983

Song L, Fang Y, Chen L, Wang J, Chen X (2021) Role of non-coding RNAs in plant immunity. Plant Commun 2(3):100180

Stief A, Altmann S, Hoffmann K, Pant BD, Scheible WR, Bäurle I (2014) Arabidopsis miR156 regulates tolerance to recurring environmental stress through SPL transcription factors. Plant Cell 26(4):1792–1807

Summanwar A, Basu U, Rahman H, Kav NN (2020) Non-coding RNAs as emerging targets for crop improvement. Plant Sci 297:110521

Sunkar R, Kapoor A, Zhu JK (2006) Posttranscriptional induction of two Cu/Zn superoxide dismutase genes in Arabidopsis is mediated by downregulation of miR398 and important for oxidative stress tolerance. Plant Cell 18(8):2051–2065

Sunkar R, Chinnusamy V, Zhu J, Zhu JK (2007) Small RNAs as big players in plant abiotic stress responses and nutrient deprivation. Trends Plant Sci 12(7):301–309

Tiwari S, Lata C (2021) Recent advancements in long noncoding RNA-mediated stress responses in rice. In: Long noncoding RNAs in plants. Academic Press, pp 63–74

Tiwari S, Jain M, Singla-Pareek SL, Bhalla PL, Singh MB, Pareek A (2023) Pokkali: a naturally evolved salt-tolerant rice shows a distinguished set of lncRNAs possibly contributing to the tolerant phenotype. Int J Mol Sci 24(14):11677

Vatov E, Gechev T (2025) Recent evidence for transgenerational adaptation resulting from stress induced changes in the cytosine methylation landscape of plants. Plant Stress 16:100812

Voinnet O (2009) Origin, biogenesis, and activity of plant microRNAs. Cell 136(4):669–687

Wang HLV, Chekanova JA (2017) Long noncoding RNAs in plants. In: Long non coding RNA biology, pp 133–154

Wang ST, Sun XL, Hoshino Y, Yu Y, Jia B, Sun ZW, Sun MZ, Duan XB, Zhu YM (2014) MicroRNA319 positively regulates cold tolerance by targeting OsPCF6 and OsTCP21 in rice (Oryza sativa L.). PLoS One 9(3):e91357

Wang, A, Hu J, Gao C, Chen G, Wang B, Lin C, Song L, Ding Y, and Zhou G (2019) Genome-wide analysis of long non-coding RNAs unveils the regulatory roles in the heat tolerance of Chinese cabbage (Brassica rapa ssp. chinensis). Scientific reports, 9(1), p.5002

Wei Y-S et al (2025) Mechanisms of Abscisic acid (ABA)-mediated plant defense responses: an updated review. Plant Stress 15:100724

Xu, J, Wang Q, Freeling M, Zhang X, Xu Y, Mao Y, Tang X, Wu F, Lan H, Cao M, and Rong T (2017) Natural antisense transcripts are significantly involved in regulation of drought stress in maize. Nucleic acids research, 45(9), pp.5126–5141

Yadav VK, Jalmi SK, Tiwari S, Kerkar S (2023) Deciphering shared attributes of plant long non-coding RNAs through a comparative computational approach. Sci Rep 13(1):15101

Yadav A, Mathan J, Dubey AK, Singh A (2024) The emerging role of non-coding RNAs (ncRNAs) in plant growth, development, and stress response signaling. Non-coding RNA 10(1):13

Yang M, Zhu P, Cheema J, Bloomer R, Mikulski P, Liu Q, Zhang Y, Dean C, Ding Y (2022) In vivo single-molecule analysis reveals COOLAIR RNA structural diversity. Nature 609(7926):394–399

Yang H, Cui Y, Feng Y, Hu Y, Liu L, Duan L (2023) Long non-coding RNAs of plants in response to abiotic stresses and their regulating roles in promoting environmental adaption. Cells 12(5):729

Yang C, Li D, Mao D, Liu X.U.E, Ji C, Li X, Zhao X, Cheng, Z, Chen C, and Zhu L (2013) Overexpression of micro RNA 319 impacts leaf morphogenesis and leads to enhanced cold tolerance in rice (O ryza sativa L.). Plant, cell & environment, 36(12), pp.2207–2218

Zhang LW, Song JB, Shu XX, Zhang Y, Yang ZM (2013) miR395 is involved in detoxification of cadmium in Brassica napus. J Hazard Mater 250:204–211

Zhang X, Dong J, Deng F, Wang W, Cheng Y, Song L, Hu M, Shen J, Xu Q, Shen F (2019) The long non-coding RNA lncRNA973 is involved in cotton response to salt stress. BMC Plant Biol 19:1–16

Zhang D, Zhang Z, Unver T, Zhang B (2021) CRISPR/Cas: a powerful tool for gene function study and crop improvement. J Adv Res 29:207–221

Zhang F, Yang J, Zhang N, Wu J, Si H (2022a) Roles of microRNAs in abiotic stress response and characteristics regulation of plant. Front Plant Sci 13:919243

Zhou M, Li D, Li Z, Hu Q., Yang C, Zhu L, and Luo H (2013) Constitutive expression of a miR319 gene alters plant development and enhances salt and drought tolerance in transgenic creeping bentgrass. Plant physiology, 161(3), pp.1375–1391

Zhang Y, Zhou Y, Zhu W, Liu J, Cheng F (2022b) Non-coding RNAs fine-tune the balance between plant growth and abiotic stress tolerance. Front Plant Sci 13:965745

Zhao Y, Zhu P, Hepworth J, Bloomer R, Antoniou-Kourounioti RL, Doughty J, Heckmann A, Xu C, Yang H, Dean C (2021) Natural temperature fluctuations promote COOLAIR regulation of FLC. Genes Dev 35(11–12):888–898

Zhao Z, Yang Y, Iqbal A, Wu Q, Zhou L (2024) Biological insights and recent advances in plant long non-coding RNA. Int J Mol Sci 25(22):11964

Chapter 9
Evolutionary Dynamics of Stress-Responsive Non-coding Regions

Priyanka Chauhan, Pratibha Verma, Ashutosh Tripathi, and Vrushali Desai

Abstract Non-coding regions play a pivotal role in plant stress responses, influencing gene regulation at epigenetic, transcriptional, and post-transcriptional levels. This chapter explores the evolutionary dynamics of stress-responsive non-coding RNAs (ncRNAs), including miRNAs, siRNAs, lncRNAs, and circRNAs, highlighting their origins, functions, and regulatory mechanisms. The chapter also discusses differences between conserved and lineage-specific regulatory sequences, emphasizing their significance in species adaptation. Recent advancements in high-throughput sequencing and functional genomics have revealed complex ncRNA-mediated regulatory networks that govern stress tolerance. Understanding these mechanisms offers potential applications in crop improvement, enhancing resilience to climate-induced stresses.Graphical Abstract

Authors Priyanka Chauhan and Pratibha Verma have equally contributed to this chapter.

P. Chauhan (✉)
School of Sciences, P P Savani University, Surat, Gujarat, India

Academy of Scientific and Innovative Research (AcSIR), Ghaziabad, Uttar Pradesh, India
e-mail: priyanka.chauhan@ppsu.ac.in

P. Verma · A. Tripathi
Academy of Scientific and Innovative Research (AcSIR), Ghaziabad, Uttar Pradesh, India

Divison of Microbial Technology, Lucknow, Uttar Pradesh, India

V. Desai
School of Sciences, P P Savani University, Surat, Gujarat, India

S. Tiwari et al. (eds.), *Decoding Non-Coding RNA*, Concepts and Strategies in Plant Sciences, https://doi.org/10.1007/978-981-96-9859-2_9

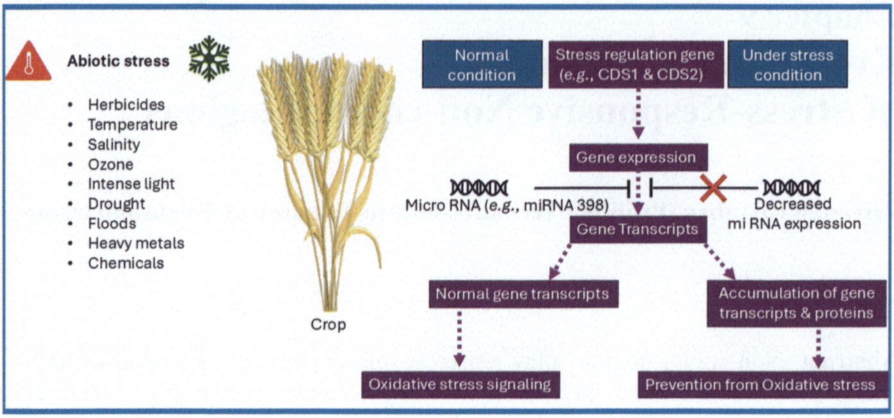

Keywords Non-coding region · Abiotic stress · Evolution · Conserved sequence · Lineage-specific elements

9.1 Introduction

More than 90% of eukaryotic genomes are transcribed, yet only about 2% of these transcripts are translated into proteins. The majority are non-coding RNAs (ncRNAs) that have little to no protein-coding potential. The rest of the transcriptome consists of ncRNA transcripts, which were once thought to be transcriptional noise due to their lack of protein-coding potential and low sequence conservation (Rai et al. 2019). Several functionally significant ncRNAs have been identified, including transfer RNAs, ribosomal RNAs, and other regulatory RNAs such as microRNAs (miRNAs), short interfering RNAs (siRNAs), long non-coding RNAs (lncRNAs), circular RNAs (circRNAs), and other non-coding RNA types. ncRNAs are further classified based on their nucleotide (nt) lengths (Yu et al. 2019). Long non-coding RNAs (lncRNAs) are transcripts longer than 200 nt that lack coding potential and form one of the largest categories of ncRNAs (Lakhotia et al. 2020). Housekeeping ncRNAs, including ribosomal RNAs, transfer RNAs, small nuclear RNAs, and small nucleolar RNAs, have several biological functions in plants (Fig. 9.1). In contrast, other ncRNAs are produced in response to external stimuli or during specific stages of development (Datta et al. 2024). Growing evidence underscores the crucial role of lncRNAs as key regulators involved in various biological processes, including development and environmental responses in plants (Zhang et al. 2023a, b). Moreover, small RNAs (sRNAs, are typically 20–24 nucleotides long, include small interfering RNAs (siRNAs) and microRNAs (miRNAs)**.** In plants, miRNAs are usually 21-nt sRNAs that guide the degradation or translational inhibition of mRNA targets, thereby repressing the expression of specific genes (Dzinyela et al. 2023).

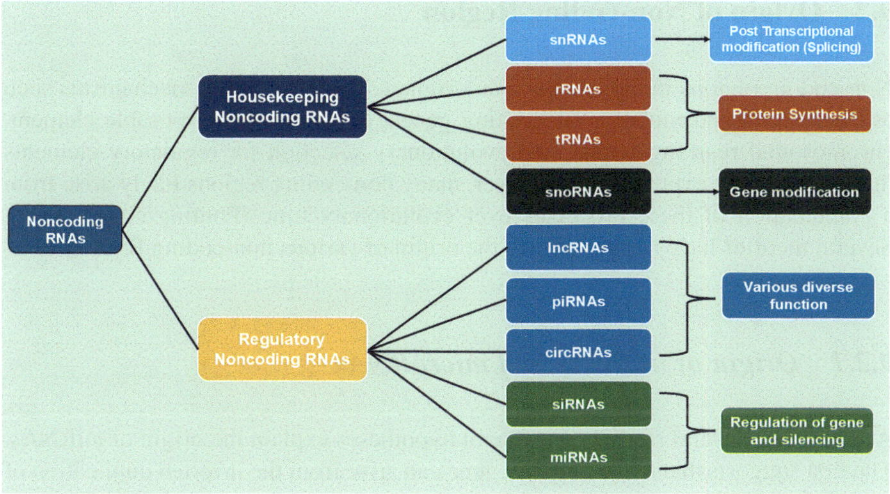

Fig. 9.1 Types of different non-coding RNAs and their function in plants

Nevertheless, high-throughput sequencing and experimental validation have demonstrated that ncRNAs are involved in diverse levels of gene regulation, including epigenetic, transcriptional, and post-transcriptional control (Cech and Steitz 2014).

As sessile organisms, plants cannot escape stress and must instead rely on avoidance and tolerance mechanisms to survive under constantly changing environmental conditions. Abiotic stresses such as cold, heat, hypoxia, drought, nutrient deficiencies, and salinity significantly impact plant growth and productivity, often leading to considerable yield losses (Zhang et al. 2022; Chauhan et al. 2021). Additionally, the growing global population, shrinking arable land due to rapid urbanization, declining soil quality, and the impacts of global warming are expected to further intensify food security challenges (Hossain et al. 2020). To cope with these harsh conditions, plants employ various gene regulatory mechanisms to restore and maintain cellular homeostasis. Recent studies have highlighted the crucial role of ncRNAs in regulating gene expression under stress conditions (Gill et al. 2024; Ding et al. 2024; Ahmed et al. 2020).

This book chapter highlights the origins of various non-coding regions in plants and their roles in stress management. Additionally, it provides a comparative analysis of conserved and lineage-specific regulatory systems in plants. Finally, the chapter discusses the adaptive significance of stress-responsive non-coding elements in different plant species.

9.2 Origin of Non-coding Region

Non-coding regions in the genome can originate through several mechanisms such as duplication and mutation of existing genes, insertion of transposable element, chromosomal rearrangements, and evolutionary selection for regulatory elements that control gene expression. Moreover, many non-coding regions likely arise from a combination of these processes over evolutionary time (Ponting et al., 2009). Several theories have been given for the origin of various non-coding RNAs.

9.2.1 Origin of miRNAs and microRNAs

According to several studies, three main hypotheses explain the origin of miRNAs. The first suggests that a new miRNA gene can arise from the inverted duplication of an existing miRNA gene or a protein-coding sequence (Hou et al. 2019). The second proposes that transcripts with inverted repeats at both ends, typically produced by transposable elements (TEs), particularly miniature inverted-repeat transposable elements (MITEs), can easily form stem-loop structures, serving as precursors for potential miRNAs (Nozawa et al. 2010, 2012; Piriyapongsa et al. 2007). The third theory posits that random hairpin formations within intronic or intergenic regions can also contribute to the emergence of new miRNA genes. According to the first hypothesis, newly formed miRNA genes retain sequence similarity to their ancestral genes but accumulate nucleotide changes over time (Tanzer and Stadler 2004; Zhang et al. 2011). Numerous miRNAs have evolved through this mechanism, allowing them to target the transcripts of their progenitor genes and regulate their expression through feedback loops. The second mechanism has led to the generation of many new miRNAs, with up to 48 identified in Arabidopsis and rice (Piriyapongsa and Jordan 2008). Brodersen et al. (2008) analyzed juvenile miRNA genes in orthologous regions between *Arabidopsis lyrata* and *Arabidopsis thaliana*, revealing that some *A. thaliana* miRNAs exhibited low sequence conservation. This suggests a lack of selection pressure on miRNA genes originating from random sequences. Similarly, miRNA genes derived from TEs showed comparable patterns (Bentwich et al. 2005). They are transcribed from miRNA genes (MIR genes) by RNA polymerase II (RNA pol II) and processed by the RNase III enzyme Dicer-like protein 1 (DCL1) (Axtell and Meyers 2018). The resulting miRNA–miRNA* duplex is then transported to the cytoplasm by HASTY (HST). In the cytoplasm, the duplex is incorporated into the RNA-induced silencing complex (RISC), which includes ARGONAUTE 1 (AGO1) proteins (Pareek et al. 2015).

9.2.2 Origin of lncRNAs

Long non-coding RNAs (lncRNAs) are transcripts longer than 200 base pairs (bp) that lack protein-coding potential. They can be classified based on their strand of origin as sense or antisense, by transcriptional orientation as divergent or convergent, and by location as intronic or intergenic (Assaf et al. 2024). Although the exact mechanisms of lncRNA origination in plants remain unclear, three potential hypotheses have been suggested based on known genomic evolution processes: (1) duplication of pre-existing lncRNAs, (2) degeneration or pseudogenization of protein-coding genes, and (3) evolution from existing transposable elements (TEs) (Hou et al. 2019). LncRNAs are transcribed by RNA polymerases II, III, IV, and V and regulate target genes through various mechanisms, including chromatin remodeling, transcriptional repression, molecular mimicry, RNA splicing, and functioning as transcriptional enhancers (Waititu et al. 2020).

9.2.3 Origin of CircRNAs

Circular RNAs (circRNAs) can originate from exons (exonic circRNAs), introns (intronic circRNAs), or intergenic regions (Liu et al. 2020). Notably, multiple circRNAs can be produced from a single gene locus through a process known as alternative circularization, and they can consist of either a single exon or multiple exons (Chen 2020). Interestingly, circRNAs are not exclusively generated from the nuclear genome; they can also arise from chloroplast and mitochondrial genomes. For instance, exonic circRNAs corresponding to mitochondrial genes such as *COX1*, *apocytochrome b*, *NAD9*, *ABCI3*, and sec-independent protein translocase have been identified in barley (Darbani et al. 2016). Additionally, 48 circRNAs were detected in the chloroplast genome and 10 in the mitochondrial genome of *Arabidopsis*, demonstrating the widespread nature of circRNA biogenesis pathways (Darbani et al. 2016; Sun et al. 2016). Although circRNAs are still not well understood, they likely represent an important class of non-coding regulatory elements in plant genomes. Furthermore, non-coding RNAs and transposable elements (TEs) influence the production of various small interfering RNAs (siRNAs), including heterochromatic siRNAs (hc-siRNAs), phased secondary siRNAs (phasiRNAs), natural antisense transcript siRNAs (NAT-siRNAs), and trans-acting siRNAs (ta-siRNAs or tasiRNAs), which are categorized as a type of phasiRNAs (Mandal et al. 2022; Yu et al. 2014). These siRNAs regulate target gene expression at both transcriptional and post-transcriptional levels. Their gene-silencing functions are either similar to or integrated with the miRNA and siRNA pathways (Lam et al. 2015).

9.2.4 Origin of SiRNAs

Small interfering RNAs (siRNAs) are categorized into three main types based on their biogenesis and mode of action: trans-acting siRNAs (tasiRNAs), heterochromatic siRNAs (hc-siRNAs), and natural antisense siRNAs (nat-siRNAs). TasiRNAs function similarly to microRNAs (miRNAs) by regulating target genes in *cis* but act at loci distinct from their site of origin. In contrast, hc-siRNAs operate in *cis* within their source regions or homologous sequences, as their origins coincide with their target sites (Zheng et al. 2009). Nat-siRNAs are further divided into *cis-nat-siRNAs* and *trans-nat-siRNAs*. *Cis-nat-siRNAs* are generated from the transcription of two RNAs from opposite strands of the same locus, whereas *trans-nat-siRNAs* are produced from transcripts originating from different loci (Jin et al. 2008). The biogenesis of siRNAs involves the formation of double-stranded RNAs (dsRNAs), the activity of Dicer-like (DCL) enzymes, and the incorporation of siRNAs into the RNA-induced silencing complex (RISC) containing ARGONAUTE (AGO) proteins, which mediate target regulation at the transcriptional or post-transcriptional level (Erdmann and Picard 2020). TasiRNA biogenesis requires the involvement of SUPPRESSOR OF GENE SILENCING 3 (SGS3), RNA-dependent RNA polymerase 6 (RDR6), and DCL4 (Gasciolli et al. 2005; Peragine et al. 2004). In contrast, nat-siRNA formation depends on SGS3, RDR6, DCL1, DCL2, and a plant-specific RNA polymerase, NRPD1A. Meanwhile, hc-siRNAs are generated through the coordinated action of RNA polymerase IV (Pol IV), DCL3, RDR2, 5S rRNA genes, and NRPD1A (Mallory and Vaucheret 2006; Borsani et al. 2005).

9.3 Role of ncRNAs in Response to Abiotic Stress in Plants

Abiotic stresses, exacerbated by climate change and human activity, hinder plant growth, challenging agriculture and the environment. In response, plants have developed mechanisms like stress perception, epigenetic modifications, and transcriptional and translational regulation. Recent studies emphasize the key roles of long non-coding RNAs (lncRNAs) in plant adaptation to these stresses (Wierzbicki et al. 2021). Several non-coding regions such as lncRNAs, miRNAs, and siRNAs are involved the abiotic stress management in different plants species (Table 9.1).

Like mRNAs, plant lncRNAs are transcribed by nuclear RNA polymerases (Pol I, II, III, IV, and V) and undergo post-transcriptional modifications, including capping, splicing, polyadenylation, and nuclear export, to ensure proper structure and function (Böhmdorfer et al. 2016) (Marquardt et al. 2014). Like protein-coding genes, they have 5′ and 3′ UTRs, introns, and exons, but their polyadenylation abundance and efficiency are generally lower than mRNAs (Cao et al. 2018).

Water is essential for plant life, and drought causing dehydration and desiccation poses a significant stress during plant establishment, negatively impacting development, yield, and survival. Recent studies have identified several drought-responsive

Table 9.1 Summary of different abiotic stress responsive non-coding RNA

S. no.	Plant	Non-coding RNA	Abiotic stress	References
1.	Arabidopsis	SRO5-P5CDH nat-siRNA	Salt	Borsani et al. (2005)
2.	Cotton	GhDNA1 lncRNAs	Drought	Tao et al. (2021)
3.	Arabidopsis	HSFB2a lncRNAs	Heat stress	Cao et al. (2021)
4.	Cotton	XH123 lncRNAs	Cold tolerance	Cao et al. (2021)
5.	Arabidopsis	CIL1 lncRNAs	Cold tolerance	Huo et al. (2022)
6.	*Populus trichocarpa*	lincRNA1795	Drought	Shuai et al. (2014)
7.	*Populus trichocarpa*	lincRNA20	Drought	Shuai et al. (2014)
8.	*Populus trichocarpa*	lincRNA2623	Drought	Shuai et al. (2014)
9.	*Populus trichocarpa*	lincRNA967	Drought	Shuai et al. (2014)
10.	*Populus trichocarpa*	lincRNA2762	Drought	Shuai et al. (2014)
11.	*Populus trichocarpa*	lincRNA1449	Drought	Shuai et al. (2014)
12.	Arabidopsis	miR160	Harmone signaling	Liu et al. (2007)
13.	Cotton	miR160, miR159	Harmone signaling	Reyes and Chua (2007)
14.	Barley	miR393	Seed dormancy	Bai et al. (2017)
15.	*Phaeodactylum tricornutum*	miR5472	Phosphate deficiency	Cruz de Carvalho et al. (2016)
16.	*Phaeodactylum tricornutum*	miR5471	Phosphate deficiency	Cruz de Carvalho et al. (2016)
17.	*Arabidopsis thaliana*	miR397	Cadmium tolerance	Ali et al. (2023)
18.	Rice	mir535	Cadmium tolerance	Yue et al. (2023)
19.	Tomato	miR398	Cadmium tolerance	Yan et al. (2023)
20.	Barley	miR156	Drought	Wang et al. (2023)
21.	*Arabidopsis thaliana*	miR159	Drought	Jiang et al. (2022)
22.	*Arabidopsis thaliana*	miR169a	Drought	Jiang et al. (2022)
23.	Wheat	miR1119	Drought	Shamloo-Dashtpagerdi et al. (2023)
24.	Arabidopsis and rice	miR393	Drought	Zhao et al. (2019)
25.	Alfalfa	miR156	Temperature stress	Matthews et al. (2019)
26.	Rapeseed	miR166e, miR319, miR3936-5p	Temperature stress	Zeng et al. (2018)
27.	Tomato	sha-miR319d	Temperature stress	Shi et al. (2019)
28.	Tea plants	miR156	Nutritional stress	Shi et al. (2019)

(continued)

Table 9.1 (continued)

S. no.	Plant	Non-coding RNA	Abiotic stress	References
29.	Peanut	miR156	Nutritional stress	Li et al. (2021)
30.	Soybean	miR169, miR395, miR397, miR399	Salt stress	Xu et al. (2013)
31.	Wheat	miR171, miR166	Salt stress	Thornburg et al. (2020)
32.	Tobacco	miR396a	Salt stress	Chen et al. (2015)

lncRNAs in species like rice (Chen et al. 2021), maize (Lv et al. 2016), wheat (Cagirici et al. 2017), tomato, cassava, switchgrass, *Eutrema salsugineum*, *rapeseed*, *Oryza rufipogon*, *Populus trichocarpa*, *Brassica juncea*, and *Cleistogenes songorica* (Yan et al. 2019; Bhatia et al. 2020). Differentially expressed lncRNAs responsive to drought stress have been identified in peanut varieties with varying drought tolerance (Ren et al. 2022), and GhDNA1 in cotton regulates drought-responsive genes by targeting AAAG DNA double strands (Tao et al. 2021). These findings highlight how lncRNAs can be either induced or suppressed by drought stress.

Temperature fluctuations limit plant distribution and crop yields, with extreme climate events, driven by global warming, posing significant challenges. LncRNAs are involved in plant responses to cold and heat stresses in species like Arabidopsis, Chinese cabbage, grapevine, banana, wheat, maize, cassava, and alfalfa (Yang et al. 2022; Qin et al. 2017; Wang et al. 2019). The antisense heat-inducible lncRNA asHSFB2a regulates heat stress in *Arabidopsis*, while the cold-responsive lncRNA XH123 improves cold tolerance in cotton seedlings (Cao et al. 2021). Additionally, cold-induced lncRNA 1 (CIL1) in *Arabidopsis* aids cold stress response by influencing hormone signaling, ROS homeostasis, and glucose metabolism (Huo et al. 2022).

miRNAs regulate plant growth and development along with abiotic stress responses by targeting functional gene either transcripts for cleavage or translation repression (Chen et al. 2009). In plants, miRNAs recognize target mRNA binding sites through near-exact sequence complementarity, leading to cleavage at the tenth nucleotide of the complementary site (Addo-Quaye et al. 2008) or translational inhibition (Chen et al. 2004). They can also inhibit translation without mRNA degradation by binding partially mismatched sequences in the 3′ UTR. Additionally, miRNAs regulate translation initiation by interfering with eukaryotic initiation factor 4E/cap and poly(A) tail functions (Rhoades et al. 2002). The function of miRNAs depends on their conserved binding sites within mRNA targets (Rhoades et al. 2002). Seed germination is influenced by the physiological state of the seed (dormancy), which is regulated by the ABA/GA (abscisic acid/gibberellins) ratio (Carrera-Castaño et al. 2020). In Arabidopsis, overexpression of miR160 leads to hyposensitivity to ABA during seed germination (Liu et al. 2013). miR160 targets ARF10 and ARF16, which regulate ABI3 expression and induce the seed dormancy (Liu et al. 2007). In cotton, the miR160/ARF axis also impacts seed size by regulating genes associated with seed development (Reyes and Chua 2007). Additionally, miR159 affects seed dormancy and germination, as its overexpression renders seeds hyposensitive to ABA and suppresses MYB33 and MYB101, positive regulators of ABA responses (Reyes and Chua 2007). Crosstalk between ABA and auxin is involved in seed germination, and miR160 may reduce ABA sensitivity by downregulating an auxin signaling component (Nonogaki 2010). Furthermore, miR393 influences seed development in barley by targeting the auxin receptors TIR1/AFBs, with both overexpression and target mimic (MIM393)-mediated inhibition affecting seed development (Bai et al. 2017).

9.4 Mechanisms of Stress Response Regulation

Functional genes regulate plant growth, development, and stress adaptation, with only about 2% of transcribed RNA encoding proteins in the eukaryotic genome (Rai et al. 2019). Advancements in functional genomics and RNA-seq technology have led to the identification of various non-coding RNAs (ncRNAs), including rRNA, tRNA, snRNA, and regulatory ncRNAs (Jha et al. 2020; Yu et al. 2019; Ariel et al. 2015). These ncRNAs are involved in various developmental processes and responses to stress. Regulatory ncRNAs in plants include microRNAs (miRNAs), small interfering RNAs (siRNAs), and long non-coding RNAs (lncRNAs, >200 nucleotides) (Li et al. 2017). MiRNAs can target lncRNAs to produce phasiRNAs, while lncRNAs can also generate miRNAs or regulate their expression (Meng et al. 2021). Advancements in molecular biology high-throughput sequencing technologies have enabled the identification of many functional genes involved in plant development and responses to abiotic stress. At the same time, the complex world of non-coding RNAs (ncRNAs) is gradually being revealed. Many ncRNAs have been the subject of several review articles in recent years (Wang et al. 2017; Ahmed et al. 2020). Gaining a deeper understanding of ncRNA-mediated regulatory networks in plant development and stress responses offers valuable insights for enhancing plant tolerance to environmental stresses. Plant ncRNA regulatory pathways are complex, with the same ncRNA regulating both development and abiotic stress responses. Despite progress, the molecular mechanisms and roles of ncRNAs and their targets are still largely unexplored. Recent studies show that manipulating ncRNAs and their targets can modify plant phenotypes and enhance stress tolerance (Singh et al. 2021). Moreover, miRNA and siRNAs influence gene expression by an extensive number of ways, such as (1) cutting of target mRNA, (2) translational suppression, and (3) transcriptional inactivation (Li et al. 2024). Furthermore, lncRNAs affect gene expression by the taking over of protein and miRNAs, controlling the stability of mRNA and translation, and modifying chromatin state (Statello et al. 2021).

Cis-regulatory elements govern the timing, amount, and spatial pattern of gene expression. These elements include enhancers, silencers, and insulators (Ijaz et al., 2020). Enhancers are DNA sequences that promote gene expression, while silencers

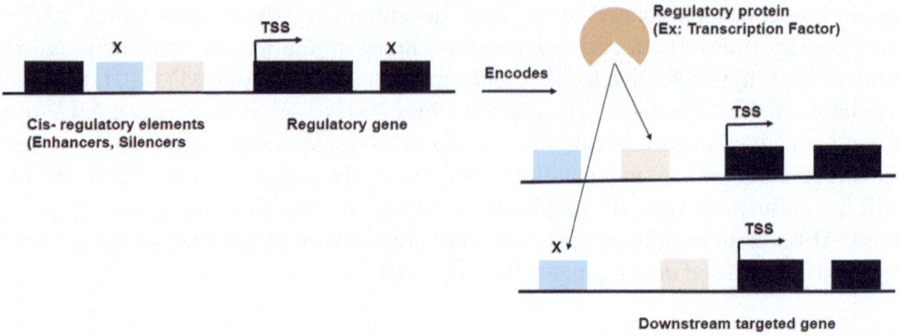

Fig. 9.2 The TSS and arrows above the DNA sequence indicate transcriptional start sites (TSSs), while solid boxes represent protein-coding exons, and shaded boxes highlight cis-regulatory elements or modules. The developmental regulatory gene encodes a transcription factor that targets the transcriptional regulatory complex of downstream genes. X marks potential DNA mutation sites, which can affect both regulatory elements and coding sequences. Mutations in upstream coding exons can influence many downstream interactions, while mutations in a single cis-regulatory module typically affect only one expression domain

suppress it. Insulators, depending on the DNA context, can either prevent enhancer or silencer interactions. Transcription factors bind to specific DNA sequences within these cis-regulatory elements and often play a key role in regulating the transcription initiation complex (Wasserman and Sandelin 2004). Cis-regulatory elements, including enhancers, are modular in nature and control gene expression across various domains (Fig. 9.2; Carrol et al. 2004). Mutations in the cis-regulatory regions of developmental genes can lead to changes in their expression patterns, impacting numerous downstream targets and forming intricate gene regulatory networks (GRNs) (Peter and Davidson 2011). As these regions often consist of multiple modular enhancers, alterations in a single module are typically restricted to affecting one expression domain of a pleiotropic protein, thereby enabling the selection of trait variants (Carroll et al. 2013).

9.5 Conserved Versus Lineage-Specific Regulatory Sequences

Decryption of evolutionary dynamics of stress-responsive non-coding regulatory sequences requires distinguishing between conserved elements, which remain functionally stable across species, and lineage-specific sequences, and emerge through evolutionary innovations to meet species-specific stress responses. These regulatory elements, including cis-regulatory elements (CREs), enhancers, promoters, and non-coding RNAs (ncRNAs), orchestrate gene expression patterns that define adaptive potential and stress response of an organism. A conserved regulatory sequence is a genomic element; due to purifying selection, it remains functionally and structurally similar across multiple taxa. There are a few criteria that are used to identify conserved sequences.

9.5.1 Conservation Across Species

9.5.1.1 Cross-Species Sequence Similarity

The comparison of sequence between multiple species allows to distinguish non-coding sequences that have been conserved in multiple species and are more likely to conservation than shared ancestry (Frazer et al. 2003). These are highly conserved elements and exhibit greater than 70% identity over a defined length (Siepel et al. 2005).

9.5.1.2 Multiple Genome Alignments

These alignments are used to calculate ancestral sequence, age-based constrained elements and conservation score between groups of genomes. Certain tools such as multiple alignment using fast Fourier transform (MAFT), clustalW, and Multiple Sequence Comparison by Log-Expectation (MUSCLE) are used to discover conservation across and between phylogenetic groups (Katoh and Standley 2013).

9.5.2 Functional Conservation

9.5.2.1 Cis-Regulatory Modules (CRMs) Alongside Preserved Transcription Factor Binding Sites (TFBSs)

Regulatory sequences including enhancers, promoters, and silencers are called cis-regulatory modules (CRMs). The promoters can be present upstream of target gene, whereas silencers and enhancers can be far away from that gene and can regulate the translation irrespective of their location and orientation. It is made up of clusters of transcriptional factors (TFs) binding sites (TFBSs) independent of their cooperative. It has transcriptional regulatory functions due to characteristic binding of cognate transcriptional factors to the binding sites harboring on it (Ni et al., 2022). For example, enhancers and promoters regulating genes related to heat shock proteins (HSPs), drought (DREB1/CBF), and oxidative stress (SOD) show evolutionary conservation (Kang et al. 2023).

9.5.2.2 Non-coding RNAs with Conserved Targets

Genomes are largely transcribed, and genomes are extensively transcribed and induce thousands of long non-coding RNAs (lncRNAs), which are generally longer than 200 nucleotides remain untranslated to functional proteins. Several microRNA (miRNAs) i.e., miR398 related to ROS modulation and long non-coding RNAs (lncRNAs) i.e., COOLAIR related to temperature-dependent flowering, are functionally present across species (Sunkar et al. 2012).

9.5.3 Phylogenetic Footprinting and Evolutionary Constraints

A bioinformatic approach, i.e., phylogenetic footprint is used to identify genomic regions present in different regions that are conserved, specifically on regulatory elements, for example CREs. The fundamental principle behind this is that genes that are functionally important involved in regulation of genes are conserved across evolution (Ijaz et al. 2020). It tends to identify non-coding regions, and ensures their unchanged functions (Frazer et al. 2003). Whereas selective constraints on syntenic regions are genomic loci that has conserved functions are syntenically maintained across all species (Budzko et al. 2023).

9.5.4 Epigenetic and Chromatin Conservation

Epigenomes are easily studied, and it only takes two measurements, separated by enough time to allow drift to build in less important genes, to estimate conservation. As random changes are frequently harmful and result in decreased fitness and loss due to negative or purifying selection, functional sections of a genome are generally more preserved (Rud et al. 2021). For example, histone alterations (H3K4me3, H3K27ac) identify species-specific conserved regulatory sequences (Villar et al. 2015).

9.5.5 Criteria for Lineage-Specificity

Gene duplication, genomic drift, and adaptation to distinct ecological niches all contribute to the emergence of a lineage-specific regulatory sequence. These components are exclusive to evolutionary lineages and do not exhibit cross-species conservation.

Absence of Homology Across Species: Regardless of present functional differences, homology in biology refers to similarities in genes or anatomical features across animals of different taxa that result from a common ancestor. According to evolutionary biology, homologous structures are those that have kept traits from a common ancestor after undergoing adaptive adaptations for distinct uses as a result of natural selection. The examples of absence of homology across species are lack of sequence similarity during comparative analysis and emergence of species-specific regulatory motifs from random mutation and positive selection (Schmitz and Bornberg-Bauer 2017; Buttler 2023).

Functional novelty in each Lineage: Genes unique to a particular lineage are frequently thought of as "novel" genes, signifying genetic novelty that has emerged within that lineage. Unique regulatory networks (LS-NCRSEs) accommodate into lineage-specific transcriptional circuits, which regulates genes in a species-dependent manner (Buttler 2023). For example, photosynthesis in maize is typically regulated by species specific sis regulatory sequences which are missing in rice (Aubry et al. 2014).

Epigenetic Distinctions in Lineage-Specific Sequences:

- Unique chromatin accessibility patterns: Regulatory elements that are missing in some species but open in others are revealed by ATAC-seq data (He et al. 2022).
- Species-specific histone modifications: Certain enhancers have distinct histone marks (for example, drought-tolerant plants have H3K27me3 repressive marks, but temperate species do not).

Researchers combine evolutionary analysis, comparative genomics, functional genomics, and bioinformatics to distinguish between conserved and lineage-specific non-coding sequences (Table 9.2).

Conclusively, strong evolutionary constraints ensure that conserved non-coding regulatory elements (CNCs), which control essential stress–response systems, are preserved across species. Evolutionary processes including mutation, duplication, and TE insertion give rise to lineage-specific non-coding regulatory elements (LS-NCREs), which allow for species-specific adaptation to environmental stress.

Table 9.2 Integrative approaches for identifying conserved versus lineage-specific regulatory elements

S. no.	Approach	Role in identifying conservation	Role in identifying lineage-specificity	References
1.	Multiple genome alignments	Detects conserved sequences across species	Identifies missing sequences in certain species	Frazer et al. (2003)
2.	Comparative Phylogenetics	Determines evolutionary conservation patterns	Tracks sequence divergence or de novo emergence	Schmitz and Bauer (2016)
3.	Functional assays (e.g., ChIP-seq, CRISPR screens, enhancer assays)	Confirms conserved regulatory functions	Tests lineage-specific regulatory activity	Villar et al. (2015)
4.	Epigenomic analysis (e.g., ATAC-seq, histone modification profiling)	Identifies conserved chromatin states	Reveals species-specific regulatory elements	He et al. (2022)

To differentiate between conserved and lineage-specific regulatory sequences, comparative genomics, functional tests, and epigenomic studies are essential.

9.6 Functional Characterization of Conserved Non-coding Sequences

In a variety of species, conserved non-coding regulatory sequences are essential for preserving basic stress–response systems. Because of their crucial regulatory roles, which guarantee a coordinated reaction to external stimuli, these sequences are frequently maintained. Their functional characterization sheds light on how organisms adjust to changing environments and preserve homeostasis.

9.6.1 Networks of Universal Stress–Response

Across several taxa, universal stress–response networks are facilitated by conserved regulatory sequences. These networks contain essential regulatory components that control important stress–response genes, such as conserved cis-regulatory motifs in promoter and enhancer regions. For instance, during thermal stress, heat shock proteins can be activated due to the high conservation of heat shock elements (HSEs) in fungi, plants, and animals (Åkerfelt et al. 2010). In a similar vein, antioxidant response elements (AREs) are extensively found in eukaryotes and are essential for defense mechanisms against oxidative stress (Ma 2013).

9.6.2 Similarities Across Species in Regulatory Systems

Numerous non-coding RNAs (ncRNAs) and transcription factor binding sites (TFBSs) are conserved across species, according to comparative genomics, suggesting that stress responses are regulated similarly. As an illustration of the evolutionary conservation of their binding motifs, the DREB (Dehydration-Responsive Element Binding) transcription factors in plants control responses to cold stress and drought in both monocots and dicots (Sakuma et al. 2006). Similar to this, conserved microRNAs like miR398 target the genes for superoxide dismutase (SOD) to control the oxidative stress response in both plants and animals (Sunkar et al. 2006).

The functional significance of these components in stress adaptation is highlighted by their cross-species conservation, underscoring the role of non-coding sequences in regulatory evolution.

9.7 Significance of Non-coding Sequences

Evolutionary adaptations that allow species to establish distinct stress–response mechanisms are represented by non-coding sequences. These sequences develop as a result of mutations, duplications, and transposable element activity, which aid in species-specific adaptations, in contrast to conserved elements, which are shared by several taxa. Knowing these components helps us understand how various creatures develop specialized coping mechanisms to deal with environmental stressors.

9.7.1 Evolutionary Developments in Stress Reactions

In adaptive evolution, lineage-specific regulatory components frequently play a crucial role, enabling species to create new stress–response pathways. These sequences include species-specific long non-coding RNAs (lncRNAs), lineage-specific transcription factor binding sites (TFBSs), and recently discovered cis-regulatory elements that alter gene expression in response to stress.

For instance, submergence tolerance, a characteristic lacking in other cereal crops, is made possible in rice (Oryza sativa) by a lineage-specific SNP mutation in the promoter region of the SUB1A gene (Xu et al. 2006). Similarly, certain plant species have evolved cold tolerance while others have not due to transposable element insertions in the DREB1 promoter (Matsui et al. 2008). These examples show how species adaptability to a variety of environmental situations is facilitated by lineage-specific regulatory sequences.

9.7.2 Contribution to Species-Specific Fitness

By improving species fitness in particular ecological niches, lineage-specific sequences give an advantage over other species. Gene expression networks are fine-tuned by such factors, enabling animals to maximize their physiological reactions to environmental stressors.

For example, in some species, lineage-specific proliferation of stress-responsive microRNAs contributes to distinct stress adaption mechanisms. According to Okuma and Kawaguchi (2021), miR2111 regulates nodulation in legumes when nitrogen is limited, a function that is absent in non-leguminous plants. Comparing *Arabidopsis lyrata* to its related *Arabidopsis thaliana*, lineage-specific transposable element insertions have altered regulatory networks in the latter, impacting drought resilience (Van de Peer et al. 2017). These evolutionary innovations demonstrate how lineage-specific sequences shape adaptive diversity and are crucial for species survival and ecological success.

9.8 Conclusion and Future Aspects

Non-coding regions play a crucial role in regulating plant responses to environmental stresses. The evolutionary dynamics of stress-responsive non-coding RNAs (ncRNAs), including miRNAs, siRNAs, lncRNAs, and circRNAs, demonstrate their significance in gene expression control, chromatin remodeling, and epigenetic regulation. These elements function in stress perception, signal transduction, and adaptive mechanisms, ensuring plant survival under adverse conditions. Evolutionary forces such as mutation, gene duplication, and transposable element activity have contributed to the diversification of these regulatory elements, enabling plants to fine-tune their responses to specific environmental pressures. Conclusively, understanding the conservation and lineage specificity of these regulatory sequences offers insights into the molecular basis of stress adaptation in plants.

Future research should focus on unravelling the complete regulatory networks involving ncRNAs and their interactions with transcription factors, chromatin modifications, and cis-regulatory elements. Advances in functional genomics, single-cell transcriptomics, and CRISPR-based gene editing will enhance our ability to characterize ncRNAs and their precise roles in plant stress tolerance. Additionally, integrating multi-omics approaches, such as epigenomics and proteomics, will provide a holistic understanding of plant adaptation mechanisms. Harnessing these insights for crop improvement through targeted gene regulation will be pivotal in developing stress-resilient crops, ensuring food security in era of climate change.

Author Contributions Conceptualization: PC; writing, reviewing, and editing: PC, PV, AT, and VD. All the authors have read and agreed to the published version of the manuscript.

Conflicts of Interest The authors declare no conflict of interest.

Funding This research has no funding.

References

Addo-Quaye C, Eshoo TW, Bartel DP, Axtell MJ (2008) Endogenous siRNA and miRNA targets identified by sequencing of the Arabidopsis degradome. Curr Biol 18(10):758–762

Ahmed W, Xia Y, Li R, Bai G, Siddique KH, Guo P (2020) Non-coding RNAs: functional roles in the regulation of stress response in Brassica crops. Genomics 112(2):1419–1424

Åkerfelt M, Morimoto RI, Sistonen L (2010) Heat shock factors: integrators of cell stress, development and lifespan. Nat Rev Mol Cell Biol 11(8):545–555

Ali S, Huang S, Zhou J, Bai Y, Liu Y, Shi L, Liu S, Hu Z, Tang Y (2023) miR397-LACs mediated cadmium stress tolerance in *Arabidopsis thaliana*. Plant Mol Biol 113(6):415–430

Ariel F, Romero-Barrios N, Jégu T, Benhamed M, Crespi M (2015) Battles and hijacks: noncoding transcription in plants. Trends Plant Sci 20(6):362–371

Assaf C, Bester Preeti, Dabas (2024) Noncoding RNA - The Dark Matter of the Genome Deciphering the Hidden Language of Long Non-Coding RNAs: Recent Findings and Challenges IntechOpen

Aubry S, Kelly S, Kümpers BM, Smith-Unna RD, Hibberd JM (2014) Deep evolutionary comparison of gene expression identifies parallel recruitment of trans-factors in two independent origins of C4 photosynthesis. PLoS Genet 10(6):e1004365
Axtell MJ, Meyers BC (2018) Revisiting criteria for plant microRNA annotation in the era of big data. Plant Cell 30(2):272–284
Bai B, Shi B, Hou N, Cao Y, Meng Y, Bian H, Zhu M, Han N (2017) microRNAs participate in gene expression regulation and phytohormone cross-talk in barley embryo during seed development and germination. BMC Plant Biol 17:1–16
Bentwich I, Avniel A, Karov Y, Aharonov R, Gilad S, Barad O, Barzilai A, Einat P, Einav U, Meiri E, Sharon E (2005) Identification of hundreds of conserved and nonconserved human microRNAs. Nat Genet 37(7):766–770
Bhatia G, Singh A, Verma D, Sharma S, Singh K (2020) Genome-wide investigation of regulatory roles of lncRNAs in response to heat and drought stress in Brassica juncea (Indian mustard). Environ Exp Bot 171:103922
Böhmdorfer G, Sethuraman S, Rowley MJ, Krzyszton M, Rothi MH, Bouzit L, Wierzbicki AT (2016) Long non-coding RNA produced by RNA polymerase V determines boundaries of heterochromatin. elife 5:e19092
Borsani O, Zhu J, Verslues PE, Sunkar R, Zhu JK (2005) Endogenous siRNAs derived from a pair of natural cis-antisense transcripts regulate salt tolerance in Arabidopsis. Cell 123(7):1279–1291
Brodersen P, Sakvarelidze-Achard L, Bruun-Rasmussen M, Dunoyer P, Yamamoto YY, Sieburth L, Voinnet O (2008) Widespread translational inhibition by plant miRNAs and siRNAs. Science 320(5880):1185–1190
Budzko L, Hoffa-Sobiech K, Jackowiak P, Figlerowicz M (2023) Engineered deaminases as a key component of DNA and RNA editing tools. Mol Ther Nucleic Acids 34:102062
Buttler CA (2023) Characterizing transposon-derived enhancers in human immunity. Doctoral dissertation. University of Colorado at Boulder
Cagirici HB, Alptekin B, Budak H (2017) RNA sequencing and co-expressed long non-coding RNA in modern and wild wheats. Sci Rep 7(1):10670
Cao H, Wahlestedt C, Kapranov P (2018) Strategies to annotate and characterize long noncoding RNAs: advantages and pitfalls. Trends Genet 34(9):704–721
Cao Z, Zhao T, Wang L, Han J, Chen J, Hao Y, Guan X (2021) The lincRNA XH123 is involved in cotton cold-stress regulation. Plant Mol Biol 106:521–531
Carrera-Castaño G, Calleja-Cabrera J, Pernas M, Gómez L, Oñate-Sánchez L (2020) An updated overview on the regulation of seed germination. Plants 9(6):703
Carroll SB, Grenier JK, Weatherbee SD (2013) From DNA to diversity: molecular genetics and the evolution of animal design. Wiley, Chichester
Cech TR, Steitz JA (2014) The noncoding RNA revolution—trashing old rules to forge new ones. Cell 157(1):77–94
Chauhan P, Verma P, Pandey S, Bhattacharya A, Tripathi A, Giri VP, Singh SP, Mishra A (2021) Endophytic microbial interaction with legume crop for developing resistance against nutrient stress. In: Microbes in land use change management. Elsevier, Amsterdam, pp 363–387
Chen X (2009) Small RNAs and their roles in plant development. Annu Rev Cell Dev Biol 25(1):21–44
Chen X (2004) A microRNA as a translational repressor of APETALA2 in Arabidopsis flower development. Science 303(5666):2022–2025
Chen LL (2020) The expanding regulatory mechanisms and cellular functions of circular RNAs. Nat Rev Mol Cell Biol 21(8):475–490
Chen L, Luan Y, Zhai J (2015) Sp-miR396a-5p acts as a stress-responsive genes regulator by conferring tolerance to abiotic stresses and susceptibility to Phytophthora nicotianae infection in transgenic tobacco. Plant Cell Rep 34:2013–2025
Chen J, Zhong Y, Qi X (2021) LncRNA TCONS_00021861 is functionally associated with drought tolerance in rice (Oryza sativa L.) via competing endogenous RNA regulation. BMC Plant Biol 21:1–12

Cruz de Carvalho MH, Sun HX, Bowler C, Chua NH (2016) Noncoding and coding transcriptome responses of a marine diatom to phosphate fluctuations. New Phytol 210(2):497–510
Darbani B, Noeparvar S, Borg S (2016) Identification of circular RNAs from the parental genes involved in multiple aspects of cellular metabolism in barley. Front Plant Sci 7:776
Datta R, Kapadia C, Danish S, Singh S (eds) (2024) Non-coding RNA in plants: modulation and stress responses. Elsevier, Chantilly
Ding T, Li W, Li F, Ren M, Wang W (2024) microRNAs: key regulators in plant responses to abiotic and biotic stresses via endogenous and cross-kingdom mechanisms. Int J Mol Sci 25(2):1154
Dzinyela R, Alhassan AR, Kiani-Pouya A, Rasouli F, Yang L, Movahedi A (2023) Role of small RNAs in plant stress response and their potential to improve crops. Crop Pasture Sci 74(12):1116–1127
Erdmann RM, Picard CL (2020) RNA-directed DNA methylation. PLoS Genet 16(10):e1009034
Frazer KA, Elnitski L, Church DM, Dubchak I, Hardison RC (2003) Cross-species sequence comparisons: a review of methods and available resources. Genome Res 13(1):1–12
Gasciolli V, Mallory AC, Bartel DP, Vaucheret H (2005) Partially redundant functions of Arabidopsis DICER-like enzymes and a role for DCL4 in producing trans-acting siRNAs. Curr Biol 15(16):1494–1500
Gill SS, Khan NA, Agarwala N, Singh K, Sunkar R, Tuteja N (2024) ncRNAs in plant development and stress responses. Plant Physiol Biochem 214:108950
He F, Tang Y, Sun Y, Wang Y, Zhu JK (2022) Evolutionary innovations in long noncoding RNAs shaping plant environmental adaptation. Nat Commun 13(1):7854
Hossain A, Krupnik TJ, Timsina J, Mahboob MG, Chaki AK, Farooq M, Bhatt R, Fahad S, Hasanuzzaman M (2020) Agricultural land degradation: processes and problems undermining future food security. In: Environment, climate, plant and vegetation growth. Springer, Cham, pp 17–61
Hou J, Lu D, Mason AS, Li B, Xiao M, An S, Fu D (2019) Non-coding RNAs and transposable elements in plant genomes: emergence, regulatory mechanisms and roles in plant development and stress responses. Planta 250:23–40
Huo C, Zhang B, Wang R (2022) Research progress on plant noncoding RNAs in response to low-temperature stress. Plant Signal Behav 17(1):2004035
Ijaz U, Pervaiz T, Ahmed T, Seemab R, Shahid M, Noman M, Nadeem M, Azeem F (2020) Plant Cis-regulatory elements: methods of identification and applications. Asian J Agric Biol 8(02)
Jha UC, Nayyar H, Jha R, Khurshid M, Zhou M, Mantri N, Siddique KH (2020) Long non-coding RNAs: emerging players regulating plant abiotic stress response and adaptation. BMC Plant Biol 20:1–20
Jiang Y, Wu X, Shi M, Yu J, Guo C (2022) The miR159-MYB33-ABI5 module regulates seed germination in Arabidopsis. Physiol Plant 174(2):e13659
Jin H, Vacic V, Girke T, Lonardi S, Zhu JK (2008) Small RNAs and the regulation of cis-natural antisense transcripts in Arabidopsis. BMC Mol Biol 9:1–13
Kang X, Zhao L, Liu X (2023) Calcium signaling and the response to heat shock in crop plants. Int J Mol Sci 25(1):324
Katoh K, Standley DM (2013) MAFFT multiple sequence alignment software version 7: improvements in performance and usability. Mol Biol Evol 30(4):772–780
Lakhotia SC, Mallick B, Roy J (2020) Non-coding RNAs: ever-expanding diversity of types and functions. In: RNA-based regulation in human health and disease. Academic Press, London, pp 5–57
Lam P, Zhao L, Eveleigh N, Yu Y, Chen X, Kunst L (2015) The exosome and trans-acting small interfering RNAs regulate cuticular wax biosynthesis during Arabidopsis inflorescence stem development. Plant Physiol 167(2):323–336
Li S, Castillo-González C, Yu B, Zhang X (2017) The functions of plant small RNA s in development and in stress responses. Plant J 90(4):654–670

Li L, Li Q, Davis KE, Patterson C, Oo S, Liu W, Liu J, Wang G, Fontana JE, Thornburg TE, Pratt IS (2021) Response of root growth and development to nitrogen and potassium deficiency as well as microRNA-mediated mechanism in peanut (Arachis hypogaea L.). Front Plant Sci 12:695234

Li Q, Wang Y, Sun Z, Li H, Liu H (2024) The biosynthesis process of small RNA and its pivotal roles in plant development. Int J Mol Sci 25(14):7680

Liu PP, Montgomery TA, Fahlgren N, Kasschau KD, Nonogaki H, Carrington JC (2007) Repression of AUXIN RESPONSE FACTOR10 by microRNA160 is critical for seed germination and post-germination stages. Plant J 52(1):133–146

Liu X, Zhang H, Zhao Y, Feng Z, Li Q, Yang HQ, Luan S, Li J, He ZH (2013) Auxin controls seed dormancy through stimulation of abscisic acid signaling by inducing ARF-mediated ABI3 activation in Arabidopsis. Proc Natl Acad Sci 110(38):15485–15490

Liu X, Hu Z, Zhou J, Tian C, Tian G, He M, Gao L, Chen L, Li T, Peng H, Zhang W (2020) Interior circular RNA. RNA Biol 17(1):87–97

Lv Y, Liang Z, Ge M, Qi W, Zhang T, Lin F, Peng Z, Zhao H (2016) Genome-wide identification and functional prediction of nitrogen-responsive intergenic and intronic long non-coding RNAs in maize (Zea mays L.). BMC Genomics 17:1–15

Ma Q (2013) Role of nrf2 in oxidative stress and toxicity. Annu Rev Pharmacol Toxicol 53(1):401–426

Mallory AC, Vaucheret H (2006) Functions of microRNAs and related small RNAs in plants. Nat Genet 38(Suppl 6):S31–S36

Mandal M, Poddar N, Kumar S (2022) Identification of novel noncoding RNAs in plants by big data analysis. In: Machine learning and systems biology in genomics and health. Singapore, Springer Nature, pp 123–145

Marquardt S, Raitskin O, Wu Z, Liu F, Sun Q, Dean C (2014) Functional consequences of splicing of the antisense transcript COOLAIR on FLC transcription. Mol Cell 54(1):156–165

Matsui A, Ishida J, Morosawa T, Mochizuki Y, Kaminuma E, Endo TA, Okamoto M, Nambara E, Nakajima M, Kawashima M, Satou M (2008) Arabidopsis transcriptome analysis under drought, cold, high-salinity and ABA treatment conditions using a tiling array. Plant Cell Physiol 49(8):1135–1149

Matthews C, Arshad M, Hannoufa A (2019) Alfalfa response to heat stress is modulated by microRNA156. Physiol Plant 165(4):830–842

Meng X, Li A, Yu B, Li S (2021) Interplay between miRNAs and lncRNAs: mode of action and biological roles in plant development and stress adaptation. Comput Struct Biotechnol J 19:2567–2574

Ni P, Wilson D, Su Z (2022) A map of cis-regulatory modules and constituent transcription factor binding sites in 80% of the mouse genome. BMC Genomics 23(1):714

Nonogaki H (2010) MicroRNA gene regulation cascades during early stages of plant development. Plant Cell Physiol 51(11):1840–1846

Nozawa M, Miura S, Nei M (2010) Origins and evolution of microRNA genes in Drosophila species. Genome Biol Evol 2:180–189

Nozawa M, Miura S, Nei M (2012) Origins and evolution of microRNA genes in plant species. Genome Biol Evol 4(3):230–239

Okuma N, Kawaguchi M (2021) Systemic optimization of legume nodulation: a shoot-derived regulator, miR2111. Front Plant Sci 12:682486

Pareek M, Yogindran S, Mukherjee SK, Rajam MV (2015) Plant MicroRNAs: biogenesis, functions, and applications. In: Plant biology and biotechnology: volume II: plant genomics and biotechnology. Springer, New Delhi, pp 639–661

Peragine A, Yoshikawa M, Wu G, Albrecht HL, Poethig RS (2004) SGS3 and SGS2/SDE1/RDR6 are required for juvenile development and the production of trans-acting siRNAs in Arabidopsis. Genes Dev 18(19):2368–2379

Peter IS, Davidson EH (2011) Evolution of gene regulatory networks controlling body plan development. Cell 144(6):970–985

Piriyapongsa J, Jordan IK (2008) Dual coding of siRNAs and miRNAs by plant transposable elements. RNA 14(5):814–821

Piriyapongsa J, Mariño-Ramirez L, Jordan IK (2007) Origin and evolution of human microRNAs from transposable elements. Genetics 176(2):1323–1337

Ponting CP, Oliver PL, Reik W (2009) Evolution and functions of long noncoding RNAs. Cell 136(4):629–641

Qin T, Zhao H, Cui P, Albesher N, Xiong L (2017) A nucleus-localized long non-coding RNA enhances drought and salt stress tolerance. Plant Physiol 175(3):1321–1336

Rai MI, Alam M, Lightfoot DA, Gurha P, Afzal AJ (2019) Classification and experimental identification of plant long non-coding RNAs. Genomics 111(5):997–1005

Ren J, Jiang C, Zhang H, Shi X, Ai X, Li R, Dong J, Wang J, Zhao X, Yu H (2022) LncRNA-mediated ceRNA networks provide novel potential biomarkers for peanut drought tolerance. Physiol Plant 174(1):e13610

Reyes JL, Chua NH (2007) ABA induction of miR159 controls transcript levels of two MYB factors during Arabidopsis seed germination. Plant J 49(4):592–606

Rhoades MW, Reinhart BJ, Lim LP, Burge CB, Bartel B, Bartel DP (2002) Prediction of plant microRNA targets. Cell 110(4):513–520

Rud D, Marjoram P, Siegmund K, Shibata D (2021) Functional human genes typically exhibit epigenetic conservation. PLoS One 16(9):e0253250

Sakuma Y, Maruyama K, Osakabe Y, Qin F, Seki M, Shinozaki K, Yamaguchi-Shinozaki K (2006) Functional analysis of an Arabidopsis transcription factor, DREB2A, involved in drought-responsive gene expression. Plant Cell 18(5):1292–1309

Schmitz JF, Bornberg-Bauer E (2016) Fact or fiction: updates on how protein-coding genes might emerge de novo from previously non-coding DNA. F1000Research 6:57

Schmitz JF, Bornberg-Bauer E (2017) Fact or fiction: updates on how protein-coding genes might emerge de novo from previously non-coding DNA. F1000Research 6:57

Shamloo-Dashtpagerdi R, Shahriari AG, Tahmasebi A, Vetukuri RR (2023) Potential role of the regulatory miR1119-MYC2 module in wheat (Triticum aestivum L.) drought tolerance. Front Plant Sci 14:1161245

Shi X, Jiang F, Wen J, Wu Z (2019) Overexpression of Solanum habrochaites microRNA319d (sha-miR319d) confers chilling and heat stress tolerance in tomato (S. lycopersicum). BMC Plant Biol 19:1–17

Shuai P, Liang D, Tang S, Zhang Z, Ye CY, Su Y, Xia X, Yin W (2014) Genome-wide identification and functional prediction of novel and drought-responsive lincRNAs in Populus trichocarpa. J Exp Bot 65(17):4975–4983

Siepel A, Bejerano G, Pedersen JS, Hinrichs AS, Hou M, Rosenbloom K, Clawson H, Spieth J, Hillier LW, Richards S, Weinstock GM (2005) Evolutionarily conserved elements in vertebrate, insect, worm, and yeast genomes. Genome Res 15(8):1034–1050

Singh P, Dutta P, Chakrabarty D (2021) miRNAs play critical roles in response to abiotic stress by modulating cross-talk of phytohormone signaling. Plant Cell Rep 40(9):1617–1630

Statello L, Guo CJ, Chen LL, Huarte M (2021) Gene regulation by long non-coding RNAs and its biological functions. Nat Rev Mol Cell Biol 22(2):96–118

Sun X, Wang L, Ding J, Wang Y, Wang J, Zhang X, Che Y, Liu Z, Zhang X, Ye J, Wang J (2016) Integrative analysis of *Arabidopsis thaliana* transcriptomics reveals intuitive splicing mechanism for circular RNA. FEBS Lett 590(20):3510–3516

Sunkar R, Kapoor A, Zhu JK (2006) Posttranscriptional induction of two cu/Zn superoxide dismutase genes in Arabidopsis is mediated by downregulation of miR398 and important for oxidative stress tolerance. Plant Cell 18(8):2051–2065

Sunkar R, Li YF, Jagadeeswaran G (2012) Functions of microRNAs in plant stress responses. Trends Plant Sci 17(4):196–203

Tanzer A, Stadler PF (2004) Molecular evolution of a microRNA cluster. J Mol Biol 339(2):327–335

Tao X, Li M, Zhao T, Feng S, Zhang H, Wang L, Han J, Gao M, Lu K, Chen Q, Zhou B (2021) Neofunctionalization of a polyploidization-activated cotton long intergenic non-coding RNA DAN1 during drought stress regulation. Plant Physiol 186(4):2152–2168

Thornburg TE, Liu J, Li Q, Xue H, Wang G, Li L, Fontana JE, Davis KE, Liu W, Zhang B, Zhang Z (2020) Potassium deficiency significantly affected plant growth and development as well as microRNA-mediated mechanism in wheat (Triticum aestivum L.). Front Plant Sci 11:1219

Van de Peer Y, Mizrachi E, Marchal K (2017) The evolutionary significance of polyploidy. Nat Rev Genet 18(7):411–424

Villar D, Berthelot C, Aldridge S, Rayner TF, Lukk M, Pignatelli M, Park TJ, Deaville R, Erichsen JT, Jasinska AJ, Turner JM (2015) Enhancer evolution across 20 mammalian species. Cell 160(3):554–566

Waititu JK, Zhang C, Liu J, Wang H (2020) Plant non-coding RNAs: origin, biogenesis, mode of action and their roles in abiotic stress. Int J Mol Sci 21(21):8401

Wang J, Meng X, Dobrovolskaya OB, Orlov YL, Chen M (2017) Non-coding RNAs and their roles in stress response in plants. Genomics, Proteomics Bioinformatics 15(5):301–312

Wang P, Dai L, Ai J, Wang Y, Ren F (2019) Identification and functional prediction of cold-related long non-coding RNA (lncRNA) in grapevine. Sci Rep 9(1):6638

Wang NH, Zhou XY, Shi SH, Zhang S, Chen ZH, Ali MA, Ahmed IM, Wang Y, Wu F (2023) An miR156-regulated nucleobase-ascorbate transporter 2 confers cadmium tolerance via enhanced anti-oxidative capacity in barley. J Adv Res 44:23–37

Wasserman WW, Sandelin A (2004) Applied bioinformatics for the identification of regulatory elements. Nat Rev Genet 5(4):276–287

Wierzbicki AT, Blevins T, Swiezewski S (2021) Long noncoding RNAs in plants. Annu Rev Plant Biol 72(1):245–271

Xu K, Xu X, Fukao T, Canlas P, Maghirang-Rodriguez R, Heuer S, Ismail AM, Bailey-Serres J, Ronald PC, Mackill DJ (2006) Sub1A is an ethylene-response-factor-like gene that confers submergence tolerance to rice. Nature 442(7103):705–708

Xu F, Liu Q, Chen L, Kuang J, Walk T, Wang J, Liao H (2013) Genome-wide identification of soybean microRNAs and their targets reveals their organ-specificity and responses to phosphate starvation. BMC Genomics 14:1–30

Yan Q, Wu F, Yan Z, Li J, Ma T, Zhang Y, Zhao Y, Wang Y, Zhang J (2019) Differential co-expression networks of long non-coding RNAs and mRNAs in Cleistogenes songorica under water stress and during recovery. BMC Plant Biol 19:1–19

Yan G, Hua Y, Jin H, Huang Q, Zhou G, Xu Y, He Y, Zhu Z (2023) Sly-miR398 participates in cadmium stress acclimation by regulating antioxidant system and cadmium transport in tomato (Solanum lycopersicum). Int J Mol Sci 24(3):1953

Yang X, Liu C, Niu X, Wang L, Li L, Yuan Q, Pei X (2022) Research on lncRNA related to drought resistance of Shanlan upland rice. BMC Genomics 23(1):336

Yu R, Jih G, Iglesias N, Moazed D (2014) Determinants of heterochromatic siRNA biogenesis and function. Mol Cell 53(2):262–276

Yu Y, Zhang Y, Chen X, Chen Y (2019) Plant noncoding RNAs: hidden players in development and stress responses. Annu Rev Cell Dev Biol 35(1):407–431

Yue E, Rong F, Liu Z, Ruan S, Lu T, Qian H (2023) Cadmium induced a non-coding RNA microRNA535 mediates Cd accumulation in rice. J Environ Sci 130:149–162

Zeng X, Xu Y, Jiang J, Zhang F, Ma L, Wu D, Wang Y, Sun W (2018) Identification of cold stress responsive microRNAs in two winter turnip rape (Brassica rapa L.) by high throughput sequencing. BMC Plant Biol 18:1–13

Zhang Y, Jiang WK, Gao LZ (2011) Evolution of microRNA genes in Oryza sativa and *Arabidopsis thaliana*: an update of the inverted duplication model. PLoS One 6(12):e28073

Zhang C, Liu C, Shi J, Wang Y, Xu C, Ye X, Liu Q, Li X, Qiao W, Yin Y, Cong Y (2022) Molecular mechanism of antibody neutralization of coxsackievirus A16. Nat Commun 13(1):7854

Zhang L, Lin T, Zhu G, Wu B, Zhang C, Zhu H (2023a) LncRNAs exert indispensable roles in orchestrating the interaction among diverse noncoding RNAs and enrich the regulatory network of plant growth and its adaptive environmental stress response. Hortic Res 10(12):uhad234

Zhang H, Liu S, Ren T, Niu M, Liu X, Liu C, Wang H, Yin W, Xia X (2023b) Crucial abiotic stress regulatory network of NF-Y transcription factor in plants. Int J Mol Sci 24(5):4426

Zhao J, Yuan S, Zhou M, Yuan N, Li Z, Hu Q, Bethea FG Jr, Liu H, Li S, Luo H (2019) Transgenic creeping bentgrass overexpressing Osa-miR393a exhibits altered plant development and improved multiple stress tolerance. Plant Biotechnol J 17(1):233–251

Zheng B, Wang Z, Li S, Yu B, Liu JY, Chen X (2009) Intergenic transcription by RNA polymerase II coordinates Pol IV and Pol V in siRNA-directed transcriptional gene silencing in Arabidopsis. Genes Dev 23(24):2850–2860

Chapter 10
Genomic Resources for Studying Stress-Responsive Non-coding Regions

Pranesh Prabhu, Samreen Abbas, Bhartendu Nath Mishra, Manoj Kumar Mishra, and Neha Verma

Abstract Non-coding regions of the genome are critical in gene regulation, particularly in reaction to environmental stress. Advances in genomic technology have made it possible to identify and characterise stress-responsive non-coding elements such as long non-coding RNAs (lncRNAs), microRNAs (miRNAs), and enhancer RNAs (eRNAs), all of which regulate gene expression under stress. This chapter gives details about genetic resources and bioinformatic methods available for studying these regulatory areas, shedding light on their functional significance. Public archives like Ensembl, NONCODE, and RNAcentral provide large datasets for non-coding RNA annotation, while computational tools like RNA fold, Target Scan, and IncRNA SNP aid in predictive analysis and functional validation. Furthermore, multi-omics integration, which includes transcriptomics, epigenomics, and proteomics, helps us comprehend stress adaptation processes. Despite advances, issues like insufficient annotation, tissue-specific expression variability, and data integration problems persist. This chapter also highlights current techniques, such as artificial intelligence-driven ncRNA prediction and CRISPR-based functional validation, which shows promise for furthering stress biology research. Understanding stress-responsive non-coding regions reveals important information about plant and animal stress adaptation, with potential implications in agriculture, medicine, and biotechnology. Researchers can better understand the complex regulatory networks that govern stress resilience by leveraging genomic resources and computational breakthroughs.

Keywords Genomic databases · Machine learning · Non-coding RNA · Omics · Stress

P. Prabhu · S. Abbas · B. N. Mishra
Department of Biotechnology, Institute of Engineering and Technology, Lucknow, Uttar Pradesh, India

M. K. Mishra · N. Verma (✉)
Department of Biotechnology, S.R Institute of Management & Technology, Lucknow, Uttar Pradesh, India

S. Tiwari et al. (eds.), *Decoding Non-Coding RNA*, Concepts and Strategies in Plant Sciences, https://doi.org/10.1007/978-981-96-9859-2_10

10.1 Introduction

In recent years, non-coding regions in plant genome have received a lot of attention, especially in terms of understanding abiotic stress responses. Non-coding regions do not produce proteins but serve critical roles in gene regulation, chromatin organisation, and stress adaptation. These sections include long non-coding RNAs (lncRNAs), microRNAs (miRNAs), small interfering RNAs (siRNAs), enhancer elements, and cis-regulatory elements, all of which help to control gene expression dynamically under stress. Drought, salt, severe temperatures, and nutrient deficits all pose significant challenges to plant survival and crop yield. Plants respond to these challenges via a highly coordinated molecular network in which non-coding DNA and RNA elements play critical regulatory roles. Advances in functional genomics, high-throughput sequencing, and multi-omics integration have shed light on the role of non-coding regions in stress responses (Lécuyer et al. 2024).

Genomic databases are very useful for researchers studying the function of non-coding RNA regions responsible for plant stress responses. These resources have detailed annotations of sequences, regulatory elements, and functional data, allowing for greater insight into the molecular mechanisms driving stress resilience. Researchers use publicly available resources, bioinformatics tools, and computer pipelines to analyse and interpret these plant genomic regions. Despite these advances, a significant hurdle remains, that is, the availability and integration of genomic resources for researching stress-responsive non-coding areas. This chapter provides an overview of the genomic resources available for studying stress-responsive non-coding regions, such as public repositories and bioinformatics tools for analysing plant non-coding elements, as well as the integration of multi-omics data to investigate regulatory mechanisms in stress adaptation. It also discusses challenges and future perspectives. Using these genomic resources, researchers can get better insights into plant stress responses and design novel crop improvement strategies, ultimately increasing plant resilience to changing environmental conditions (Roychowdhury et al. 2023).

The rapid improvements in bioinformatics and multi-omics technology have transformed plant stress adaptation research, allowing for a better knowledge of non-coding areas and their regulatory functions. High-throughput sequencing methods, such as RNA-seq and short RNA sequencing, have aided in the identification of stress-responsive non-coding RNAs (ncRNAs), which include microRNAs (miRNAs) and long non-coding RNAs. Chromatin immune-precipitation sequencing (ChIP-seq) and the assay for transposase-accessible chromatin sequencing (ATAC-seq) have shed light on the epigenetic control of stress-related genes and cis-regulatory regions that influence transcriptional responses under stress circumstances (Gupta et al. 2023). Furthermore, machine learning (ML) and artificial intelligence (AI) techniques have been included into bioinformatics pipelines to anticipate functional non-coding components, hence enhancing annotation accuracy and functional characterisation. Multi-omics integration, which combines genomes, transcriptomics, epigenomics, and metabolomics, enables the identification of essential stress-responsive regulatory networks, hence improving the ability to identify critical non-coding elements involved in stress adaption. The combination of these computational tools with high-throughput datasets has significantly increased the ability to analyse, interpret, and exploit non-coding areas for crop improvement, opening the way for the production of stress-tolerant plant types (Vincent et al. 2022).

10.2 Public Repositories and Bioinformatic Tools

10.2.1 Genomic Databases for Non-coding Regions

Plant stress-responsive non-coding regions must be identified, classified, and functionally analysed using strong genomic databases that include high-quality sequencing data, transcriptional regulators, and functional annotations. These databases are crucial for researchers looking to understand the molecular networks that control abiotic stress adaption. The most important genomic repositories are described in detail below, illustrated in Fig. 10.1, and their features are summarised in Table 10.1.

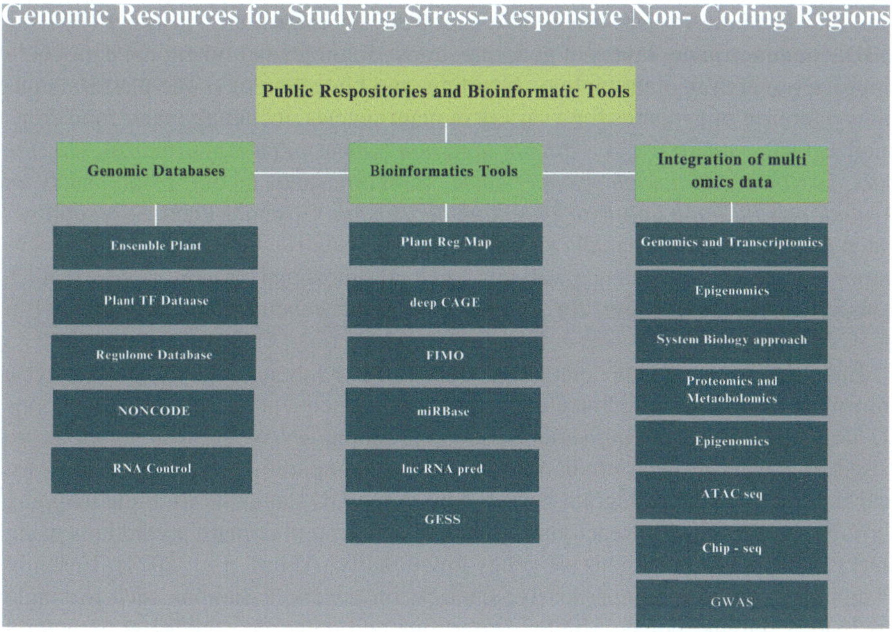

Fig. 10.1 List of genomic resources for studying stress-responsive non-coding regions

Table 10.1 Genomic database highlighting its features and applications with website links

Database	Key features	Application in stress research	Website
Ensembl Plants	Genome annotations, cis-regulatory elements, cross-species comparisons	Identifying stress-responsive regulatory sequences and conserved ncRNAs	https://plants.ensembl.org/
Plant TFDB	Transcription factor classification, TF-target interactions, regulatory networks	Studying TF-mediated gene regulation under abiotic stress	http://planttfdb.gao-lab.org/
Regulome DB	Enhancer regions, transcription factor binding sites, histone modification marks	Mapping stress-responsive enhancers and promoter elements	https://regulomedb.org/
NONCODE	lncRNA, miRNA, siRNA annotations, expression profiles, functional predictions	Investigating stress-responsive lncRNAs and miRNAs	http://www.noncode.org/
RNAcentral	Aggregated ncRNA repository, functional annotations, sequence search tools	Exploring stress-responsive ncRNA interactions and regulatory pathways	https://rnacentral.org/

10.2.1.1 Ensembl Plants: A Comprehensive Genome Annotation Platform

Ensembl Plants is a freely available genomic database that provides high-quality genome annotations, comparative genomics insights, and regulatory element data for a diverse range of plant species. Ensembl Plants, a component of the wider Ensembl Genome Project run by the European Bioinformatics Institute (EMBL-EBI), combines many layers of genomic, transcriptomic, and functional datasets to improve research in plant biology, breeding, and biotechnology. The platform contains reference of genomes for a variety of plant species, including major food crops such as *Oryza sativa* (rice), *Triticum aestivum* (wheat), *Zea mays* (maize), *Glycine max* (soybean), *Solanum lycopersicum* (tomato), and model plants such as *Arabidopsis thaliana* and *Brachypodium distachyon*. Ensembl Plants is a sophisticated genome browser that allows researchers to visualise gene architectures, regulatory components, sequence variants, and chromosomal organisation, giving a comprehensive foundation for researching plant genetics and evolution (Liu et al. 2022a).

Ensembl Plants genome annotation workflow is a key feature that offers precise information about gene architecture such as coding sequences, untranslated regions (UTRs), alternative splicing variants, and non-coding RNAs. These annotations are based on advanced gene prediction algorithms, comparative sequencing analysis, and experimental data integration. Gene Ontology (GO) annotations, route mapping through the KEGG and Reactome databases, and protein domain predictions using InterPro all help to characterise genes functionally (Abbas et al. 2021). Ensembl Plants also provides comprehensive datasets on genetic variations such as single nucleotide polymorphisms (SNPs), insertions and deletions (INDELs), structural variations, and copy number variations (CNVs), which are required for genome-wide association studies (GWAS) and marker-assisted breeding programs. This enables researchers to investigate the genetic variety of plant species and find significant loci linked to agronomically essential features like drought tolerance, disease resistance, and yield improvement (Wang et al. 2023a).

Beyond genome annotation, Ensembl Plants contributes significantly to comparative genomics by offering ortholog and paralog predictions, synteny analysis, and whole-genome alignments. These characteristics aid evolutionary research by allowing researchers to track gene duplications, conservation, and divergence among plant species. The platform uses several sequence alignment algorithms to identify conserved gene clusters and transcriptional regulatory networks, offering information on the mechanisms that drive plant adaptability and domestication (Simopoulos et al. 2018). Ensembl Plants uses phylogenetic data to rebuild evolutionary relationships between plant lineages, providing vital insights into the functional development of major gene families such as transcription factors, resistance genes, and metabolic enzymes.

Ensembl Plants also makes major contributions to regulatory genomics by annotating promoters, enhancers, transcription factor binding sites, and chromatin accessibility regions. These regulatory components are essential for understanding gene expression patterns in various environmental situations and stress responses. The platform uses datasets from techniques like chromatin immunoprecipitation sequencing (ChIP-seq) and assay for transposase-accessible chromatin sequencing (ATAC-seq) to find cis-regulatory regions that affect gene expression. This information is especially useful for synthetic biology and genetic engineering applications, where precise gene regulation is required to create stress-resistant crops (Wang et al. 2023b). Furthermore, Ensembl Plants provides access to transcriptome information, such as RNA sequencing (RNA-seq) data, allowing researchers to compare gene expression levels across tissues, developmental stages, and experimental treatments. These datasets help to further functional genomics research by highlighting differentially expressed genes implicated in essential physiological processes like photosynthesis, secondary metabolism, and stress adaption.

One of Ensembl Plants' most powerful uses is agricultural genomics and crop enhancement. The platform enables precision breeding strategies to generate high-yielding and climate-resilient crop varieties by combining multi-omics data such as genomes, transcriptomics, epigenomics, and proteomics. Genome editing tools, such as CRISPR-Cas9, rely on correct genome annotations to target specific genes for change, and Ensembl Plants offers the reference data required to properly build guide RNA (Zhu et al. 2016). Furthermore, by linking genetic differences to phenotypic features, the database facilitates marker-assisted selection (MAS), allowing breeders to speed up the development of new cultivars with improved agronomic traits. The availability of high-quality reference genomes also allows for pan-genome investigations, in which researchers compare numerous accessions of a species to discover core and variable genes, revealing genetic elements behind phenotypic plasticity and adaptation to different environments (Simopoulos et al. 2019).

Ensembl Plants is more than simply a database; it is a full analytical platform that provides plant researchers with a variety of bioinformatics tools and resources. The Ensembl Genome Browser allows users to study gene structures, search for sequence similarity using BLAST/BLAT tools, obtain genomic variations using the Variant Effect Predictor (VEP), and download massive datasets for offline analysis. The platform also provides APIs and programmatic access, enabling integration with custom workflows for large-scale genomic studies. Researchers in plant breeding, evolutionary biology, and functional genomics can use these techniques to propose novel ideas and evaluate results through experimental methods.

10.2.1.2 Plant Transcription Factor Database (Plant TFDB): A Repository for Transcription Factors and Regulatory Elements

The Plant Transcription Factor Database (Plant TFDB) is a specialised bioinformatics resource that provides detailed information about transcription factors (TFs) and transcriptional regulators in plants. Because transcription factors play an important role in gene regulation by controlling the expression of target genes in response to environmental stimuli, developmental cues, and physiological conditions, Plant TFDB is an invaluable resource for understanding the complex regulatory networks that govern plant growth, development, stress adaptation, and secondary metabolism (Jin et al. 2017). Plant TFDB, which is hosted and managed by Peking University's Centre for Bioinformatics, has been updated several times to assure the inclusion of high-quality genomic and functional annotations for numerous plant species. The database includes curated information on transcription factor families, DNA-binding motifs, gene expression patterns, protein–protein interactions, and functional genomics data, making it an invaluable resource for both basic and applied plant research.

One of the key goals of Plant TFDB is to systematically identify and classify transcription factors in plants using domain architecture and sequence homology. Transcription factors are classified into several families, including MYB, bHLH, AP2/ERF, WRKY, NAC, and MADS-box, each of which has a different role in plant growth and stress response. The classification follows standard guidelines established by the Plant Transcription Factor Database Initiative (PlnTFDB) and other reputable sources (Zhang et al. 2018). Each identified TF is connected to reference genomes from well-annotated species, such as *Arabidopsis thaliana*, crops such as *Oryza sativa* (rice) and *Triticum aestivum* (wheat), and commercially important plants like *Solanum lycopersicum* (tomato) and *Glycine max* (soybean). Plant TFDB provides genome-wide transcription factor annotations across several species, allowing researchers to investigate the evolution, conservation, and diversification of regulatory elements in plants.

Plant TFDB's integration of DNA-binding motif analysis and promoter prediction techniques enables researchers to find cis-regulatory regions linked to transcription factor binding sites. These characteristics are especially relevant for understanding the regulatory systems that underpin plant responses to abiotic stresses (such as drought, salinity, and temperature changes) and biotic interactions (such as pathogen defence and symbiotic relationships) (Chen et al. 2015) The database contains experimentally validated TF–target interactions, demonstrating gene regulatory networks involved in important biological processes. Furthermore, the co-expression analysis module assists in discovering functionally similar transcription factors that may collaborate to regulate complicated signalling networks. This information is backed by ChIP-seq datasets, DAP-seq data, and RNA-seq expression profiles, allowing researchers to investigate tissue-specific expression and dynamic regulatory connections under a variety of environmental settings.

One of Plant TFDB's key features is its user-friendly interface and sophisticated bioinformatics tools, which allow researchers to easily search, browse, and retrieve transcription factor-related information. Users can search for TFs by species, gene ID, family categorisation, sequence similarity, or functional annotations. The Gene Ontology (GO) enrichment analysis feature sheds light on the biological roles of transcription factors, connecting them to pathways such as hormone signalling (auxins, gibberellins, abscisic acid), secondary metabolite production, and stress responses (Li et al. 2024).

Plant TFDB provides a foundation for advanced gene regulation research in the age of functional genomics and systems biology, enabling genome editing technologies like CRISPR-Cas9 and synthetic biology techniques. By combining regulatory data with transcriptome, proteome, and epigenome databases, researchers can develop precision breeding strategies and create stress-resistant, high-yield crop types. The database has been extensively used in research on transcriptional control of blooming time, root architecture alterations, drought tolerance, and plant–microbe interactions, indicating its broad usefulness across various fields of plant science (Baillo et al. 2019). As the research field of plant functional genomics advances, future upgrades to Plant TFDB are projected to contain single-cell transcriptomics, spatial transcriptomics, and long-read sequencing data, providing even more thorough insights into spatio-temporal gene regulation. In addition, using machine learning and artificial intelligence (AI)-based predictions for TF binding motifs and gene regulatory networks will improve Plant TFDB's accuracy and predictive capabilities. These developments will strengthen its position as a premier resource for plant gene control research, allowing for discoveries that will have an influence on crop genetics, biotechnology, and sustainable agriculture.

10.2.1.3 Regulome DB: A Database for Functional Regulatory Sequences

Regulome DB, which was originally established to annotate functional regulatory elements in the human genome, is now being investigated for its potential use in investigating non-coding regulatory regions in plant genomes. Stress-responsive non-coding genes such as long non-coding RNAs (lncRNAs), microRNAs (miRNAs), and cis-regulatory elements (CREs) must be identified and functionally characterised in order to understand plant responses to abiotic and biotic stresses. These non-coding elements regulate gene expression via interacting with transcription factors, chromatin modifiers, and signalling pathways. Regulome DB provides a framework for integrating multiple omics datasets, such as chromatin accessibility

(ATAC-seq), transcription factor binding (ChIP-seq), and DNA methylation profiles, to discover regulatory regions implicated in plant stress adaption.

One of Regulome DB's most important contributions to plant stress biology is its capacity to detect and rate genomic variations based on their expected regulatory influence. Regulome DB uses a scoring system similar to that used in human genetics to award regulatory scores to non-coding variations based on their relationship with transcription factor binding sites, histone modifications, and enhancer activity (Boyle et al. 2012). In plants, this approach can be used to prioritise single nucleotide polymorphisms (SNPs) in non-coding areas that contribute to stress tolerance. Drought-responsive SNPs in the promoter regions of essential genes involved in abscisic acid (ABA) signalling, for example, can be prioritised according to their likelihood of affecting transcription factor binding.

In addition to SNP prioritisation, Regulome DB allows for the investigation of epigenetic alterations and chromatin accessibility in stress-responsive non-coding genes. DNA methylation, histone changes, and chromatin remodelling all play important roles in controlling plant stress responses by inhibiting or activating stress-inducible genes. Regulome DB, by combining methylation-sensitive sequencing datasets with chromatin accessibility profiles (ATAC-seq), can aid in the identification of epigenetic markers associated with transcriptionally active or repressed regulatory regions in stress-exposed plants. This is especially valuable for studying long-term stress memory in plants, where specific regulatory elements undergo stable epigenetic changes, allowing for faster adaptation to repeated stress events.

As plant functional genomics advances, Regulome DB's future applications in stress biology are likely to include machine learning-based predictions of non-coding regulatory elements, integration with single-cell transcriptomic data, and improved annotation of lncRNA and miRNA target sites. The capacity to investigate long-distance chromatin connections using techniques such as Hi-C and Capture-C will improve Regulome DB's predictive potential in detecting stress-responsive enhancer–promoter relationships. Furthermore, as more plant genomes are fully annotated, comparative regulatory genomics using Regulome DB will provide deeper insights into conserved and species-specific regulatory networks that drive stress adaption. Finally, by improving our understanding of plant non-coding gene regulation, RegulomeDB will help to develop crop improvement techniques for resilience to climate change, drought, salinity, and disease attacks.

10.2.1.4 NONCODE: A Comprehensive Database for Non-coding RNAs

NONCODE is a comprehensive database dedicated to the annotation and functional characterisation of non-coding RNAs (ncRNAs), including long non-coding RNAs (lncRNAs), microRNAs (miRNAs), circular RNAs (circRNAs), and small nucleolar RNAs (snoRNAs). Because ncRNAs play an important role in post-transcriptional regulation, chromatin remodelling, and stress adaption in plants, NONCODE is an invaluable resource for studying their functional mechanisms in response to environmental stress. Stress-responsive ncRNAs in plants affect gene expression by epigenetic changes, RNA interference, and transcription factor interaction, influencing a variety of stress response pathways including drought resistance, salinity tolerance, heat stress, and pathogen defence. The incorporation of high-throughput sequencing information into NONCODE enables researchers to detect differentially expressed ncRNAs under diverse stress situations, resulting in a better understanding of their regulatory activities. By connecting ncRNA expression profiles to transcriptional networks, NONCODE gives insights into how non-coding components contribute to plant resilience and response to environmental changes (Petrov et al. 2017).

One of the most important uses of NONCODE in stress biology is the identification and characterisation of lncRNAs implicated in plant stress responses. lncRNAs function as transcriptional regulators, molecular scaffolds, or decoys, modulating stress-related signalling pathways through interactions with transcription factors, chromatin modifiers, and RNA-binding proteins. Several drought-induced lncRNAs have been found in rice (*Oryza sativa*), regulating root architecture and water uptake efficiency via interactions with auxin-responsive genes. Similarly, in *Arabidopsis thaliana*, lncRNAs such as COOLAIR and APOLO have been found to mediate temperature-dependent flowering time regulation, showing how lncRNAs integrate stress signals with developmental processes. NONCODE facilitates the functional annotation of these lncRNAs by integrating RNA sequencing (RNA-seq), ribosome profiling, and chromatin immunoprecipitation sequencing (ChIP-seq) data, providing researchers with a comprehensive view of lncRNA-mediated stress regulatory networks (Zhao et al. 2021).

Beyond lncRNAs, NONCODE is an important resource for investigating stress-responsive miRNAs, which play a critical role in post-transcriptional gene silencing. Plant miRNAs control stress adaption by targeting mRNAs that encode transcription factors, transporters, and stress-protective proteins, thereby fine-tuning gene expression in adverse conditions. For example, miR398 and miR397 are known to modulate oxidative stress responses in Arabidopsis by regulating the expression of superoxide dismutase (SOD) and laccase genes, respectively. Similarly, miR528 in rice has been linked to salinity and drought stress tolerance by regulating antioxidant pathways (Fang et al. 2014). NONCODE offers genome-wide miRNA annotation, expression profiling, and functional predictions, allowing researchers to discover new miRNAs associated with various stress conditions. NONCODE advances our understanding of ncRNA-mediated stress regulatory circuits in plants by combining small RNA sequencing datasets, making it easier to uncover stress-responsive miRNA–target relationships (Xie et al. 2014).

NONCODE's ability to combine multi-omics data, including as transcriptomics, epigenomics, and metabolomics, to develop predictive models of ncRNA activities is critical for the future of plant stress biology research. As emerging technologies like single-cell RNA sequencing (scRNA-seq) and long-read sequencing advance, NONCODE will become increasingly important in understanding the dynamic regulation of ncRNAs under stress. Furthermore, combining deep learning and AI-based methodologies will improve the database's prediction capacities, allowing for more accurate identification of ncRNA-driven regulatory networks in plants.

This will help crop development initiatives improve stress resilience while also providing new targets for genetic engineering and synthetic biology applications. Finally, by increasing our understanding of stress-responsive ncRNAs, NONCODE will make a significant contribution to sustainable agriculture and food security in the face of climate change (Xiyuan et al. 2017).

10.2.1.5 RNAcentral: A Centralised Platform for Non-coding RNA Sequences

RNAcentral is a comprehensive, freely accessible database that integrates and unifies non-coding RNA (ncRNA) sequences from a number of sources, including primary databases such as Ensembl, RefSeq, and Rfam (The RNAcentral Consortium 2021). It acts as a primary repository for a wide range of ncRNA types, including long non-coding RNAs (lncRNAs), microRNAs (miRNAs), transfer RNAs (tRNAs), small nucleolar RNAs (snoRNAs), and circular RNAs (circ RNAs), all of which play important roles in gene regulation. Stress-responsive plant ncRNAs are especially interesting because they help plants adapt to a variety of environmental difficulties such as drought, salinity, severe temperatures, and pathogen infections. One of RNAcentral's significant benefits in the investigation of stress-responsive lncRNAs is the integration of transcriptomic and genomic information, which makes it easier to identify differentially expressed lncRNAs (DE-lncRNAs) under stress circumstances. IncRNAs have been shown to modify chromatin states, operate as protein interaction scaffolds, and act as molecular decoys to regulate transcription factors, all of which influence stress responses on numerous levels (Sweeney et al. 2021).

For example, in Arabidopsis thaliana, the lncRNA COLDAIR regulates vernalisation-mediated blooming by altering the chromatin state of FLC, a major flowering repressor, under cold stress conditions. Similarly, in Oryza sativa, the lncRNA Drought-Induced Long Noncoding RNA1 (DRL1) has been shown to influence drought stress responses by modulating ABA signalling. By allowing researchers to investigate cross-species conservation and expression patterns of lncRNAs, RNAcentral aids in the discovery of novel regulatory mechanisms used by plants to cope with abiotic stresses (Allmer 2022).

In addition to lncRNAs, miRNAs constitute a crucial family of ncRNAs accessible in RNAcentral that play important roles in plant stress response. miRNAs work by post-transcriptionally controlling gene expression by mRNA degradation or translational inhibition, allowing plants to quickly adapt their cellular metabolism in response to environmental stress. Several miRNAs, including miR398, miR396, and miR159, have been widely investigated for their roles in oxidative stress tolerance, drought responses, and pathogen defence. For example, miR398 controls superoxide dismutase (SOD) enzymes, which aid plants in managing oxidative stress during drought and salinity stress. Similarly, miR159 has been linked to the regulation of MYB transcription factors involved in stress-induced leaf senescence. RNAcentral offers a centralised platform for accessing functional annotations, sequence alignments, and expression profiles of miRNAs, facilitating in-depth analysis of stress-induced miRNA–mRNA regulatory networks across various plant species (Jannesar et al. 2020).

As plant stress biology research progresses, the integration of RNAcentral with functional genomics, epigenomics, and multi-omics datasets will be critical in expanding our understanding of ncRNA-mediated gene regulation. RNAcentral's future upgrades are intended to incorporate single-cell RNA sequencing (scRNA-seq) datasets, improved annotation of plant-specific ncRNAs, and advanced bioinformatics techniques for predicting ncRNA activities (Bhatia et al. 2017). Furthermore, new technologies like long-read sequencing and deep-learning-based functional predictions will improve our ability to characterise previously unannotated ncRNAs involved in stress reactions. The significance of RNAcentral in integrating ncRNA data to plant stress adaptation mechanisms may provide useful insights for producing climate-resilient crops via genetic engineering, precision breeding, and synthetic biology approaches (Song et al. 2019). RNAcentral is poised to become a valuable resource for researchers seeking to understand plant stress-responsive non-coding gene regulatory networks by constantly increasing its data integration and analysis capabilities.

10.2.2 Bioinformatics Tools for Non-coding Region Analysis

Advanced bioinformatics tools are required for the functional characterisation of stress-responsive non-coding areas in plants, including sequence analysis, motif finding, regulatory element mapping, and non-coding RNA prediction. These methods are critical in identifying transcription factor binding sites (TFBS), predicting enhancer elements, analysing long non-coding RNAs (lncRNAs), and integrating multi-omics datasets. Some of the most widely used bioinformatics tools for analysing non-coding regions in plant stress studies are listed below and summarised in Table 10.2.

10.2.2.1 Plant Reg Map: Regulatory Element Prediction for Plants

PlantRegMap (Plant Regulatory Map) is a comprehensive bioinformatics resource that integrates transcription factors (TFs), regulatory elements, and gene co-expression networks to help researchers explore gene regulation in plants (Tian et al. 2020). This database sheds light on plant transcriptional regulatory interactions, particularly those involving non-coding RNAs (ncRNAs) such as long non-coding RNAs (lncRNAs) and microRNAs (miRNAs), which play critical roles in stress adaption. Plant life is dependent on the precise regulation of gene expression mediated by ncRNAs and their associated transcription factors in the face of diverse abiotic conditions (e.g., drought, salt, temperature changes) and biotic challenges (e.g., pathogen infections). PlantRegMap allows researchers to investigate

Table 10.2 Bioinformatics tools highlighting its function and applications with website links

Tool name	Function	Application in stress research	Website
PlantRegMap	Regulatory element prediction for transcription factor binding sites (TFBS)	Identifies cis-regulatory elements related to drought and salt stress	http://plantregmap.gao-lab.org/
deepCAGE	Enhancer identification using deep learning on CAGE data	Maps stress-responsive enhancer elements controlling gene expression	https://deepcage.org/
FIMO (MEME Suite)	Motif discovery and scanning in non-coding DNA	Discovers novel transcription factor motifs regulating stress responses	http://meme-suite.org/tools/fimo
miRBase	MicroRNA (miRNA) annotation and target prediction	Detects miRNAs involved in plant responses to abiotic stress conditions	https://www.mirbase.org/
LncRNApred	Prediction and classification of plant long non-coding RNAs (lncRNAs)	Classifies stress-responsive lncRNAs and their regulatory interactions	https://lncrnapred.org/
GESS	Genome-wide small RNA (sRNA) target analysis	Analyses sRNA-mediated gene regulation in plant stress adaptation	https://gess.org/

stress-responsive ncRNA regulatory networks by integrating multi-omics data like as ChIP-seq (chromatin immunoprecipitation sequencing), ATAC-seq (chromatin accessibility data), and RNA-seq (transcriptomic data). This capability is especially useful for investigating regulatory mechanisms in important crops such as rice (*Oryza sativa*), wheat (*Triticum aestivum*), maize (*Zea mays*), and Arabidopsis (*Arabidopsis thaliana*), where ncRNA-mediated regulation is critical to plant resilience to environmental challenges (Tian et al. 2020).

A major strength of PlantRegMap is its large-scale transcription factor (TF) binding site prediction capability, which allows researchers to identify cis-regulatory elements (CREs) and transcription factor binding motifs in promoters of stress-responsive genes. Since non-coding RNAs interact with transcription factors to regulate gene expression, this feature helps in uncovering TF–ncRNA interactions that contribute to plant stress responses. For example, lncRNAs can function as molecular sponges, sequestering stress-related transcription factors and altering their availability to downstream target genes.

PlantRegMap is important for understanding stress-responsive miRNA regulatory networks in addition to lncRNAs. MicroRNAs are important post-transcriptional regulators that control gene expression by targeting messenger RNAs (mRNAs) for destruction or translational repression in response to environmental stress. Several plant miRNAs have been widely researched for their roles in stress adaption, including miR398, which regulates superoxide dismutase (SOD) genes to manage oxidative stress in plants, and miR393, which modifies auxin signalling pathways during abiotic stress conditions. PlantRegMap uses co-expression networks and promoter motif analysis to give a functional annotation of miRNA-mediated stress response pathways, assisting researchers in discovering novel miRNAs involved in biotic and abiotic stress adaption (Kulkarni and Vandepoele 2020).

As functional genomics improves, PlantRegMap is likely to become increasingly important in integrating single-cell regulatory data, epigenetic changes, and 3D genome architectural information. Future improvements are likely to include improved machine learning models for predicting transcription factor binding sites, which will allow for more accurate detection of stress-induced regulatory interactions. Furthermore, comparative regulatory genomics using PlantRegMap will shed light on the evolutionary conservation of ncRNA-mediated stress responses across plant species, aiding crop improvement programs aimed at increasing resilience to climate change and environmental stress. PlantRegMap, which provides a high-resolution map of plant ncRNA regulatory networks, is poised to become a valuable resource for understanding and developing stress-resistant crops through gene-editing and synthetic biology approaches. This will not only promote fundamental plant science studies, but also have real-world applications in sustainable agriculture and food security (Kulkarni and Vandepoele 2020).

10.2.2.2 deepCAGE: Enhancer Identification in Non-coding Regions

deepCAGE (Cap Analysis Gene Expression) is a high-throughput sequencing approach that maps transcription start sites (TSSs) and analyses promoter architecture at the single-nucleotide level. This method is critical for finding stress-responsive plant non-coding genes, such as long non-coding RNAs (lncRNAs), microRNAs (miRNAs), and enhancer RNAs (eRNAs), which modulate gene expression under abiotic and biotic stress training. Researchers can learn about the regulation of non-coding RNAs (ncRNAs) and their function in stress adaption by combining deepCAGE-derived TSS data with transcriptome and epigenomic datasets (Liu et al. 2022b).

DeepCAGE's principal application in plant stress biology is to identify promoter areas and alternative TSS usage in non-coding RNAs. Many plant lncRNAs and miRNAs exhibit stress-induced transcriptional activation, and deepCAGE allows for precise identification of their transcription start sites (TSSs) under various stress conditions. Similarly, deepCAGE research in rice revealed that heat stress produces unique transcription start sites in stress-sensitive miRNA precursors, demonstrating that ncRNA transcription is extremely flexible and responsive to environmental signals. DeepCAGE sheds light on how non-coding RNAs regulate gene expression in response to environmental changes by detecting stress-dependent changes in promoter activity (Roy et al. 2015).

In addition to finding novel stress-induced ncRNAs, deepCAGE is useful for characterising enhancer RNAs (eRNAs) and cis-regulatory regions that aid in stress adaption. Enhancer RNAs, a type of transcriptionally active ncRNA, are crucial for attracting transcription factors and chromatin modifiers to activate stress-related genes. Similarly, in maize, deepCAGE data was utilised to identify heat-responsive enhancers that regulate stress-protective pathways, indicating how non-coding components organise plant resistance in harsh environmental conditions. Integrating deepCAGE data with ChIP-seq and ATAC-seq datasets allows researchers to find stress-specific enhancer–promoter connections, resulting in a functional roadmap of ncRNA-mediated stress. As plant functional genomics advances, deepCAGE is predicted to be a critical tool for integrating single-cell transcriptional profiling, epigenetic changes, and 3D chromatin interactions.

10.2.2.3 FIMO (MEME Suite): Motif Discovery in Non-coding DNA

FIMO (Find Individual Motif Occurrences) is a powerful bioinformatics tool that allows researchers to scan DNA sequences for matches to known motif matrices, revealing important information about cis-regulatory elements (CREs) and transcription factor (TF) binding sites in non-coding regions of the genome. FIMO is very beneficial in discovering functional promoter motifs, enhancers, and regulatory regions that mediate gene expression under abiotic and biotic stress conditions. Plants respond to environmental stresses including drought, salt, temperature variations, and pathogen attacks by activating or repressing stress-responsive genes, a process that relies significantly on transcription factors binding to non-coding DNA regions.

One of the most important uses of FIMO is the identification of stress-responsive cis-regulatory elements (CREs) in non-coding areas, which serve as binding sites for TFs that regulate gene expression during stress (Buske et al. 2010). For example, studies in rice (*Oryza sativa*) have utilised FIMO to look for DREB (Dehydration-Responsive Element Binding) motifs in the promoters of drought-induced lncRNAs, uncovering novel regulatory components that aid in drought adaptation.

In the future, FIMO is expected to be an essential tool for integrating machine learning-based motif discovery, single-cell transcriptomics, and 3D chromatin architecture data, furthering our understanding of non-coding gene regulation in plant stress responses (Franco-Zorrilla et al. 2014). Future advances are likely to improve FIMO's ability to detect tissue- and condition-specific TF binding sites, allowing for more precise functional annotations of stress-responsive non-coding RNAs.

10.2.2.4 miRBase: MicroRNA Prediction and Functional Annotation

miRBase is a public database that serves as the principal repository for microRNA (miRNA) sequences, annotations, and functional information from a variety of species, including plants. MiRNAs play an important role in post-transcriptional gene regulation; therefore they are engaged in plant responses to a variety of abiotic and biotic stressors, including drought, salinity, heat, cold, oxidative stress, and pathogen attack. miRBase includes genome-wide miRNA annotations, sequence alignments, projected target genes, and expression profiles, making it an invaluable resource for researching stress-responsive plant miRNAs. miRBase allows researchers to find, classify, and functionally annotate new miRNAs that govern plant stress adaption by combining data from next-generation sequencing (NGS) studies, RNA-seq datasets, and computational predictions (Wang et al. 2019). One of the most important applications of miRBase in plant stress biology is the identification of stress-inducible miRNAs and their target genes, which mediate vital physiological responses to environmental stresses.

Researchers may use miRBase to obtain, compare, and analyse the expression levels of stress-responsive miRNAs in different plant species, allowing them to identify evolutionarily conserved and species-specific regulatory mechanisms (Wang et al. 2019). This comparative method is especially useful when researching major crop species such as rice (*Oryza sativa*), wheat (*Triticum aestivum*), maize (*Zea mays*), and soybean (*Glycine max*), where miRNA-based regulation is critical for yield stability and environmental adaptation (Tripathi et al. 2015).

Beyond miRNA identification, miRBase allows for the investigation of miRNA–mRNA interaction networks, which provide mechanistic insights into how miRNAs regulate gene expression under stress situations. Many abiotic stress-responsive miRNAs function as negative regulators, suppressing target mRNAs encoding transcription factors (TFs), stress-related proteins, and hormone signalling components. In wheat, for example, miR156 regulates SPL (SQUAMOSA promoter-binding-like) transcription factors that are important in drought and heat stress adaptation (Ravichandran et al. 2019). Similarly, miR172 regulates flowering time in maize by targeting AP2-like transcription factors in response to environmental stimuli. Researchers can use miRBase in conjunction with functional genomics methodologies like degradome sequencing, CLIP-seq, and gene co-expression networks to experimentally confirm miRNA–target interactions and find novel regulatory mechanisms for stress resilience.

miRBase's importance in stress-responsive plant miRNA research is projected to grow in the future as single-cell transcriptomics, long-read sequencing technologies, and artificial intelligence-based miRNA discovery methodologies progress. Future upgrades are projected to improve miRBase's skills in predicting miRNA function, finding miRNA isoforms (isomiRs), and combining multi-omics data for a more comprehensive knowledge of plant stress adaption (Kozomara et al. 2019).

10.2.2.5 LncRNApred: Prediction of Long Non-coding RNAs (lncRNAs)

LncRNApred is a sophisticated computational method for identifying and classifying long non-coding RNAs (lncRNAs), a critical class of non-coding RNAs that govern plant responses to environmental stressors. Unlike protein-coding RNAs, lncRNAs use a range of processes, including chromatin remodelling, transcriptional control, and post-transcriptional alterations. LncRNApred uses machine learning techniques, sequence feature analysis, and evolutionary conservation data to identify lncRNAs from coding RNAs, allowing for large-scale identification of lncRNAs across plant species. Given the growing evidence that lncRNAs play a role in plant responses to abiotic and biotic stresses such as drought, salinity, heat, cold, and pathogen infection, LncRNApred is crucial for understanding the regulatory networks that allow plants to adapt to stress (Pian et al. 2016).

LncRNApred, which analyses expression patterns, epigenetic changes, and sequence motifs, provides a complete platform for investigating the functional functions of lncRNAs in stress adaption (Liu et al. 2019). In addition to finding stress-responsive lncRNAs, LncRNApred is an effective method for predicting functional relationships between lncRNAs and other regulatory molecules, such as microRNAs (miRNAs) and transcription factors (TFs) (Liu et al. 2015).

10.2.2.6 GESS: Genome-Wide Analysis of Small RNA Target

The Global Energy Score for short RNAs (GESS) is a computational framework for identifying and characterising short RNAs (sRNAs), including microRNAs (miRNAs) and small interfering RNAs (siRNAs), which influence plant responses to abiotic and biotic stress situations. Stress-responsive non-coding RNAs (ncRNAs) play an important role in post-transcriptional gene regulation in plants. They target mRNAs for destruction or translational repression, influencing stress tolerance and adaptation. GESS uses thermodynamic binding principles to determine the energetic stability of sRNA–mRNA interactions, allowing researchers to predict functional sRNAs that regulate stress-responsive genes (Yilmazel et al. 2014). Given the increasing importance of sRNAs in plant stress biology, GESS is an invaluable tool for discovering novel ncRNA-mediated regulatory pathways that contribute to plant resilience under environmental challenges such as drought, salinity, temperature changes, and pathogen infections.

The primary application of GESS in plant stress research is to identify stress-responsive miRNAs and siRNAs involved in transcriptional and post-transcriptional regulation. Many plant microRNAs, including miR398, miR166, and miR396, have been extensively studied for their involvement in oxidative stress tolerance, drought adaptation, and hormone signalling pathways. For example, miR398 is known to modulate superoxide dismutase (SOD) genes, which aid in the reduction of oxidative damage induced by drought and salinity stress (Fang et al. 2014). GESS enables researchers to analyse global energy scores of miRNA–mRNA interactions, allowing for more exact predictions of stress-induced miRNA targets and their downstream effects. GESS supports the functional annotation of stress-responsive ncRNAs by combining high-throughput sequencing datasets like as RNA-seq and degradome sequencing, making it a valuable computational tool for large-scale plant transcriptomics studies.

GESS is useful in understanding siRNA-mediated stress responses, notably in the RNA-directed DNA methylation (RdDM) and gene silencing pathway. Small interfering RNAs have critical roles in epigenetic alterations, transposon silencing, and chromatin remodelling, all of which are required for plant adaptation to environmental stressors. For example, in rice (Oryza sativa) and wheat (Triticum aestivum), GESS has been used to discover siRNAs that cause DNA methylation in stress-responsive genes, altering their expression in harsh conditions (Adamson et al. 2012). GESS-assisted investigations in Arabidopsis thaliana have also found siRNA clusters important in pathogen defence responses, underscoring the importance of ncRNAs in immunity and stress-induced epigenetic regulation. Integrating GESS predictions with methylome sequencing and histone modification data allows researchers to uncover the epigenetic activities of stress-responsive short RNAs, leading to new insights into mechanism of plant adaptation.

10.2.3 Integration of Multi-omics Data

The study of stress-responsive non-coding regions in plants necessitates a thorough multi-omics approach to find regulatory networks that control abiotic stress adaptation. By combining genomes, transcriptomics, epigenomics, proteomics, and metabolomics, researchers can discover functional links between non-coding RNA elements, gene expression variations, chromatin alterations, and physiological responses to stress. This section investigates how distinct omics layers help to understanding plant stress adaptation, using specific case studies and real-world examples to demonstrate their relevance.

10.2.3.1 Genomics and Transcriptomics: Identifying Stress-Responsive Non-coding Elements

Genomic sequencing lays the groundwork for finding stress-responsive non-coding areas, such as cis-regulatory elements, enhancer regions, and different types of non-coding RNAs. Long-read sequencing technologies, such as PacBio and Oxford Nanopore, have made it easier to annotate long non-coding RNAs (lncRNAs), short RNAs (sRNAs), and regulatory motifs in plant genomes. Transcriptomics, which uses RNA sequencing (RNA-seq) and short RNA sequencing (sRNA-seq), enables researchers to characterise the expression patterns of ncRNAs in response to stress situations. This allows for the discovery of stress-induced lncRNAs, miRNAs, and other regulatory components (Lakhotia 2012). A multi-omics research of tomato (*Solanum lycopersicum*) under salt stress found that miR166 and miR396 regulate genes involved in osmotic balance and ion homeostasis.

10.2.3.2 Genomics and Transcriptomics: Identifying Stress-Responsive Non-coding Elements

Epigenomic alterations such as DNA methylation, histone modifications, and chromatin accessibility all play important roles in regulating stress-responsive gene expression. Researchers can use techniques like as ChIP-seq (Chromatin Immunoprecipitation Sequencing), ATAC-seq (Assay for Transposase-Accessible Chromatin Sequencing), and Whole-Genome Bisulfite Sequencing (WGBS) to investigate how stress alters the epigenetic landscape and affects non-coding regulatory elements. Stress-induced hypomethylation in promoter regions activates stress-related genes, whereas hypermethylation inhibits non-essential functions during stress adaption. Influence chromatin accessibility, making regulatory sites more accessible to transcription factors under stress, demonstrates that stress-responsive enhancers undergo structural modifications, allowing transcription factors to bind and control gene expression (Lakhotia 2012).

10.2.3.3 Proteomics and Metabolomics: Linking Non-coding RNAs to Functional Responses

Proteomics and metabolomics fill the gap between gene regulation and functional phenotypic responses. Because many ncRNAs have post-transcriptional functions, understanding stress adaption requires an understanding of their impact on protein expression and metabolism. Proteomics (Mass Spectrometry-based analysis) enables researchers to quantify changes in protein abundance under stress situations. Metabolomics (LC-MS and GC-MS) detects changes in primary and secondary metabolites, exposing important metabolic pathways involved in stress tolerance (Satrio et al. 2024). A proteomic and metabolomic examination of winter wheat (*Triticum aestivum*) revealed that lncRNAs influence sugar metabolism genes, resulting in increased accumulation of raffinose and sucrose, which protect cells against freezing damage.

10.2.3.4 Systems Biology Approaches for Multi-omics Integration

To properly utilise multi-omics data, systems biology methodologies are used to combine genomic, transcriptomic, epigenomic, proteomic, and metabolomic data. Gene Regulatory Network (GRN) Analysis Connects stress-responsive ncRNAs to transcription factors, target genes, and epigenetic changes. Machine Learning in Multiomics: AI-powered methods (such as Random Forest and Deep Learning) anticipate functional stress-responsive non-coding areas. Visualisation tools (e.g., Cytoscape, STRING) can help map ncRNA interactions in stress circuits. A network-based multi-omics study in rice (Oryza sativa) discovered a miRNA–TF–mRNA interaction hub that controls cold stress adaptation. The combination of RNA-seq, ChIP-seq, and proteomics demonstrated that miR319 inhibits TCP transcription factors, allowing the plant to regulate photosynthetic efficiency and membrane fluidity at low temperatures.

10.2.3.5 Epigenomics: Histone Modification and DNA Methylation in Stress-Responsive Genes

Epigenomics controls plant responses to abiotic stress by affecting gene expression, chromatin structure, and transcription factor accessibility. Two key epigenetic mechanisms—histone alterations and DNA methylation—are essential for stress-responsive gene regulation. These changes allow plants to fine-tune their genetic responses to changing environmental conditions like drought, salt, and temperature stress.

Histone Modifications and Their Role in Stress Adaptation

Histone changes including methylation, acetylation, phosphorylation, and ubiquitination change chromatin structure and affect the accessibility of stress-responsive genes. Studies using chromatin immunoprecipitation followed by sequencing (ChIP-seq) found important histone marks related with stress adaptation: H3K4me3 and H3K27ac. These marks are associated with active transcription and are more prevalent in stress-induced genes, ensuring rapid activation under stress circumstances. H3K27me3 is a restrictive histone mark that normally silences stress-related genes but is deleted during stress, allowing genes to be activated. H3K9me2 has a function in heterochromatin formation and stress memory, allowing plants to respond more quickly to recurring stress episodes. In Arabidopsis, H3K27me3 demethylation activates heat shock transcription factors, which improves thermotolerance by preserving heat stress memory.

DNA Methylation and Stress Responses

DNA methylation involves the addition of methyl groups (CH3) to cytosine bases, primarily in CG, CHG, and CHH sequence contexts. DNA methylation is a key epigenetic modification that suppresses gene expression in stress-responsive regulatory regions. Hypomethylation in promoter regions leads to the activation of stress-responsive genes. Hypermethylation is often observed in repetitive sequences and transposable elements (TEs) to prevent genome instability under stress conditions. Stress-induced DNA demethylation has been observed in drought-responsive genes, allowing rapid adaptation to water-deficient conditions. In Oryza sativa, drought stress induced global DNA hypomethylation, particularly in genes related to ABA signalling and root development. This resulted in an increased expression of drought-adaptive genes, improving water uptake and survival.

10.2.3.6 ChIP-seq: Identifying Regulatory Protein-Binding Sites

Chromatin immunoprecipitation and sequencing (ChIP-seq) is a high-throughput approach for mapping DNA–protein interactions in plant genomes. It is commonly used to detect transcription factor binding sites, histone alterations, and chromatin-associated proteins that control stress-responsive genes and non-coding RNAs.

The Mechanism of ChIP-Seq in Stress Research

ChIP-seq includes crosslinking proteins and DNA, followed by immunoprecipitation with antibodies that target transcription factors (TFs), histone modifications, or chromatin-associated proteins. The collected DNA fragments are subsequently sequenced to determine their chromosomal locations, revealing details about stress-responsive regulatory networks (Jaini et al. 2015).

- Identification of Stress-Responsive Tissue Factor ChIP-Seq enables the identification of TFs that bind to promoters, enhancers, and ncRNA loci under stress circumstances.
- Histone Modification Mapping assists in identifying active and repressive chromatin states that regulate stress gene expression.
- Enhancer Discovery identifies stress-activated enhancers and superenhancers that influence gene regulation.

10.2.3.7 ATAC-Seq: Chromatin Accessibility Data to Explore Non-coding Elements

ATAC-seq (Assay for Transposase-Accessible Chromatin Sequencing) is a next-generation sequencing approach that allows scientists to map open chromatin areas and uncover stress-responsive regulatory elements in non-coding DNA. Unlike ChIP-seq, which requires prior knowledge of target proteins, ATAC-seq provides a global picture of accessible chromatin, making it excellent for identifying new regulatory elements (Zhang et al. 2022).

ATAC-Seq in Stress-Responsive Chromatin Studies

- Identification of Open Chromatin Regions reveals enhancer and promoter accessibility under stress conditions.
- Mapping Stress-Specific Chromatin Remodelling identifies dynamic chromatin changes in response to abiotic stress.
- Enhancer–Promoter Interactions help define regulatory networks influencing ncRNA-mediated stress responses.

10.2.3.8 Genome-Wide Association Studies (GWASs): Identifying SNPs in Non-coding Regions Related to Stress Tolerance

Genome-wide association studies (GWASs) are an effective method for discovering single nucleotide polymorphisms (SNPs) associated with stress tolerance. GWAS, unlike traditional gene-mapping methodologies, examines genetic differences across many populations, correlating SNP markers to agronomically significant stress-adaptive features (Ma et al. 2021).

GWAS and Stress-Responsive Non-coding Regions

- SNPs in Promoter and Enhancer Regions: GWAS has identified regulatory variants that impact gene expression during stress adaptation.
- miRNA and lncRNA Polymorphisms: Variations in ncRNA loci influence post-transcriptional regulation under stress.
- Epigenetic Variants in Stress Genes: Methylation-sensitive SNPs (mSNPs) influence chromatin accessibility and gene expression patterns.

10.3 Challenges and Future Perspectives

The study of stress-responsive non-coding areas in plants has received a lot of attention because of their function in gene regulation, chromatin remodelling, and stress adaption. Despite breakthroughs in genomics, epigenomics, and multi-omics technologies, there are still significant hurdles to completely characterising and utilising non-coding areas for crop development. One of the most difficult issues in plant non-coding RNA research is the inadequate functional annotation of non-coding components. While protein-coding genes have been widely explored, a large proportion of plant genomes are made up of non-coding DNA whose function is unknown. Poorly characterised lncRNAs and miRNAs: Although hundreds of long non-coding RNAs (lncRNAs) and microRNAs (miRNAs) have been discovered using RNA-seq and computational methods, only a tiny subset have been experimentally confirmed. Unlike protein-coding genes, non-coding regions frequently have poor sequence conservation, making functional predictions challenging. Functional genomics advances, such as CRISPR-based perturbation screens and loss-of-function experiments, can aid in the identification of stress-responsive non-coding areas in plants. The integration of genomes, transcriptomics, epigenomics, and metabolomics data is crucial for understanding ncRNA-mediated gene regulation, but it poses numerous problems. Multi-omics investigations generate vast amounts of high-dimensional data, making it challenging to extract useful biological insights. Different sequencing methodologies and analytical tools frequently result in conflicting ncRNA annotations and data interpretations. Machine learning-based techniques and AI-driven integrative multi-omics pipelines can help to harmonise data and reconstruct stress-responsive regulatory networks.

Long-term epigenetic tracking using third-generation sequencing technologies and chromatin profiling (ChIP-seq, ATAC-seq) can improve our understanding of epigenetic stability in plant stress responses. CRISPR-based genome editing has revolutionised plant genetics, but its application in modifying stress-responsive non-coding regions remains challenging. Editing regulatory elements, enhancers, and ncRNA loci can result in unintended disruptions to gene regulatory networks. Unlike coding genes, cis-regulatory elements work context-dependently, making it impossible to predict how their changes affect plant stress response. CRISPR-dCas9-based epigenome editing is a viable option since it allows for precise DNA methylation and histone changes without affecting DNA sequences. Despite these challenges, newer technology and research trends are creating new opportunities for using non-coding regions to improve plant stress tolerance. AI-enabled bioinformatics tools have revolutionised ncRNA identification, functional annotation, and target prediction. Most ncRNA research use bulk tissue RNA-seq, which averages gene expression across several cell types. However, stress responses frequently use cell-type-specific regulation systems. Single-cell RNA-seq (scRNA-seq) allows for the investigation of ncRNA expression patterns at the single-cell level, revealing cell-specific stress adaption pathways. Single-cell ATAC-seq reveals alterations in chromatin accessibility in stress-responsive ncRNA sites. Combining single-cell transcriptomics and epigenomics will help us better understand how individual cells regulate stress adaptation via ncRNAs. Researchers can generate stress-tolerant crop varieties by targeting stress-responsive ncRNAs using genome editing and synthetic biology methods. Programmable ncRNAs could be designed to activate stress-response pathways on demand. CRISPR-Cas13 can precisely edit ncRNA sequences without affecting genomic DNA. Using ncRNAs to fine-tune hormone pathways may give crops greater resilience to shifting climates. Field studies with genetically altered miRNAs and lncRNAs could transform sustainable agriculture and food security. The study of stress-responsive non-coding areas is at a promising stage, providing new insights into gene regulation, stress adaptation, and crop development. While obstacles such as functional annotation deficiencies, epigenetic variability, and genome editing complications remain, the combination of AI, single-cell technologies, and CRISPR-based tools is likely to solve them. Future study will focus on decoding regulating ncRNA networks at a systems level, ultimately leading to the development of climate resilience.

10.4 Conclusion

The investigation of stress-responsive non-coding areas in plants has revolutionised our understanding of gene regulation, chromatin remodelling, and stress adaptation processes. These non-coding components, which include long non-coding RNAs (lncRNAs), microRNAs (miRNAs), enhancer regions, and transcription factor binding sites, play critical roles in influencing plant responses to abiotic stressors such as drought, salinity, severe temperatures, and nutritional constraints. Recent advances in high-throughput sequencing, multi-omics integration, and genome editing technologies have revealed new details on how these regulatory elements regulate plant stress tolerance. Despite tremendous advances, hurdles remain in functionally annotating non-coding RNAs, integrating complicated multi-omics datasets, and using CRISPR-based genome editing to non-coding areas. However, cutting-edge technologies like AI-driven ncRNA prediction, single-cell omics, and synthetic biology techniques are projected to solve these challenges, ushering in a new era of precision agriculture. Moving forward, CRISPR-based RNA editing, ncRNA-guided stress tolerance mechanisms, and synthetic regulatory circuits have the potential to revolutionise crop engineering. The production of stress-tolerant crops through targeted ncRNA alterations has enormous potential for maintaining food security in the face of climate change. Researchers can better understand the functional complexity of non-coding elements by using integrative systems biology methodologies, ultimately leading to more sustainable agriculture practices and increased global food output.

References

Abbas F, Ke Y, Zhou Y, Yu Y, Waseem M, Ashraf U, Wang C, Wang X, Li X, Yue Y, Yu R, Fan Y (2021) Genome-wide analysis reveals the potential role of MYB transcription factors in floral scent formation in Hedychium coronarium. Front Plant Sci 12:623742. https://doi.org/10.3389/FPLS.2021.623742/PDF

Adamson B, Smogorzewska A, Sigoillot FD, King RW, Elledge SJ (2012) A genome-wide homologous recombination screen identifies the RNA-binding protein RBMX as a component of the DNA-damage response. Nat Cell Biol 14(3):318–328. https://doi.org/10.1038/ncb2426

Allmer J (2022) Noncoding RNA databases. Curr Pharm Biotechnol 24(7):825–831. https://doi.org/10.2174/1570180819666220520124511/CITE/REFWORKS

Baillo EH, Kimotho RN, Zhang Z, Xu P (2019) Transcription factors associated with abiotic and biotic stress tolerance and their potential for crops improvement. Genes 10(10):771. https://doi.org/10.3390/GENES10100771

Bhatia G, Goyal N, Sharma S, Upadhyay SK, Singh K (2017) Present scenario of long non-coding RNAs in plants. Non-Coding RNA 3(2):16. https://doi.org/10.3390/NCRNA3020016

Boyle AP, Hong EL, Hariharan M. Cheng Y, Schaub, MA, Kasowski M, & Snyder M (2012) Annotation of functional variation in personal genomes using RegulomeDB. Genome research, 22(9), 1790–1797

Chen ZY, Guo XJ, Chen ZX, Chen WY, Liu DC, Zheng YL, Liu YX, Wei YM, Wang JR (2015) Genome-wide characterization of developmental stage- and tissue-specific transcription factors in wheat. BMC Genomics 16(1):1–15. https://doi.org/10.1186/S12864-015-1313-Y/FIGURES/3

Gupta S, Kaur R, Sharma T, Bhardwaj A, Sharma S, Sohal JS, Singh SV (2023) Multi-omics approaches for understanding stressor-induced physiological changes in plants: an updated overview. Physiol Mol Plant Pathol 126:102047. https://doi.org/10.1016/J.PMPP.2023.102047

Fang J, et al (2014) Conformational analysis of processivity clamps in solution demonstrates that tertiary structure does not correlate with protein dynamics. Structure 22(4):572–581

Fabian A. Buske , Mikael Bodén , Denis C. Bauer , Timothy L. Bailey Author Notes Bioinformatics, Volume 26, Issue 7, April 2010, Pages 860–866, https://doi.org/10.1093/bioinformatics/btq049

Xie C, Yuan J, Li H, Li M, Zhao G, Bu D, Zhu W, Wu W, Chen R, Zhao Y (2014) NONCODEv4: exploring the world of long non-coding RNA genes. Nucleic Acids Res 42(D1):D98–D103. https://doi.org/10.1093/NAR/GKT1222

Xiyuan L, Dechao B, Liang S, Yang W, Shuangsang F, Hui L, Haitao L, Chunlong L, Wenzheng F, Runsheng C, Yi Z (2017) Using the NONCODE database resource. Curr Protoc Bioinformatics 58(1):12.16.1–12.16.19. https://doi.org/10.1002/CPBI.25

Yilmazel B, Hu Y, Sigoillot F, Smith JA, Shamu CE, Perrimon N, Mohr SE (2014) Online GESS: prediction of miRNA-like off-target effects in large-scale RNAi screen data by seed region analysis. BMC Bioinformatics 15(1):1–6. https://doi.org/10.1186/1471-2105-15-192/TABLES/1

Zhang X, Yao C, Fu S, Xuan H, Wen S, Liu C, Li F, Liu A, Bi S, Zhang S, Li S (2018) Stress2TF: a manually curated database of TF regulation in plant response to stress. Gene 638:36–40. https://doi.org/10.1016/J.GENE.2017.09.067

Zhang Z, Lin L, Chen H, Ye W, Dong S, Zheng X, Wang Y (2022) ATAC-seq reveals the landscape of open chromatin and cis-regulatory elements in the Phytophthora sojae genome. Mol Plant-Microbe Interact 35(4):301–310. https://doi.org/10.1094/MPMI-11-21-0291-TA/ASSET/IMAGES/LARGE/MPMI-11-21-0291-TAF5-1648652321022.JPEG

Zhao L, Wang J, Li Y, Song T, Wu Y, Fang S, Bu D, Li H, Sun L, Pei D, Zheng Y, Huang J, Xu M, Chen R, Zhao Y, He S (2021) NONCODEV6: an updated database dedicated to long non-coding RNA annotation in both animals and plants. Nucleic Acids Res 49(D1):D165–D171. https://doi.org/10.1093/NAR/GKAA1046

Zhu S, Li W, Liu J, Chen CH, Liao Q, Xu P, Xu H, Xiao T, Cao Z, Peng J, Yuan P, Brown M, Liu XS, Wei W (2016) Genome-scale deletion screening of human long non-coding RNAs using a paired-guide RNA CRISPR–Cas9 library. Nat Biotechnol 34(12):1279–1286. https://doi.org/10.1038/nbt.3715

Chapter 11
Utilization of Non-coding Regions for Enhancing Stress Resilience in Crop Species

Santosh Kumar Singh, Neethu Asokan, S. Jeevan, Parul Saxena, and Bhawna Kalra

Abstract Plants are facing extreme challenges due to alterations in the natural environmental conditions and the rise of pollutants in the surrounding atmosphere. They have several adaptive mechanisms to acclimatize to the environment. But the survival capacity of plants is challenged due to several stressed conditions as drought, salinity, high temperature, pesticide, and heavy metal pollution. The need to develop crops that are resilient and capable of facing environmental changes is rising. Such crops can help in establishing a sustainable agricultural setup in harsher climatic conditions. It is essential to understand the regulatory mechanism of genes in plants for the identification of the specific genetic regions that can be harnessed to develop stress-resistant crops. The coding regions of the DNA are being extensively studied for the specific functional proteins and related cellular functions. But the non-coding regions are less investigated. The non-coding regions in plant genomes have a significant role in gene regulation patterns. Many non-coding regions are known to control the regulation of responsive genes under stress conditions. The current chapter explains the types of non-coding regions and their significant role in stress responses in plants. The methodological approach to harness the potential regulatory functions of non-coding regions using different tools and techniques is explained in the chapter. The chapter explains various case studies related to the improvement of stress-resilient crops. The challenges and ethical concerns of

S. K. Singh (✉) · N. Asokan · S. Jeevan
Department of Life Sciences, Sri Sathya Sai University for Human Excellence, Kalaburagi, Karnataka, India
e-mail: santosh.s@sssuhe.ac.in; neethu.asokan@sssuhe.ac.in; jeevan.s@ssslsg.org; jeevan@sssuhe.ac.in

P. Saxena
A*STAR, Agency for Science, Technology and Research, Singapore, Singapore
e-mail: mailtoparul@gmail.com

B. Kalra
Faculty of Agricultural Sciences, Department of Genetics and Plant Breeding, SGT University, Gurugram, Haryana, India
e-mail: bhawna_fasc@sgtuniversity.org

S. Tiwari et al. (eds.), *Decoding Non-Coding RNA*, Concepts and Strategies in Plant Sciences, https://doi.org/10.1007/978-981-96-9859-2_11

gene modulations are elaborated in the chapter, as well as the ideas of overcoming challenges and limitations. This chapter includes a comprehensive analysis of regulatory components of the non-coding region of plant genome and explains the importance of such regions in producing resilient crops to address the food security and challenges to sustainable farming systems.

Keywords Non-coding regions · Stress resilience · ncRNAs · Enhancers · Promoters · Genome editing · Sequencing techniques · Food security

11.1 Introduction

The plant's defence mechanism and acclimatization potential are facing serious challenges due to fluctuating environmental and climatic conditions. Human activities and uncontrolled developments are raising concerns about the food security issues for the world's growing population. It has become important to evaluate methods and strategies to develop sustainable agricultural practices. The different stress components that exert maximum pressure on plants' adaptive abilities should be addressed to enable successful farming practices and increase food production. Several scientists are trying to develop more adaptive crops that are capable of surviving in abiotic and biotic stress conditions. The survival potential of plants lies in their genome structure. Certain DNA regions can encode stress-responsive proteins under specific stresses (Nakashima and Yamaguchi-Shinozak 2006). The coding region of DNA is extensively studied for its expression pattern and encoded proteins. The study of the non-coding region is equally important since many nucleotide sequences in non-coding regions have regulators for the stress-responsive genes. It is significant to understand the importance of such genes and their possible role in the development of stress-resilient crops.

11.1.1 Scope and Importance of the Studies Related to Non-coding Regions

Non-coding regions or introns are very crucial in controlling the expression pattern of many genes. Most non-coding regions of DNA are known for their inability to encode for proteins as compared to the coding regions of the DNA. But, their roles in complex genomic expression and regulation are notable (Fu 2014). Non-coding regions include several key regulatory elements as enhancers, insulators, operators, promoters, introns, and non-coding RNAs (ncRANs). The promoters are the major regulatory elements that transduce signals to operators and ribosome binding sites (RBS) and, ultimately to the structural genes. Introns are interfering sequences that control the partial or full expression of genes by suitable splicing methods in the expression of specific genes. Gene expression is strictly controlled at transcription as well as post-transcriptional levels. Introns have a significant role in the regulation mechanism. Many non-coding RNAs and micro-RNAs are already reported to have an important role in post-transcriptional gene regulations (Bhogireddy et al. 2021). Such controls become more significant and prominent during the genetic behaviour of cells under stress conditions.

Several studies have reported the scope and role of non-coding regions (NCRs) in stress conditions such as drought, salinity, heavy metal accumulation, pesticide toxicity, etc. It is highly important to understand the type of NCRs and the scope to utilize them in the development of stress-resilient crops (Bashir and Husaini 2025). Advancement in sequencing techniques and functional genomics can provide valuable understanding into the regulatory mechanism of NCRs and their role in plant survival, improving crop production. Several potential applications of NCRs can be identified and applied to sustainable agricultural development. Non-coding region-mediated regulations of stress-responsive genes may develop the ability of survival in plants, especially in adverse climatic and soil conditions. It can increase the crop production to many folds and might be suitable to farmers on the scale of cost & benefit ratio. The stress resilient crops may not require much of external inputs as fertilizers, pesticides, nematicides, etc. and economic losses to the farmers could be prevented. It is essential to understand the molecular mechanism by which non-coding regions regulate the gene expression of stress-responsive genes, for suitable application in the field using Biotechnology and Gene Engineering (Patil et al. 2024). The integration of plant breeding techniques into molecular and gene engineering techniques may enable the development of stress resilience in crops. It will lead to sustainable and economical farming approaches contributing to food security and reducing the stress-load on crops.

11.1.2 Components of the Study

The chapter includes the study and explanation of several components of non-coding genes related to research in the current scenario. It includes studies about different types of NCRs and their role in the regulation of stress-responsive genes. The authors have explained the mechanism of gene regulation in stress-related genes with emphasis on non-coding regions. The identification techniques and functional role of several NCRs are also explained in the current chapter. Several gene manipulation approaches and case studies that would help understand the impact of NCRs in improving stress resilience are explained with suitable examples. Gene regulation is a complex process, and several challenges are expected while practicing gene manipulations. The limitations and potential opportunities of using NCRs are explained by the authors, suggesting future research directions. The ethical issues, regulatory and integrative approach of traditional breeding, and biotechnology-based methods are explained with suitable examples.

11.2 Non-coding Regions (NCRs)

Non-coding sequences are able to influence a large group of structural genes by regulating its activity. Their role is observed to be significant with respect to their contribution in adoptive responses of plants to critical environmental changes. These regions also support plant growth and development. Their role in defence against several pathogenic organisms is also established.

11.2.1 Type and Functions of NCRs

There are several non-coding regions that are able to influence genes related to many biological functions (Fig. 11.1). They can be activators, regulators, binding sites for activator proteins, insulator sequences, and many more. Some examples are listed as follows:

(a) Promoters are the binding sites for the key enzyme "RNA Polymerase." These sequences are upstream of structural genes and help in the binding of several transcription factors (especially in eukaryotes) to promote and enhance the rate of transcription.
(b) Enhancers are nucleotide sequences that increase the frequency and rate of transcription of genes. These sequences are suitable binding sites for many transcription factors and transcription activator proteins (TAPs), which help in the speedy transcription of mRNA of specific genes in different adaptive responses.

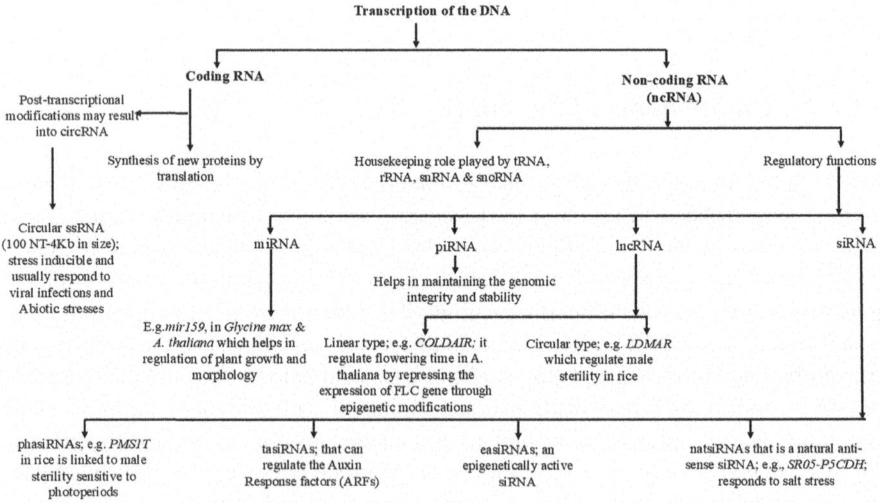

Fig. 11.1 Different types of NCRs and their role (Hamilton and Baulcombe 1999; Holoch and Moazed 2015; Aravin et al. 2008; Chen et al. 2020; Majumdar 2024)

(c) Non-coding sequences or Introns are usually removed by splicing using a complex spliceosome during post-transcriptional modifications. These nucleotide sequences play important role in producing altered proteins by modulated gene expression patterns with the help of alternative splicing methods. It helps in the synthesis of new proteins, sometimes stress-responsive proteins, that can help in the acclimatization of plants.

(d) Some RNA molecules have only regulatory functions and do not encode any proteins. These non-coding or ncRNAs are known to perform various functions. Such as chromatin re-modelling during epigenetic regulations may be done by long non-coding RNAs (lncRNAs). Similarly, enhancers, silencers, and insulators are intergenic regions of DNA that do not encode any protein but can alter the gene expression pattern in favour of plants. Micro-RNAs are small complementary sequences of mRNA encoded by structural genes. They are able to silence or neutralize the translation mechanism due to their complementarity to sense RNAs of many genes. Transposons or jumping genes can carry regulatory sequences and partial structural genes to different regions of DNA and can make necessary changes in gene expression (Kakutani 2002). The synergistic effects of NCRs are very important in plants' adaptive responses and help them in regulating several structural genes.

The major role of non-coding regions is to maintain genomic integrity and stability. Despite their inability to encode for proteins, they can alter the gene expression pattern and help in regulating structural genes. They have the potential of influencing the production of diverse proteins by alternate splicing type of mechanisms and modify the chromatic structure changing its accessibility to the transcription machinery. The long non-coding RNAs play an important role in DNA methylation and alterations of histone proteins (Bhardwaj et al. 2020). DNA methylation, acetylation, and histone modifications are important steps of epigenetic changes in DNA (Niederhuth and Schmitz 2017). It has also been reported that the NCRs can contribute to DNA repair mechanisms during stress-mediated DNA damage. It is reported that 98% of the human genome is unable to encode proteins but has an important role in controlling the gene expression during developmental and stress conditions in plants and animals. It is also reported that the complex regulatory networks are mostly made up of these NCRs and are important components of the genomic structure of an organism.

11.2.2 The Concept of Stress Resilience in Plants

Plants adapt to environmental conditions through morphological, physiological, and molecular adaptations at cellular and tissue levels. They have several enzymatic and non-enzymatic antioxidants, which play different roles in combating oxidative stress conditions. Many osmoprotectants are reported to be produced in plants under specific stress conditions like drought. Stress-responsive genes play a vital role in maintaining the plant's adaptation potential. The coordinated action of NCRs ensures the precise plant response to specific stress and aids them to adapt to extreme cold or heat, salinity/alkalinity conditions, and other stresses.

Drought stress has a serious influence on growth and development of plants. Plants have developed physiological methods to limit the rate of transpiration and accumulation of osmoprotectants as trehalose and proline. The stress-responsive genes may encode dehydrins and aquaporins to cope with the stress. Similarly, under extreme saline conditions, salts may be accumulated in specialized cells or salt glands. Also, the higher transcription rate of salt-responsive genes encoding osmolyte production and ion channel regulating genes such as *SOS1* and *HKT1* are reported in many plants. Heat Shock Proteins (*HSPs*) are produced in many plants under extreme high-temperature conditions. Plants are able to produce anti-freeze proteins, too, under very low temperature conditions to avoid cellular injuries to cells. Under nutrient deficiency conditions, the mobilization of nutrient reserves and related enzymatic activities are strictly under the control of stress-responsive genes. Stresses like metal toxicity and radiation may cause damage at cellular and molecular levels. The free radicals produced in such stress conditions may cause severe oxidative stress. Plants use enzymatic antioxidants such as Catalase, Peroxidase, Superoxide Dismutase, IAA oxidase, etc. and non-enzymatic antioxidants as proline, ascorbate, flavonoids, tocopherol, etc. to combat the stress and reduce the cellular oxidative stress. Some stress-responsive genes encode for metal-binding proteins as metallothionein and phytochelatin to detoxify the heavy metals.

Plants can fight and survive the pathogens and pests by activating their hypersensitive responses and triggering the molecular signalling mechanism (Jiang and Cui 2023). It is observed and reported that salicylic acid, jasmonic acid, and ethylene help plants to survive attacks from biological pests and pathogens. It is observed that the structural genes related to many pathogenesis-related proteins (PR-proteins) and antimicrobial peptides are upregulated by enhancers and transcription activator proteins (TAPs). Many secondary compounds are synthesized in the Jasmonic acid signalling pathway to combat the attack of pests on plants. The plant's defence responses could be elevated in specific stress conditions to help plants boost their survival capacity and increase overall productivity. It is important to understand the role of non-coding regulators in plant defences to enable researchers to produce stress-resilient crops.

11.3 Development of Stress Resilience in Plants at the Molecular Level

As mentioned above, the plant defence systems are a complex network of genetic, physiological, and molecular signals that assist plants in adapting to different stress conditions. Resilient crops are able to withstand different stresses with the help of stress-responsive proteins and other secondary compounds. Some key examples are as follows:

(a) Stress-responsive genes encode stress-combating compounds such as HSPs, aquaporins, dehydrins, late embryogenesis abundant (LEA) proteins, etc. (Hand et al. 2011).
(b) Regulatory genes encode transcription factors and other regulatory proteins to alter the expression of stress-responsive genes. Some examples are DREB (dehydration-responsive element-binding) proteins, NAC (NAM, ATAF, and CUC) transcription factors, and MYB transcription factors (Dubouzet et al. 2003).
(c) Genes involved in signalling pathways may not have a direct role in combating the oxidative stress in cells, but many of them enable molecular switches to receive the stress signals and transduction of responses. Some examples are genes encoding for receptor-like kinases (RLKs), mitogen-activated protein kinases (MAPKs), calcium-dependent protein kinases (CDPKs), etc. (Ye et al. 2017).

Some stress signals can activate the transcription factors and enable their binding to promoter regions of DNA and, in turn, switch on the structural genes in response to specific stress conditions. For example, DREB proteins are activated only in response to drought stress. The activated proteins bind to the promoter region of drought-responsive genes and trigger the transcription and translation of suitable proteins. Similarly, ncRNAs, microRNAs, and lncRNAs play an important role during post-transcriptional regulation and alter the mRNA stability, affecting the rate of translation (Imaduwage and Hewadikaram 2024). miRNAs are complementary to sense RNA and can degrade the mRNA, reducing the synthesis of selective proteins. For example, anti-sense RNA mediated degradation of channel proteins in cell membrane allows the closure of transport channel of water during water-stress condition. Translational repression may be important in such conditions and can be achieved by non-coding regions of DNA. Certain miRNAs are induced to regulate the genes involved in stress responses. For example, miR398 is overexpressed during free radical-induced oxidative stress. It helps in reducing the overexpression of copper/zinc superoxide dismutase (CSD) genes, which helps in managing reactive oxygen species levels.

The regulatory sequences may also control the synthesis of proteins involved in post-translational modifications. Such proteins can modify regular proteins by ubiquitination and phosphorylation like mechanisms and activate or degrade proteins involved in stress responses. For example, activation of MAP-kinases by phosphorylation, post translation, may activate downstream signalling mechanism and in turn can activate several stress responsive genes to produce osmoprotectants/antioxidants as proline, trehalose, ascorbate, flavonoid, glutathione, phenolics, and several enzymes as Catalase, SOD, etc. During high-temperature conditions, histone acetyltransferases (HATs) are assembled on the genes encoding different types of HSPs. It mediates over the expression of genes and synthesis of heat shock proteins, making plants more adaptive to heat stress. Different protein isoforms can be made with the help of alternate splicing regulated by non-coding regions of the DNA such as dehydration-responsive Responsive Element binding protein 2A (DREB2A) protein.

It is suggested that the integrative approach to involve traditional plant breeding techniques with genomics, proteomics, and gene engineering may help in the production of stress-resilient crops. High throughput sequencing and genome editing techniques can enable researchers to manipulate the key genes and their encoded products in transgenic crops to mitigate the stress conditions. It will certainly increase productivity and might address the food security issues.

11.4 Using NCRs in Crop Improvement!

The desirable traits in crops could be modified using molecular and gene manipulation approaches of genomics & proteomics techniques. Several gene editing techniques such as CRISPR/Cas can help in gene editing in precise gene locations to get desired traits expressed in transgenic crops (Bravo-Vázquez et al. 2024). It can boost the plant's survival capacity and help develop natural defence mechanisms against a wide range of stresses. Manipulation of many non-coding regions may fine-tune the expression of stress-responsive genes and increase the overall growth of plants by expression of genes for agronomic traits. Researchers are trying to utilize these regulatory sequences to elevate crop's resilience and stress responses. We need to understand different innovative approaches and techniques proposed for developing stress-resistant/tolerant crops and improving total crop yield.

11.4.1 Identification and Characterization of NCRs Using Sequencing Approaches

It is crucial to understand and characterize the non-coding regions (NCRs) for their utilization in regulating the overexpression or underexpression of genes of interest. Its potential applications can be established and used in the development of stress-resilient crops. RNA sequencing is the best technique to identify and sequence the non-coding regions and characterize their role in specific gene regulation mechanisms. RNA-sequencing of circular RNAs, miRNAs, long non-coding RNAs, or regulatory sequences upstream or downstream of structural genes is used to study the role of specific NCR for its control over the expression of stress-responsive genes (Meng et al. 2010).

11.4.1.1 RNA-Sequencing: RNA Sequencing Is Done in the Following Steps

(a) Isolation of RNA molecules from desired cells or tissues using suitable protocols such as the TRIzol reagent method (Chomczynski and Sacchi 1987).

(b) Purified and quantified RNA molecules must be subjected to fragmentation by using enzymatic or chemical degradation methods. Enzyme RNase III can cleave the dsRNA and yield 5′-P and 3′-OH groups. Similarly, RNase A can cleave ssRNA at the 3′ end of cytosine and uracil. Enzymatic methods provide precise control over the size of RNA fragments generated, despite of enzymatic bias, sometimes. Chemical degradation can be done using divalent cations as Mg^{+2} and Zn^{+2} at high temperature conditions (94 °C) in specific buffer solutions. Cations can break the phosphodiester bonds. This is a simple method and yields uniform RNA fragments. Though the over-fragmentation may be a slight problem sometimes.
(c) Synthesis of cDNA using reverse transcriptase enzyme. Based on the length of the RNA template and secondary structures, different types of reverse transcriptase can be used in the synthesis of complementary DNA. Moloney Murine Leukaemia Virus Reverse Transcriptase (M-MLV RT) is useful for long RNA templates. Avian Myeloblastosis Virus Reverse Transcriptase (AMV RT) is used with RNA templates that have many secondary structures.
(d) Clonal amplification of the DNA using PCR and suitable primers.
(e) Amplified DNA fragments are subjected to suitable sequencer as High Throughput Seq. technique and sequence reads are prepared (Bray et al. 2016; Patro et al. 2017).
(f) RAW data can be used as an input for computational tools for further analysis. Using Bioinformatics tools and sequence alignment services, the sequence reads can be prepared for billions of bases. Character to character alignment should be done and the number of "reads" aligned to specific genes must be prepared. The total number of "Reads" for each gene can be counted and compared with available DNA data banks. RNA seq. can help in understanding the gene expression pattern of NCRs and might be used further for elevating the expression of desired traits in crops.

11.4.1.2 Chromatin Immunoprecipitation Sequencing (ChIP-Seq)

Chromatin Immunoprecipitation Sequencing (ChIP-seq) can be used to observe the protein-DNA interactions in complex chromatin structure. With the help of this technique, we can find the RNA binding sites for specific proteins. Different steps of this technique are shown in Fig. 11.2.

11.4.2 Computational Approaches and Bioinformatics Tools

It is significant to study the bioinformatics approaches to understand the role of data banks and tools to analyse, characterize, and identify the non-coding regions of the DNA. These tools help in providing and connecting with related transcriptomic and proteomic data. Several tools and algorithms are used in bioinformatics to identify

The first step is to cross-link the proteins to DNA using a formaldehyde reagent

↓

Ultrasonication or enzymatic digestion for chromatic shearing or fragmentation (Range: 200-500 bp)

↓

Incubation: Specific antibodies (to target specific proteins) + fragments of chromatin

↓

Protein A or G beads are used to capture the antibody-protein-DNA complex.

↓

Sample heating to reverse the cross link and release the DNA: Protein complex

↓

Purify the DNA, add sequencing adapters, and subject to sequencing (a high-throughput sequencing platform such as Illumina sequencing technique

↓

Analysis of the sequencing data using bioinformatics tools for sequence alignment and identification of significant protein binding regions

Fig. 11.2 Mechanism of sequencing in Chromatin Immunoprecipitation technique

and study the functional genomics and proteomic data of ncRNAs (Gao et al. 2015). In this approach, the first process is to analyse the alignment of RNA seq. reads with suitable reference genome, which in turn could be selected from a specific database. The second step is to assemble the transcripts and annotation of known and novel ncRNAs (Veneziano et al. 2015). Some examples of common tools used for this purpose are DARIO, CPSS, ncPRO-seq, etc. The next step is to link the probable functions of ncRNAs with available novel ncRNA species and categorize them based on the specific gene and related proteins (Sun et al. 2013, 2012). Some examples of bioinformatics tools, techniques, and data banks are listed below:

(a) TopHat and Cufflinks tools are used to align and assemble the RNA transcripts.
(b) miRDeep is a tool to detect, identify, and quantify the novel miRNAs from RNA-Seq data.
(c) lncRNA-MFDL is a machine learning-based tool that can help in differentiating the long non-coding RNAs from protein-coding RNAs.

(d) The ncPRO-seq tool can help in the detection and identification of small and novel ncRNA.
(e) ChIP-seq is used to identify the interaction between chromatic and ncRNA. It is significant to understand the regulatory role played by the non-coding RNAs.
(f) RNAcentral and miRbase are huge databases for ncRNA and miRNA, respectively.

The above-mentioned techniques, tools, and databases play a significant role in the identification and sequence alignment of non-coding RNA molecules. They can be used for functional analysis of NCRs and their probable use in the production of stress-resilient crops. The major problem of addressing the differentiation of coding and non-coding regions of the DNA could be addressed using such a wide range of tools. The study of parameters as conserved sequences, novel non-coding regions, and expression patterns may also help in mitigating this problem. Even machine learning algorithms can be used to get accurate ncRNA data. But experimental validation of data obtained from bioinformatics tools is compulsory to avoid any undesirable results. ChIP-seq, RIP-seq, and CLIP-seq can be used to study and assess the interaction of ncRNAs with proteins and chromatin molecules. It can be validated with the functional outcomes (Choudhuri 2010; Musacchia et al. 2015).

11.4.3 Molecular and Genomic Editing Approaches

CRISPR/Cas9 genome editing is used to edit unique sites in non-coding regions to study their functions. The guide RNA molecules are specifically designed for NCR regions, and enzyme-mediated double-strand break creates mutations. The resulting mutations can be compared with the wild type of genomic sequences for altered functions. It helps to detect and confirm the role of each single gene in non-coding regions. The role of promoters, enhancers, operators, etc. can be assessed by this molecular biology-based approach. Similarly, the reporter assays are used to study the role and function of enhancers and promoter sequences. In this approach, NCR regions are cloned upstream of a reporter gene and inserted into a cell. The expression pattern of the reporter gene can give clues to the regulatory activity of NCR genes. Luciferase and GFP are the major reporter genes used for stress-related studies.

Open chromatic regions are indicators of higher gene expression activities. Because they provide space for the binding of suitable transcription machinery. Such regions can be identified by ATAC-Seq. method. Specific adapters are inserted into open chromatic regions with the help of transposase enzyme. Further, these regions are sequenced and mapped to identify the regulatory sequences of NCRs. The Bisulphite Sequencing approach is used to study the DNA methylation pattern of NCRs (Grehl et al. 2020). This technique can help in understanding the epigenetic regulation of NCRs and the specific role they play in stress resilience during stress conditions.

11.5 Applications in Crop Improvement and Case Studies

The non-coding regions are not transcribed and translated to form proteins in most cases, but they have significant regulatory functions and control the expression of several stress-responsive genes. Many researchers are investigating the possibility of potential applications of such NCRs in crop improvement and the development of better agronomic traits. Long non-coding RNAs (lncRNAs) and miRNAs are reported to control several traits related to seed size, length of tillers, and flowering (Zhu and Wang 2012; Waheed and Zeng 2020). One lncRNA that is popularly known to be associated with the higher seed weight in rice is *LAIR*. It is observed that the overexpression of *LAIR* may result in increased grain yield with improved vigour. Another long-non-coding RNA is MIS-SHAPEN ENDOSPERM (MISSEN), which is a regulator of genes related to seed shape. Its underexpression may cause increased endosperm content and seed size (Wang et al. 2018; Zhou et al. 2021). Some more examples of roles of non-coding regions of genome are listed as below:

(a) The NCRs as Cis response elements (CREs) have the potential to regulate traits such as panicle size and rice yield by regulating *OsREM20* gene (Wu et al. 2021).
(b) *The SICLV3* gene can be edited in its CRE region of promoter sequence and is reported to regulate the fruit size in tomato (Li et al. 2018).
(c) miRNAs encoded from NCR of DNA are observed to be edited in the conservative sequences of genes as *miR156d* and *miR396e* to regulate the plant structure, grain size, and nitrogen use efficiency in plants (Miao et al. 2019; Zhang et al. 2021; Jiang et al. 2022).
(d) The core sequence of lncRNA in the gene *SVR* in rice and *GARR2* in maize are reported to be edited to regulate the seed vigour and GA3 contents, respectively (Gao et al. 2022; Li et al. 2022).
(e) *The SQUAMOSA* promoter binding protein-like family (*SPL*) is reported to regulate plant grain developmental process. *SPL* family genes as *OsSPL13* are observed to promote cell size, grain length and overall yield in rice (Si et al. 2016; Wang et al. 2012).
(f) Many lncRNAs are reported to respond to various stresses. Drought-induced RNA (DRIR) is overexpressed during drought stress and regulates the expression of Abscisic acid signal transduction pathways (Qin et al. 2017).

Some more successful applications and case studies of using NCRs in enhancing crop yield are reported in many studies (Bashir and Husaini 2025). For example, drought tolerance in rice is reported to be increased by activating stress-responsive enhancers. Such enhancers were identified by using ChIP-Seq. and ATAC-Seq. techniques in rice, and its activation was attained by CRISPR/Cas9-mediated genome editing. This experiment proved the activation of drought-responsive genes. The plants were reported to yield higher even in reduced water availability. Another example is reported about the role of *miR398* in *Arabidopsis* spp. Overexpression of this gene was reported to be an inhibitor signal for the copper/zinc superoxide dismutase (*CSD*) genes (Frohn et al. 2024). It resulted in the low level of free radicals and reactive oxygen species and helped plants to overcome the oxidative stress. lncRNAs DRINC (Drought induced lncRNA) in Maize is reported to exhibit over-expression during drought stress conditions and upregulate the expression of stress-responsive genes.

Several non-coding regions are able to regulate genes directly or indirectly involved in stress responses of plants. It makes them highly significant in breeding techniques. They can be used for their potential impact on plant growth and survival potential, and many agronomic traits with improved performance can be developed in crops using suitable gene editing and gene manipulation in regulatory regions of NCRs. The success stories of using NCRs in the development of stress-resilient crops prove the potential of such regulatory regions in plant genome. These regions can be manipulated using suitable tools and techniques of Biotechnology, Proteomics, and genomics for crop improvement and higher productivity. It might be very helpful in addressing the food security issues, especially in the major environmental challenges posed to plants.

11.6 Challenges and Limitations

Though the potential applications of NCR in the development of stress-resilient crops are undoubtedly significant, certain limitations pose great challenges to its application in the field. Identification and characterization of non-coding regions is complex and requires skilled persons and advanced laboratory facilities. The differentiation of coding and non-coding regions is also a big challenge. Since under various stress conditions, the signal transduction may vary and involve multiple components, it is hard to determine the exact regulatory role of a certain NCR in DNA. Even if we can manage the sequencing and identification process, the interpretation of data can be done only by highly skilled researchers. Prior knowledge of the entire gene of interest, which is a target of gene editing, is required for precise application of techniques as CRISPR/Cas9. There is always a chance of accidental editing at non-targets or off-targets, resulting in undesired consequences of genomic manipulations. It is important to adhere to the guidelines and regulations related to gene manipulation and production of GMOs. A prior risk assessment and consideration of ethical issues must be taken care of for safe applications in the field. Some limitations and challenges are listed as under:

(a) Technical challenges include the lack of extensive knowledge of NCRs since these nucleotide sequences comprise more than 96% of the genomic DNA region. This makes their identification more challenging. Since multiple NCRs might work in coordination to regulate a group of genes, the functions in different stress conditions may vary and the study of functionality is more complex. The characterization of these regions requires advanced computational approaches and experimental validation.

(b) The epigenetic regulations as DNA methylation, acetylation, histone modifications, etc. vary across tissues and cellular stages. They also vary in different environmental conditions. The exposure of chromatic regions to transcription machinery may be dependent on several factors. Such variations may complicate the establishment of certain roles of specific NCRs associated with stress resilience (Abdulraheem et al. 2024; Fujimoto et al. 2012).
(c) Lack of noticeable phenotypic changes due to the involvement of NCRs in multiple regulatory functions makes it difficult to validate its role.
(d) Non-targets might be edited leading to genomic modifications responsible for genomic instability. The accuracy of data is a bigger challenge during such processes.
(e) The filed validation of laboratory findings is also a big challenge in fluctuating environmental conditions. Even after successful validation, the modified crops for their genetic traits have to be passed by complex legal and regulatory measures and approvals in specific countries.

Challenges and limitations can be resolved by using advanced tools and techniques available in Molecular Biology, Gene Engineering, and Bioinformatics. The outcomes are so much significant that these challenges are worth facing and resolving. There will be a good possibility to harness the potential of non-coding regions to develop stress-resilient crops and maintain the food security for growing populations.

11.7 Brighter Future of Stress-Resilient Crops: Summary of the Study

11.7.1 Future Prospects

The identification and characterization of non-coding regions of plant genome are becoming increasingly easier with the advancement of sequencing techniques and gene manipulations tools. Advancements in the technique, such as "Single-cell RNA sequencing" and Long Read Sequencing as PacBio, Oxford Nanopore, etc. can enable researchers to better understand and have deep insights into the non-coding transcriptome and functional genomics of NCRs. Integrated approaches that use genomics, proteomics, transcriptomics, and bioinformatics may facilitate the complete understanding of regulatory networks of NCRs and unique stress-specific interactions of non-coding regions with other regulatory signals. Validation of functions of different non-coding regions with precise gene editing and its integration with the mechanism of epigenetic modification in the chromatin region can help in understanding the plant's response to stress conditions and resilience (Chao et al. 2022; Zhang et al. 2023).

The novel stress resilient alleles and its regulatory network may be identified with the help of epigenetic variations in wild plant species. Genomic and epigenomic studies must be thoroughly done to uncover the genetic signals and regulatory mechanisms behind stress tolerance in plants. The crops resilient to stresses must be developed with an emphasis on species tolerant to a wide range of stress conditions. A combination of non-coding regions of genome-based improvement and traditional agronomic practices as crop rotation, smart farming, soil health management, etc. will enhance the overall potential of crops to stresses and help in increasing the productivity. Such stress-resilient crops may mitigate the adverse impact of climate modulations on agricultural production.

11.7.2 Summary and Conclusion

Developing stress-resilient crops is essential to enhance crop production and feed the growing population. The climatic fluctuations are creating a lot of pressure on the potential survival capacity of crops under various abiotic and biotic stresses. The impact of stresses is affecting the productivity of crops and raising food security concerns. Several stresses, such as salinity, meal toxicity, pesticide toxicity, UV radiations, etc. are extensively studied at the physiological and molecular level. The studies reveal that the plant's adaptive responses to specific stress conditions are primarily regulated at the molecular level. The genes encode several stress-responsive proteins and other secondary compounds that help in mitigating the effect of stresses. But the coding regions of DNA are only 2% of the total genomic load. It means the importance of non-coding regions (NCRs) in the genome becomes meaningful to understand the impact of regulatory signals on downstream structural genes. Different types of non-coding regions are reported as miRNA, lncRNAs, enhancers, promoters, ribosome binding sites, activator protein binding sites, etc. Each non-coding region may have a unique impact on stress-responsive genes. Directly or indirectly, the overexpression or underexpression of such non-coding regions and related ncRNAs may change the expression pattern of structural genes (Chen et al. 2023).

It is important to identify, characterize, and differentiate the coding regions from non-coding regions. Several tools and techniques of molecular biology, biotechnology, genomics, and proteomics are reported for the identification and characterization of these NCRs. The data interpretation with the help of bioinformatics tools and databases is equally important to correlate and validate the laboratory data and establish the functionality of non-coding regions. ncRNAs, miRNAs, and lncRNAs can modulate and change the gene expression at both transcriptional and post-transcriptional levels. Many ncRNAs have a role in alternative splicing methods to produce altered proteins. Several genes are reported to have regions being regulated by non-coding regions against oxidative stress, drought stress, salinity, etc.

Genomic editing techniques and their integration with epigenetic changes in chromatin have helped in better understanding the regulatory role of non-coding regions. The targeted modifications of unique nucleotide sequences in non-coding regions as promoters may help in the binding of transcriptional activators or enhancer proteins, which can activate or inhibit the stress-responsive genes (Chen et al. 2006). Several techniques are explained in the chapter to identify and characterize the non-coding regions, such as RNA-Seq., ChIP-Seq., bioinformatics tools, CRISPR/Cas9 genome editing, reporter assays, etc. Some success stories and case studies are explained in the chapter, highlighting the unique target gene sequences and related improvements in agronomic traits of particular crops. It is reported that the overexpression of miRNAs and lncRNAs, such as *miR398* and *DRINC,* may increase stress resilience in crops like *Arabidopsis* and maize. However, the authors have also highlighted the challenges of using NCRs to produce stress-resistant crops due to the complexity of the genome. There are significant risks of making gene manipulations in off-targets, producing consequences like gene pollution. It is important to make a proper assessment and validation of the data before field trials. It is also important to understand the rules and regulations associated with the application of genetically altered crops. Ethical issues must be addressed, and proper guidelines should be followed before making any genetic alteration in the natural genome. The advancements in the tools and techniques can help in reducing the errors during experiments, and authors are optimistic about the future of stress-resilient crop production using gene editing and modulation in non-coding regions. The food security issues could be resolved, and a growing population may be fed properly with the development and utilization of stress-resilient crops.

References

Abdulraheem MI, Xiong Y, Moshood AY, Cadenas-Pliego G, Zhang H, Hu J (2024) Mechanisms of plant epigenetic regulation in response to plant stress: recent discoveries and implications. Plants 13:163

Aravin AA, Sachidanandam R, Bourc'his D, Schaefer C, Pezic D, Toth KF, Bestor T, Hannon GJ (2008) A piRNA pathway primed by individual transposons is linked to de novo DNA methylation in mice. Mol Cell 31(6):785–799

Bashir T, Husaini AM (2025) Non-coding RNAs and their role in plants: prospective omics-tools for improving growth, development and stress tolerance in field crops. Mol Biol Rep 52:249

Bhardwaj V et al (2020) Role of histone modifications in stress responses and adaptation in plants. J Exp Bot 71(14):4301–4314

Bhogireddy S, Mangrauthia SK, Kumar R et al (2021) Regulatory non-coding RNAs: a new frontier in regulation of plant biology. Funct Integr Genomics 21:313–330

Bravo-Vázquez LA, Méndez-García A, Chamu-García V et al (2024) The applications of CRISPR/Cas-mediated microRNA and lncRNA editing in plant biology: shaping the future of plant non-coding RNA research. Planta 259:32

Bray N, Pimentel H, Melsted P et al (2016) Near-optimal probabilistic RNA-seq quantification. Nat Biotechnol 34:525–527

Chao H, Hu Y, Zhao L, Xin S, Ni Q, Zhang P, Chen M (2022) Biogenesis, functions, interactions, and resources of non-coding RNAs in plants. Int J Mol Sci 23:3695

Chen Z, Zhang H, Jablonowski D, Zhou X, Ren X, Hong X et al (2006) Mutations in *ABO1/ELO2*, a subunit of holo-elongator, increase abscisic acid sensitivity and drought tolerance in Arabidopsis thaliana. Mol Cell Biol 26:6902–6912

Chen L, Zhu QH, Kaufmann K (2020) Long non-coding RNAs in plants: emerging modulators of gene activity in development and stress responses. Planta 252:92
Chen YH, Lu J, Yang X, Huang LC, Zhang CQ, Liu QQ, Li QF (2023) Gene editing of non-coding regulatory DNA and its application in crop improvement. J Exp Bot 74(19):6158–6175
Chomczynski P, Sacchi N (1987) Single-step method of RNA isolation by acid guanidinium thiocyanate-phenol-chloroform extraction. Anal Biochem 162(1):156–159
Choudhuri S (2010) Small noncoding RNAs: biogenesis, function, and emerging significance in toxicology. J Biochem Mol Toxicol 24(3):195–216
Dubouzet JG, Sakuma Y, Ito Y, Kasuga M, Dubouzet EG, Miura S, Seki M, Shinozaki K, Yamaguchi-Shinozaki K (2003) OsDREB genes in rice, *Oryza sativa* L., encode transcription activators that function in drought-, high salt- and cold-responsive gene expression. Plant J 33:751–763
Frohn S et al (2024) Evolutionary conserved and divergent responses to copper zinc superoxide dismutase inhibition in plants. Plant Cell Environ 1:16
Fu XD (2014) Non-coding RNA: a new frontier in regulatory biology. Natl Sci Rev 1(2):190–204
Fujimoto R, Sasaki T, Ishikawa R, Osabe K, Kawanabe T, Dennis ES (2012) Molecular mechanisms of epigenetic variation in plants. Int J Mol Sci 13(8):9900–9922
Gao Y, Wang J, Zhao F (2015) CIRI: an efficient and unbiased algorithm for de novo circular RNA identification. Genome Biol 16(1):4
Gao Q, Liu J, Weng H, Yuan X, Xiao W, Wang H (2022) A long non coding RNA derived from lncRNA–mRNA networks modulates seed vigor. Int J Mol Sci 23:9472
Grehl C, Wagner M, Lemnian I, Glaser B, Grosse I (2020) Performance of mapping approaches for whole-genome bisulfite sequencing data in crop plants. Front Plant Sci 11:176
Hamilton AJ, Baulcombe DC (1999) A species of small antisense RNA in posttranscriptional gene silencing in plants. Science 286(5441):950–952
Hand SC, Menze MA, Toner M, Boswell L, Moore D (2011) LEA proteins during water stress: not just for plants anymore. Annu Rev Physiol 73:115–134
Holoch D, Moazed D (2015) RNA mediated epigenetic regulation of gene expression. Nature 16:71–84
Imaduwage I, Hewadikaram M (2024) Predicted roles of long non-coding RNAs in abiotic stress tolerance responses of plants. Mol Hortic 4:20
Jiang N, Cui J (2023) Editorial: Non-coding RNA assisted plant response to stresses. Front Plant Sci 14:1276427
Jiang Y et al (2022) The miR159-MYB33-ABI5 module regulates seed germination in *Arabidopsis*. Physiol Plant 174(2):e13659
Kakutani T (2002) Transposon silencing and gene regulation in plants. Curr Opin Plant Biol 5(1):1–6
Li T, Yang X, Yu Y, Si X, Zhai X, Zhang H, Dong W, Gao C, Xu C (2018) Domestication of wild tomato is accelerated by genome editing. Nat Biotechnol 36:1160–1163
Li W, Chen Y, Wang Y, Zhao J, Wang Y (2022) Gypsy retrotransposon derived maize lncRNA GARR2 modulates gibberellin response. Plant J 110:1433–1446
Majumdar A (2024) Non-coding RNAs and their role in enhancing virus resistance in plants. Eur J Plant Pathol 171:1–16
Meng Y et al (2010) MicroRNA-mediated signalling involved in plant root development. Biochem Biophys Res Commun 393(3):345–349
Miao C, Wang Z, Zhang L, Yao J, Hua K, Liu X, Shi H, Zhu JK (2019) The grain yield modulator miR156 regulates seed dormancy through the gibberellin pathway in rice. Nat Commun 10:3822
Musacchia F, Basu S, Petrosino G (2015) Annocript: a flexible pipeline for the annotation of transcriptomes able to identify putative long noncoding RNAs. Bioinformatics 31(13):2199–2201
Nakashima K, Yamaguchi-Shinozak K (2006) Regulons involved in osmotic stress-responsive and cold stress-responsive gene expression in plants. Physiol Plant 126:62–71
Niederhuth CE, Schmitz RJ (2017) Putting DNA methylation in context: from genomes to gene expression in plants. Biochim Biophys Acta 1860:149–156

12.1 Introduction

Around 90% of the genome is actively transcribed into RNA. Despite this, only 1–2% of this transcribed RNA is believed to encode potential protein-coding genes, while the other 98–99% of RNAs are either not well characterized or do not encode any functional proteins. This portion is often referred as "dark matter," "transcriptional noise," "junk DNA," or simply an "experimental artifact." These RNAs are classified as non-coding RNAs (ncRNAs) (Doolittle 2013; Li and Liu 2019). The genetic framework of plants comprises both coding and non-coding regions, which collectively regulate physiological functions and responses to environmental stimuli. Historically, plant genomics research has concentrated on coding sequences that directly facilitate protein synthesis. However, non-coding regions, which were previously dismissed as insignificant, are now recognized as essential regulators of gene expression and cellular activities. These non-coding regions, which include promoters, enhancers, microRNAs, long non-coding RNAs (lncRNAs), and intronic sequences, have been found to play pivotal roles in controlling stress tolerance mechanisms in plant (Domínguez-Rosas et al. 2023; Kaikkonen et al. 2011; Waititu et al. 2020).

Stress tolerance is vital for plants to thrive under fluctuating environmental factors like drought, salinity, high temperatures, and cold. This adaptive ability is not solely influenced by protein-coding genes but is profoundly shaped by the complex regulatory network orchestrated by non-coding regions (Gelaw and Sanan-Mishra 2021; Patra et al. 2023; Yadav et al. 2024). Non-coding RNAs (ncRNAs), which are functional RNA molecules that do not encode proteins, have become crucial regulators in various biological processes. These ncRNAs—including microRNAs (miRNAs), small interfering RNAs (siRNAs), long non-coding RNAs (lncRNAs), and circular RNAs (circRNAs)—play a vital role in regulating gene expression at both transcriptional and post-transcriptional levels. For example, miRNAs and siRNAs mediate gene silencing through RNA interference (RNAi), impacting plant development, stress responses, and defense mechanisms. lncRNAs participate in chromatin remodeling, transcriptional regulation, and protein interactions, while circRNAs act as molecular sponges for miRNAs, regulating their availability. Together, ncRNAs enable plants to adapt to both abiotic and biotic stresses, control flowering time, and fine-tune hormone signaling pathways (Hajieghrari and Farrokhi 2022; Jiang et al. 2023; Mirdar Mansuri et al. 2022; Zhang et al. 2022).

One of the greatest challenges in current plant genomics research is unraveling the complexities of non-coding regions and their precise roles in stress responses. Unlike coding sequences, the functional elements within non-coding regions are often difficult to pinpoint, their activities subtle, and their effects sometimes context-dependent (Amaral et al. 2013). Many of these regions regulate the expression of stress-responsive genes, acting as switches or amplifiers in response to stress signals. For instance, stress-induced transcription factors and signaling pathways are often regulated at the level of non-coding RNAs, which can modulate their expression through diverse mechanisms such as RNA interference, epigenetic modifications, and chromatin remodeling. This makes understanding the role of non-coding regions not only challenging but crucial for the development of stress-tolerant crops in the face of climate change (Jin et al. 2023; Wang et al. 2017; Yu et al. 2019).

Despite these challenges, the research community has made significant strides in deciphering the contribution of non-coding regions to plant stress tolerance. Advances in high-throughput sequencing technologies, such as RNA-Seq and ChIP-Seq, have allowed for more comprehensive analyses of the transcriptomes and epigenomes, uncovering new insights into the regulatory networks controlled by non-coding regions. These technologies have enabled the identification of non-coding RNAs (ncRNAs) that play key roles in mediating plant responses to abiotic stress (Micheel et al. 2021; Yu et al. 2019). Additionally, the integration of bioinformatics tools and computational models has facilitated the prediction and functional annotation of non-coding regions, although challenges remain in validating their exact biological functions. As the field continues to evolve, several future directions hold promise for advancing our understanding of non-coding regions and their roles in stress tolerance (Gong et al. 2021). One area of particular interest is the study of epigenetic regulation in non-coding regions, including the role of DNA methylation, histone modification, and chromatin remodeling in controlling gene expression during stress responses. This could lead to the identification of novel epigenetic markers that could be used to predict and enhance stress tolerance in crops (Abdulraheem et al. 2024; Le et al. 2020). Furthermore, as our understanding of plant stress physiology becomes more nuanced, the integration of multi-omics approaches—combining genomics, transcriptomics, proteomics, and metabolomics—will likely provide a more holistic view of the regulatory networks that govern stress responses (Ali et al. 2022; Roychowdhury et al. 2023; Yang et al. 2021).

Another exciting frontier is the potential for non-coding regions to serve as targets for genetic engineering and molecular breeding. With advancements in CRISPR-Cas technologies, precise editing of non-coding regions to enhance stress tolerance may become a feasible strategy. Furthermore, the exploration of natural genetic diversity in non-coding regions across different plant species offers a unique opportunity for the discovery of novel regulatory elements that confer resilience to various stresses (Simna and Han 2022; Shanmugam et al. 2017; Waititu et al. 2020; Wu et al. 2020).

In this chapter, we will delve into the current challenges faced in non-coding region research, particularly focusing on the complexities of their identification, annotation, and functional validation. We will explore the promising future directions, including the integration of cutting-edge technologies, the role of epigenetics, and the potential for applying this knowledge in crop improvement. Ultimately, the goal is to highlight how a deeper understanding of non-coding regions can contribute to the development of stress-tolerant crops, an essential aspect of ensuring food security in an increasingly unpredictable climate. Understanding ncRNA-mediated regulation provides valuable insights into plant biology and offers potential applications in crop improvement for stress tolerance and enhanced productivity.

12.2 Challenges in Studying Non-coding RNAs (ncRNAs)

Plants face a wide range of abiotic and biotic stresses, including drought, salinity, heat, cold, and pathogen attacks. To survive and adapt to these stresses, plants have evolved complex signaling pathways, many of which are regulated by ncRNAs. In response to abiotic stresses such as drought, heat, and salinity, ncRNAs regulate key stress-responsive genes involved in protecting cellular integrity, maintaining osmotic balance, and activating protective proteins such as heat shock proteins and antioxidants (Atkinson and Urwin 2012; Lal et al. 2023; Zhu 2016). For instance, miRNAs are known to regulate the expression of genes involved in the response to drought by targeting key transcription factors like DREB2 (dehydration-responsive element-binding protein 2), which controls the expression of drought-inducible genes (Singh et al. 2023; Zhang et al. 2022). Similarly, lncRNAs have been shown to regulate osmotic stress responses by interacting with the gene expression machinery (Kęska et al. 2021; Yang et al. 2023). Non-coding RNAs are also critical in mediating plant defense responses to biotic stresses, including pathogen attacks (bacteria, fungi, and viruses). miRNAs and siRNAs regulate the expression of defense-related genes such as those encoding pathogenesis-related proteins, receptor-like kinases, and components of the salicylic acid (SA) and jasmonic acid (JA) signaling pathways (Carthew and Sontheimer 2009; Khraiwesh et al. 2011). Additionally, lncRNAs are involved in regulating the immune response by controlling the expression of immune receptors and defense genes. By fine-tuning these defense mechanisms, ncRNAs help plants mount effective responses to biotic stress (Dresselhaus and Hückelhoven 2018; Traubenik et al. 2024). Non-coding RNAs are involved in the intricate cross-talk between different stress response pathways. For example, miRNAs can regulate the interplay between drought and salinity stress pathways, while lncRNAs may integrate responses to both biotic and abiotic stresses (Wang et al. 2023). Understanding how ncRNAs orchestrate these complex interactions will be essential for developing strategies to enhance stress tolerance in crops (Ben Rejeb et al. 2014; He et al. 2020). However, studying them presents several challenges due to their diverse nature and complex regulatory mechanisms.

12.2.1 Identification and Characterization

Unlike protein-coding genes, non-coding RNAs (ncRNAs) lack open reading frames (ORFs), which are typically used to identify genes in genome annotation pipelines. This absence makes it challenging to detect ncRNAs using traditional gene prediction approaches that rely on ORF-based criteria. Instead, their identification requires specialized computational tools and experimental techniques to confirm their existence and biological function (Adjeroh et al. 2024; Mattick et al. 2023; Zhao et al. 2020).

Long non-coding RNAs (lncRNAs) often exhibit poor sequence conservation across species, making it challenging to determine their evolutionary significance and functional relevance. Unlike protein-coding genes, which are generally well-conserved due to selective pressure on their amino acid sequences, lncRNAs evolve rapidly, often displaying low levels of sequence similarity even among closely related species. This characteristic poses difficulties in identifying orthologous lncRNAs and in inferring their biological roles through comparative genomics (Housman and Ulitsky 2016). In contrast to small non-coding RNAs like microRNAs (miRNAs) and small interfering RNAs (siRNAs), which are highly conserved due to their critical roles in gene regulation, lncRNAs tend to be lineage-specific or even species-specific. Their lack of strong sequence conservation suggests that their function is often dependent on secondary structure, chromatin interactions, or specific genomic loci rather than strict sequence homology (Statello et al. 2020; Zhou et al. 2023). This means that functional annotation of lncRNAs often requires in-depth, species-specific validation through experimental approaches. Despite their poor sequence conservation, certain lncRNAs show conserved functional elements, such as structural motifs or syntenic relationships with nearby protein-coding genes. Some studies suggest that conservation at the level of promoter regions, splicing patterns, or tertiary RNA structures may contribute to lncRNA function across species. Additionally, emerging evidence indicates that while primary sequences of lncRNAs may diverge, their roles in regulatory networks and biological pathways may be preserved (Johnsson et al. 2013; Szcześniak et al. 2021; Mattick et al. 2023).

12.2.2 Distinguishing Between Correlation and Causation in ncRNA-Stress Response Relationships

Many non-coding RNAs (ncRNAs) exhibit differential expression patterns under stress conditions, suggesting potential involvement in stress responses and adaptation mechanisms. However, merely observing changes in their expression levels does not necessarily confirm a direct functional role in stress tolerance. Establishing a causal relationship between an ncRNA and stress response is challenging, as expression changes can result from secondary effects rather than direct regulatory functions (Amaral et al. 2013; Miguel et al. 2020). A major issue in distinguishing correlation from causation is the complexity of gene regulatory networks. Stress-induced alterations in ncRNA levels may occur as a downstream consequence of other molecular events rather than being a direct driver of stress adaptation. For instance, ncRNAs might be co-expressed with stress-responsive genes without actively contributing to their regulation. Similarly, the observed differential expression could be part of a general transcriptional shift rather than a specific functional response Mirdar Mansuri et al. 2022; Wu et al. 2022).

To determine whether an ncRNA plays a direct role in stress response, rigorous genetic and molecular studies are required. Functional validation approaches such as loss-of-function and gain-of-function experiments can help establish causality. Techniques such as CRISPR interference (CRISPRi) and RNA interference (RNAi) can be used to knock down ncRNA expression, while overexpression studies can assess whether increased ncRNA levels enhance stress tolerance (Ben Rejeb et al. 2014; He et al. 2018). If stress-related phenotypic changes are observed upon these manipulations, it provides stronger evidence of functional involvement. Additionally, ncRNA-target interaction studies are crucial for understanding their mechanistic role in stress response. Methods such as RNA pulldown assays, crosslinking immunoprecipitation (CLIP), and chromatin isolation by RNA purification (ChIRP) can help identify direct molecular targets of ncRNAs. If an ncRNA is found to interact with stress-related transcription factors, signaling molecules, or chromatin modifiers, it strengthens the case for a direct regulatory role. Moreover, comparative studies in multiple species or environmental conditions can help determine whether an ncRNA consistently contributes to stress tolerance rather than being an incidental product of stress-induced transcriptional changes (Dresselhaus and Hückelhoven 2018; Pandey et al. 2017; Zhang et al. 2023; Zhu 2016).

12.2.3 Limitations in Proving ncRNA Functionality Under Chemically Induced Abiotic Stress Tolerance

Chemically induced abiotic stress conditions are widely used in laboratory experiments to study non-coding RNAs (ncRNAs) involvement in stress tolerance. These controlled setups enable researchers to apply specific stressors—such as salinity (NaCl), drought (polyethylene glycol, PEG), heavy metals (Cd, Hg, or As), oxidative stress (hydrogen peroxide, H_2O_2), and temperature stress (heat shock)—in a reproducible manner. However, while thesessss methods provide valuable insights into molecular responses, they often fail to fully replicate the complexity of real environmental stress conditions (Imaduwage and Hewadikaram 2024, Urquiaga et al. 2021). This limitation makes it challenging to validate the actual functional role of lncRNAs in natural stress responses. One of the key issues is that chemically induced stress often lacks the dynamic fluctuations and multifactorial interactions present in natural environments. In the field, plants and organisms experience varying degrees of stress over time, often influenced by factors such as temperature changes, soil composition, microbial interactions, and competing biological pathways. These conditions create a more complex regulatory landscape, where lncRNA expression and function may differ significantly from observations made under artificial stress treatments in the laboratory (Roychowdhury et al. 2023).

Additionally, chemical treatments can introduce off-target effects that may confound experimental results. For instance, high concentrations of NaCl used to simulate salinity stress might induce osmotic stress responses that are not directly comparable to natural soil salinity conditions, where factors like ion composition, water availability, and microbial interactions also play crucial roles (Acosta-Motos et al. 2017). Similarly, PEG-induced drought stress primarily affects osmotic balance but does not fully mimic the physiological and morphological changes caused by natural drought conditions, such as fluctuating water availability and long-term adaptation mechanisms (Graham et al. 2020; Naeem et al. 2020).

12.2.4 Complexity of ncRNA-Mediated Regulatory Pathways

Non-coding RNAs (ncRNAs) interact with multiple molecular components, including DNA, RNA, and proteins, forming intricate regulatory networks. Unlike protein-coding genes that function primarily through their translated products, ncRNAs exert their effects through diverse mechanisms such as chromatin remodeling, transcriptional regulation, post-transcriptional modifications, and protein-RNA interactions (Kaikkonen et al. 2011). The complexity of these regulatory pathways makes it challenging to fully decipher their functional roles and molecular mechanisms. One of the key challenges in studying ncRNA-mediated regulation is the multifunctionality and context-dependent roles of ncRNAs. A single ncRNA can interact with multiple targets, influencing different cellular processes depending on cell type, developmental stage, or environmental conditions (Mattick and Makunin 2006). For instance, long non-coding RNAs (lncRNAs) can function as molecular scaffolds, bringing together different proteins to form ribonucleoprotein complexes, or as decoys, sequestering transcription factors and preventing them from binding to DNA. Similarly, microRNAs (miRNAs) regulate gene expression by binding to complementary mRNA sequences, leading to translational repression or degradation (Dayal et al. 2024). Furthermore, ncRNAs often participate in feedback loops and cross-talk between different regulatory pathways, adding another layer of complexity. For example, competing endogenous RNA (ceRNA) networks involve lncRNAs, pseudogenes, and circular RNAs (circRNAs) acting as miRNA sponges, indirectly regulating protein-coding gene expression. This interconnected network complicates the identification of direct regulatory relationships and requires systematic approaches to untangle these interactions (Fu 2014; Statello et al. 2020). Despite advancements, challenges remain in fully elucidating ncRNA-mediated regulatory networks, particularly due to the context-specific nature of ncRNA functions and the lack of conserved sequence elements in many ncRNAs. To overcome these challenges, a combination of experimental and computational strategies is essential to build a comprehensive understanding of ncRNA biology and its role in gene regulation, stress response, and disease pathways (Cajal et al. 2019; Kaikkonen et al. 2011).

12.2.5 Lack of Comprehensive Databases for Non-coding RNAs

Unlike protein-coding genes, which have been extensively annotated and cataloged in well-established databases such as GenBank, UniProt, and Ensembl, non-coding RNA (ncRNA) databases are still developing. The limited availability of comprehensive, well-curated databases for non-coding regions significantly hampers comparative genomics, functional predictions, and large-scale ncRNA research, particularly in plant species (Amaral et al. 2010). Several resources, such as ENCODE (Encyclopedia of DNA Elements), NONCODE, LNCipedia, and RNAcentral, provide valuable data on non-coding RNAs (Chaudhary and Banerjee 2024; Sweeney et al. 2021; Zhao et al. 2021). However, these databases are heavily focused on model organisms like humans, mice, and a few well-studied plant species (e.g., Arabidopsis and rice), leaving many agriculturally and ecologically important crops underrepresented. This limited coverage poses challenges for researchers aiming to study the evolutionary conservation and functional roles of ncRNAs across diverse species (Allmer 2022; Fritah et al. 2014; Mattick et al. 2023).

One of the primary issues is the lack of high-quality ncRNA annotations in non-model plants. Unlike protein-coding genes, which can often be predicted based on conserved sequence motifs and open reading frames (ORFs), ncRNA identification requires specialized computational approaches, such as RNA secondary structure predictions, co-expression analysis, and experimental validation. The absence of reference datasets for many species makes it difficult to apply these methods universally. Furthermore, many existing ncRNA databases suffer from inconsistent annotation standards, incomplete functional characterization, and limited integration with other genomic resources. For instance, while some databases provide transcript-level information, they may lack associated metadata, such as ncRNA-protein interactions, expression profiles across different tissues, or responses to environmental stressors. The lack of such integrated data restricts the ability to perform comparative genomic analyses and hinders efforts to link ncRNAs to specific biological functions (Fritah et al. 2014; Xu et al. 2022).

To address these limitations, expanded sequencing efforts and improved database curation are essential. Large-scale transcriptomic initiatives, such as PlantENCODE, the 1000 Plant Transcriptome Project, and the Genome 10 K project, can help generate the necessary data to enhance ncRNA annotation. Additionally, machine learning and deep-learning approaches are being developed to predict ncRNA functions based on existing datasets, which could aid in filling the knowledge gaps in underrepresented species (Khalid et al. 2023; Noviello et al. 2020). While progress has been made in developing ncRNA databases, significant gaps remain, particularly for plant species and crops. Expanding ncRNA annotations, standardizing data integration, and improving functional validation efforts are crucial to overcome these challenges and advancing non-coding RNA research in both fundamental and applied sciences (Orro and Trombetti 2023; Tan et al. 2022).

12.2.6 Difficulty in Experimental Validation

Validating the functions of non-coding RNAs in stress responses necessitates advanced techniques, including chromatin immunoprecipitation sequencing (ChIP-Seq), assay for transposase-accessible chromatin using sequencing (ATAC-Seq), and CRISPR interference (CRISPRi). These methods are resource-intensive and not always feasible for large-scale studies (Nickerson and Momen-Heravi 2024; Yadav et al. 2024). Many of these techniques require expensive reagents, specialized equipment (e.g., high-throughput sequencers, cell sorters), and bioinformatics expertise for data analysis (Adjeroh et al. 2024). Experimental protocols optimized for model organisms (e.g., Arabidopsis, rice, mice) may not be directly applicable to non-model species, including crops with complex genomes (Ning et al. 2019). Context-Specific Functions: ncRNAs often act in a condition-dependent manner, meaning their functions may change under different environmental or stress conditions, requiring multiple validation experiments (Sánchez and Huarte 2013). Unlike protein-coding genes, which can be studied using enzyme activity or protein–protein interaction assays, ncRNAs typically lack direct functional readouts, making it harder to confirm their biological roles (Winkle et al. 2021).

12.2.7 Epigenetic Modifications and Their Role in Non-coding RNA Regulation

Epigenetic modifications, including DNA methylation, histone alterations, and chromatin remodeling, regulate gene expression and the function of non-coding RNAs (ncRNAs). These modifications control genome accessibility, affecting both protein-coding genes and ncRNAs like long non-coding RNAs (lncRNAs), microRNAs (miRNAs), and small interfering RNAs (siRNAs). However, understanding their role in ncRNA regulation under environmental stress is challenging due to their dynamic and reversible nature. DNA methylation, particularly in CpG islands, often silences genes by methylating cytosine residues. In ncRNAs, this silencing can occur at promoters or regulatory regions. Histone modifications, such as acetylation and methylation, influence chromatin structure. Acetylation is linked to gene activation, while histone methylation can either repress or activate gene expression, depending on the mark. For example, H3K27me3 represses lncRNA expression, while H3K4me3 promotes the transcription of stress-related ncRNAs. Chromatin accessibility, influenced by these modifications, determines how tightly DNA is packed and how accessible ncRNA promoters are to the transcriptional machinery. ATP-dependent chromatin remodeling complexes can reposition nucleosomes, regulating ncRNA expression in response to environmental stress (Mattick et al. 2023; Peschansky and Wahlestedt 2014; Wei et al. 2017).

One of the primary challenges in studying epigenetic modifications is their dynamic and reversible nature. Epigenetic marks can change in response to various environmental cues, including abiotic stresses such as drought, salinity, and temperature extremes, but these changes are often temporary or context-dependent. For example, histone acetylation levels might increase in response to heat stress but return to baseline levels after the stress subsides, making it difficult to determine the lasting effects of such modifications (Abdulraheem et al. 2024; Baluska et al. 2023). Epigenetic modifications can be highly tissue- and stress-specific, complicating the interpretation of global epigenomic data. For example, a specific histone modification might be induced in response to drought stress in roots but not in leaves. Similarly, DNA methylation changes may differ between plant species or genotypes in response to the same stressor. This tissue and stress-specificity require careful experimental design, using high-resolution techniques to capture these variations in different cell types and under various stress conditions (Thiebaut et al. 2019; Ramakrishnan et al. 2022). To gain a comprehensive understanding of the role of epigenetic modifications in ncRNA regulation under stress, it is essential to integrate epigenomic data with functional genomic information (e.g., transcriptomics, proteomics, and metabolomics). However, the integration of these datasets is challenging due to the complexity and size of the data. Epigenomic data provide information on the chromatin state, while transcriptomic data reveal which genes and ncRNAs are actively transcribed. Combining these datasets requires sophisticated bioinformatic tools and machine learning models to identify causal relationships between epigenetic marks and ncRNA expression (Baluska et al. 2023; Gudsnuk and Champagne 2012).

12.2.8 Genome-Wide Association Studies (GWAS) Limitations

Genome-Wide Association Studies (GWAS) have played a crucial role in identifying genetic loci linked with complex traits, such as stress tolerance in plants and other organisms. By linking genetic variants to phenotypic traits, GWAS has provided valuable insights into the genomic basis of stress adaptation. However, when it comes to non-coding regions of the genome, the effectiveness of GWAS is limited due to several challenges related to the identification and functional interpretation of non-coding variants (Giral et al. 2018; Uffelmann et al. 2021). While GWAS can successfully pinpoint genetic loci associated with stress tolerance, many of the significant variants identified often lie within non-coding regions of the genome, including promoters, enhancers, introns, and non-coding RNAs (such as lncRNAs, miRNAs, and small RNAs). These variants can affect gene regulation by altering transcription factor binding sites, chromatin structure, or ncRNA expression, but their exact regulatory roles are often difficult to determine without further experimental validation. For example, a Single Nucleotide Polymorphism (Single Nucleotide Polymorphism) in a gene's promoter may lead to differential gene expression, but whether the Single Nucleotide Polymorphism directly affects the binding of a transcription factor or influences chromatin accessibility cannot be determined from GWAS data alone (Schipper and Posthuma 2022; Vitsios et al.

2021). Non-coding regions of the genome are less well-annotated compared to protein-coding regions, and their functional roles are still not fully understood. This makes it challenging to interpret the significance of variants in these regions. The lack of reference datasets and standardized functional annotations for non-coding regions across species further complicates this process (Kircher and Ludwig 2022; Peña-Martínez and Rodríguez-Martínez 2024).

Non-coding regions often function in complex regulatory networks that involve multiple interacting components, including DNA, RNA, and proteins. The regulatory impact of a variant in a non-coding region may not be easily deciphered without understanding the context of other molecular interactions. Epigenetic modifications, such as DNA methylation and histone modifications, can influence the activity of non-coding regulatory elements, adding another layer of complexity to the interpretation of GWAS findings. These interactions often require multiomics approaches, combining genomic, transcriptomic, and epigenomic data, to capture the full impact of non-coding variants on gene regulation (Morova et al. 2023; Palos et al. 2022; Pesole et al. 2019; Zhang et al. 2018; Zhong et al. 2022).

Statistical power in GWAS studies can be limited, especially in complex traits like stress tolerance, which may involve contributions from many small-effect variants across the genome. This can lead to false positives, particularly in non-coding regions where variants may not have a clear functional effect. Population structure, environmental variables, and genetic linkage can further confound the results, leading to misinterpretations of the role of non-coding variants in stress adaptation (Tibbs Cortes et al. 2021; Uffelmann et al. 2023). While GWAS can identify genetic associations with stress tolerance, the next step is to validate the functional significance of non-coding variants. Since stress responses are dynamic and context-dependent, it is important to study non-coding variants under different environmental conditions and across various tissues to confirm their functional relevance. Field-based or in vivo studies that measure the phenotypic outcomes of variant-induced modifications in non-coding regions can provide real-world validation of the regulatory roles of these variants in stress tolerance (Cano-Gamez and Trynka 2020; Chen and Li 2021; Morova et al. 2023; Schipper and Posthuma 2022).

12.3 Future Directions and Opportunities in Non-coding RNA (ncRNA) Research for Plant Stress Tolerance

The rapid acceleration in the discovery of novel non-coding RNAs (ncRNAs) in plants has unveiled an intricate and dynamic regulatory network that governs plant stress responses. However, despite these advances, our understanding of ncRNA diversity and their precise functional roles remains incomplete. As environmental challenges such as climate change, soil salinity, drought, and pathogen attacks continue to threaten global food security, unlocking the full potential of ncRNAs in plant stress tolerance has become a critical frontier in agricultural and molecular biology research. A fundamental step toward leveraging ncRNAs for stress tolerance is the systematic identification and classification of ncRNAs across diverse plant species and stress conditions. Various ncRNA classes, including long non-coding RNAs (lncRNAs), small RNAs (miRNAs, siRNAs), circular RNAs (circRNAs), and other newly discovered variants, must be cataloged comprehensively. This effort will require large-scale high-throughput sequencing technologies, coupled with advanced computational approaches, to establish expression profiles and functional annotations for these ncRNAs under different stress conditions (Jin et al. 2023; Zhang et al. 2022; Zhao et al. 2024). Integration of multi-omics data, including transcriptomics, proteomics, and metabolomics, will be instrumental in delineating the regulatory roles of ncRNAs. Computational tools and machine learning algorithms can further aid in predicting ncRNA functions and interactions with stress-related genes, uncovering novel pathways that contribute to plant resilience (Fu 2014; Mathews et al. 2007; Sato et al. 2021). While some ncRNAs have been linked to plant stress tolerance, their precise molecular mechanisms often remain elusive. To fully exploit their regulatory potential, deeper functional studies are necessary. Investigating target genes of stress-responsive miRNAs, or characterizing chromatin remodeling activities regulated by lncRNAs, could provide groundbreaking insights into plant stress adaptation (Li et al. 2022; Jannesar et al. 2020).

Emerging biotechnological tools such as CRISPR/Cas9, RNA interference (RNAi), and overexpression systems provide powerful means to validate the roles of individual ncRNAs. These technologies can be leveraged to selectively knock out, enhance, or fine-tune ncRNA expression, thereby elucidating their functional contributions to stress responses (Gui et al. 2022; Jacobs et al. 2015; Tao et al. 2023). Additionally, stress-specific ncRNA studies should span various abiotic and biotic stressors, including drought, salinity, heat, cold, and pathogen-induced stress, as different stress conditions may activate distinct ncRNA regulatory pathways (Di et al. 2014; Ma et al. 2022). Spray-induced gene silencing (SIGS) presents an eco-friendly approach using synthetic dsRNA to suppress pathogenic factors and enhance abiotic stress resilience (Li and Liu 2019; Tiwari, and Lata 2021; Xie et al. 2015). Artificial microRNAs (amiRNAs) enable targeted gene silencing, improving drought tolerance in crops like potato and rice. Short tandem target mimics (STTMs) and virus-induced gene silencing (VIGS) serve as additional tools to modulate miRNA functions in response to abiotic stresses. RNA interference (RNAi) techniques and miPEPs, which enhance miRNA accumulation, further expand the toolkit for studying ncRNA-mediated stress responses (Lämke and Bäurle 2017; Ponting et al. 2009). Advances in sequencing and bioinformatics have identified many ncRNAs, but their roles in plant stress adaptation remain largely unexplored. While sRNA–lncRNA interactions in disease resistance are better understood, abiotic stress networks need further study. Future studies should focus on ncRNA roles in plant-–microbe interactions under stress conditions (Bhogireddy et al. 2021; Jacobs et al. 2015; Liu et al. 2017; Traubenik et al. 2024).

While ncRNA-based modifications can enhance stress tolerance, they often come at the cost of reduced growth and yield. Uncoupling stress resistance from growth inhibition is a critical challenge before widespread application. The use of stress-inducible promoters may mitigate negative effects associated with constitutive expression of defense genes. Additionally, controlled laboratory conditions do not fully replicate field environments, necessitating extensive multi-location trials to validate the stability and effectiveness of ncRNA-mediated stress tolerance. Another crucial avenue is ncRNA-mediated transgenerational stress memory, which could pave the way for developing crops with improved resilience and yield (Jampala et al. 2021; Ma et al. 2013; Roulé et al. 2022; Xu et al. 2019; Zhou et al. 2024). The interaction between ncRNA expression and epigenetic modifications is another promising research avenue. Future research should aim to decode the complex relationship between stress-induced epigenetic changes and ncRNA-mediated regulation. Integrating epigenetic profiling with transcriptomic and ncRNA datasets will offer a more holistic view of gene regulation under stress, paving the way for targeted epigenome editing strategies to enhance plant resilience (Cao et al. 2024; Huang and Jin 2022; Magar et al. 2024; Megha et al. 2015).

The development of synthetic ncRNAs that mimic or inhibit endogenous ncRNAs could further revolutionize plant stress management. These approaches, combined with traditional breeding techniques, could accelerate the development of crops capable of thriving in challenging environments (Dangelmaier and Lal 2021; Ding et al. 2019; Huo et al. 2021; Hüttenhofer and Vogel 2006). The vast and complex landscape of ncRNA regulatory networks necessitates a systems biology approach, integrating multi-omics data to build comprehensive models of ncRNA-mediated stress responses. Artificial intelligence (AI) and machine learning (ML) are poised to transform ncRNA research by accelerating the discovery and functional annotation of novel ncRNAs. AI-driven algorithms can analyze large datasets of ncRNA sequences and phenotypic traits to predict functions, identify stress-related ncRNA biomarkers, and uncover hidden regulatory circuits. Additionally, AI tools can predict interactions between ncRNAs and their target genes or proteins, facilitating the identification of key regulatory nodes in stress tolerance pathways (Achawanantakun et al. 2015; Chowdhary et al. 2021; Huang et al. 2024; Kunz et al. 2020).

As ncRNA research advances, the development of ncRNA-based biotechnological tools offers exciting possibilities for enhancing crop performance under stress. Small RNA-based gene silencing or activation systems could be tailored to regulate stress-responsive genes. Moreover, ncRNA-based biosensors could be designed to monitor plant stress responses in real time, providing a valuable tool for precision agriculture (Huang et al. 2024; Lott et al. 2020; Ravasi et al. 2006; Song and Zhang 2017). By integrating these innovative strategies, it is possible to engineer crops with superior resilience to environmental stressors, contributing to global food security in the face of climate change.

Key Future Directions

- Understanding ncRNA coordination in multiple stress responses
- Uncoupling stress resistance from growth inhibition
- Efficient identification of stress-responsive ncRNAs
- Investigating ncRNA roles in transgenerational stress memory
- Discovering optimal stress-inducible promoters

- Decoding hormone-ncRNA interactions in stress adaptation
- Accelerating functional lncRNA identification for crop improvement

By addressing these research questions, ncRNA studies can lead to transformative advancements in plant biotechnology, ultimately enhancing stress tolerance and agricultural productivity.

12.4 Conclusion

The future of ncRNA research in plant stress tolerance is teeming with possibilities. The convergence of high-throughput sequencing, genome-editing technologies, multi-omics integration, and AI-driven analyses will propel our understanding of ncRNA functions to new heights. By harnessing the regulatory potential of ncRNAs, we can develop novel strategies to enhance crop resilience, ensuring agricultural sustainability in an increasingly unpredictable climate. Moving forward, interdisciplinary collaboration and continued investment in ncRNA research will be key to unlocking groundbreaking solutions for stress-adapted crop development.

References

Abdulraheem MI, Xiong Y, Moshood AY, Cadenas-Pliego G, Zhang H, Hu J (2024) Mechanisms of plant epigenetic regulation in response to plant stress: recent discoveries and implications. Plants 13(2):163. https://doi.org/10.3390/PLANTS13020163

Achawanantakun R, Chen J, Sun Y, Zhang Y (2015) LncRNA-ID: long non-coding RNA IDentification using balanced random forests. Bioinformatics 31(24):3897–3905. https://doi.org/10.1093/BIOINFORMATICS/BTV480

Acosta-Motos JR, Ortuño MF, Bernal-Vicente A, Diaz-Vivancos P, Sanchez-Blanco MJ, Hernandez JA (2017) Plant responses to salt stress: adaptive mechanisms. Agronomy 7(1):18. https://doi.org/10.3390/AGRONOMY7010018

Adjeroh DA, Zhou X, Paschoal AR, Dimitrova N, Derevyanchuk EG, Shkurat TP, Loeb JA, Martinez I, Lipovich L (2024) Challenges in LncRNA biology: views and opinions. Non-Coding RNA 10(4):43. https://doi.org/10.3390/NCRNA10040043

Ali A, Altaf MT, Nadeem MA, Karaköy T, Shah AN, Azeem H, Baloch FS, Baran N, Hussain T, Duangpan S, Aasim M, Boo KH, Abdelsalam NR, Hasan ME, Chung YS (2022) Recent advancement in OMICS approaches to enhance abiotic stress tolerance in legumes. Front Plant Sci 13:952759. https://doi.org/10.3389/FPLS.2022.952759/PDF

Allmer J (2022) Noncoding RNA databases. Curr Pharm Biotechnol 24(7):825–831. https://doi.org/10.2174/1570180819666220520124511

Amaral PP, Clark MB, Gascoigne DK, Dinger ME, Mattick JS (2010) lncRNAdb: a reference database for long noncoding RNAs. Nucleic Acids Res 39(Database issue):D146. https://doi.org/10.1093/NAR/GKQ1138

Amaral PP, Dinger ME, Mattick JS (2013) Non-coding RNAs in homeostasis, disease and stress responses: an evolutionary perspective. Brief Funct Genomics 12(3):254–278. https://doi.org/10.1093/BFGP/ELT016

Atkinson NJ, Urwin PE (2012) The interaction of plant biotic and abiotic stresses: from genes to the field. J Exp Bot 63(10):3523–3544. https://doi.org/10.1093/JXB/ERS100

Baluska F, Souza GM, Kovalchuk I (2023) Role of epigenetic factors in response to stress and establishment of somatic memory of stress exposure in plants. Plants 12(21):3667. https://doi.org/10.3390/PLANTS12213667

Ben Rejeb I, Pastor V, Mauch-Mani B (2014) Plant responses to simultaneous biotic and abiotic stress: molecular mechanisms. Plants (Basel) 3(4):458–475. https://doi.org/10.3390/PLANTS3040458

Bhogireddy S, Mangrauthia SK, Kumar R, Pandey AK, Singh S, Jain A, Budak H, Varshney RK, Kudapa H (2021) Regulatory non-coding RNAs: a new frontier in regulation of plant biology. Funct Integr Genomics 21(3–4):313. https://doi.org/10.1007/S10142-021-00787-8

Cajal SR, Segura MF, Hümmer S (2019) Interplay between ncRNAs and cellular communication: A proposal for understanding cell-specific signaling pathways. Front Genet 10(APR):446580. https://doi.org/10.3389/FGENE.2019.00281/PDF

Cano-Gamez E, Trynka G (2020) From GWAS to function: using functional genomics to identify the mechanisms underlying complex diseases. Front Genet 11:505357. https://doi.org/10.3389/FGENE.2020.00424/PDF

Cao W, Yang L, Zhuang M, Lv H, Wang Y, Zhang Y, Ji J (2024) Plant non-coding RNAs: The new frontier for the regulation of plant development and adaptation to stress. Plant Physiol Biochem 208:108435. https://doi.org/10.1016/J.PLAPHY.2024.108435

Carthew RW, Sontheimer EJ (2009) Origins and mechanisms of miRNAs and siRNAs. Cell 136(4):642. https://doi.org/10.1016/J.CELL.2009.01.035

Chaudhary U, Banerjee S (2024) Decoding the non-coding: tools and databases unveiling the hidden world of "junk" RNAs for innovative therapeutic exploration. ACS Pharmacol Transl Sci 7(7):1901–1915. https://doi.org/10.1021/ACSPTSCI.3C00388/ASSET/IMAGES/MEDIUM/PT3C00388_0009.GIF

Chen L, Li MJ (2021) Editorial: deciphering non-coding regulatory variants: computational and functional validation. Front Bioeng Biotechnol 9. https://doi.org/10.3389/FBIOE.2021.769614

Chowdhary A, Satagopam V, Schneider R (2021) Long non-coding RNAs: mechanisms, experimental, and computational approaches in identification, characterization, and their biomarker potential in cancer. Front Genet 12:649619. https://doi.org/10.3389/FGENE.2021.649619/PDF

Dangelmaier EA, Lal A (2021) Experimental validation of the noncoding potential for lncRNAs. Methods Mol Biol 2348:221–230. https://doi.org/10.1007/978-1-0716-1581-2_15

Dayal S, Chaubey D, Joshi DC, Ranmale S, Pillai B (2024) Noncoding RNAs: emerging regulators of behavioral complexity. Wiley Interdiscip Rev RNA 15(3):e1847. https://doi.org/10.1002/WRNA.1847

Di C, Yuan J, Wu Y, Li J, Lin H, Hu L, Zhang T, Qi Y, Gerstein MB, Guo Y, Lu ZJ (2014) Characterization of stress-responsive lncRNAs in Arabidopsis thaliana by integrating expression, epigenetic and structural features. Plant J 80(5):848–861. https://doi.org/10.1111/TPJ.12679

Ding Y, Shi Y, Yang S (2019) Advances and challenges in uncovering cold tolerance regulatory mechanisms in plants. New Phytol 222(4):1690–1704. https://doi.org/10.1111/NPH.15696

Domínguez-Rosas E, Hernández-Oñate MÁ, Fernandez-Valverde SL, Tiznado-Hernández ME (2023) Plant long non-coding RNAs: identification and analysis to unveil their physiological functions. Front Plant Sci 14:1275399. https://doi.org/10.3389/FPLS.2023.1275399/PDF

Doolittle WF (2013) Is junk DNA bunk? A critique of ENCODE. Proc Natl Acad Sci 110(14):5294–5300. https://doi.org/10.1073/PNAS.1221376110

Dresselhaus T, Hückelhoven R (2018) Biotic and abiotic stress responses in crop plants. Agronomy 8(11):267. https://doi.org/10.3390/AGRONOMY8110267

Fritah S, Niclou SP, Azuaje F (2014) Databases for lncRNAs: a comparative evaluation of emerging tools. RNA 20(11):1655. https://doi.org/10.1261/RNA.044040.113

Fu XD (2014) Non-coding RNA: a new frontier in regulatory biology. Natl Sci Rev 1(2):190–204. https://doi.org/10.1093/NSR/NWU008

Gelaw TA, Sanan-Mishra N (2021) Non-coding RNAs in response to drought stress. Int J Mol Sci 22(22):12519. https://doi.org/10.3390/IJMS222212519

Giral H, Landmesser U, Kratzer A (2018) Into the wild: GWAS exploration of non-coding RNAs. Front Cardiovasc Med 5:412556. https://doi.org/10.3389/FCVM.2018.00181/PDF

Gong Y, Zhu W, Sun M, Shi L (2021) Bioinformatics analysis of long non-coding RNA and related diseases: an overview. Front Genet 12:813873. https://doi.org/10.3389/FGENE.2021.813873/BIBTEX

Graham N, Patil GB, Bubeck DM, Dobert RC, Glenn KC, Gutsche AT, Kumar S, Lindbo JA, Maas L, May GD, Vega-Sanchez ME, Stupar RM, Morrell PL (2020) Plant genome editing and the relevance of off-target changes. Plant Physiol 183(4):1453. https://doi.org/10.1104/PP.19.01194

Gudsnuk K, Champagne FA (2012) Epigenetic influence of stress and the social environment. ILAR J 53(3–4):279. https://doi.org/10.1093/ILAR.53.3-4.279

Gui Q, Yang Z, Chen C, Yang F, Wang S, Dong R (2022) Identification and characterization of long noncoding RNAs involved in the aluminum stress response in Medicago truncatula via genome-wide analysis. Front Plant Sci 13:1017869. https://doi.org/10.3389/FPLS.2022.1017869/BIBTEX

Hajieghrari B, Farrokhi N (2022) Plant RNA-mediated gene regulatory network. Genomics 114(1):409–442. https://doi.org/10.1016/J.YGENO.2021.12.020

He M, He CQ, Ding NZ (2018) Abiotic stresses: general defenses of land plants and chances for engineering multistress tolerance. Front Plant Sci 871:419713. https://doi.org/10.3389/FPLS.2018.01771/PDF

He X, Guo S, Wang Y, Wang L, Shu S, Sun J (2020) Systematic identification and analysis of heat-stress-responsive lncRNAs, circRNAs and miRNAs with associated co-expression and ceRNA networks in cucumber (Cucumis sativus L.). Physiol Plant 168(3):736–754. https://doi.org/10.1111/PPL.12997

Housman G, Ulitsky I (2016) Methods for distinguishing between protein-coding and long noncoding RNAs and the elusive biological purpose of translation of long noncoding RNAs. Biochim Biophys Acta 1859(1):31–40. https://doi.org/10.1016/J.BBAGRM.2015.07.017

Huang CY, Jin H (2022) Coordinated epigenetic regulation in plants: a potent managerial tool to conquer biotic stress. Front Plant Sci 12:795274. https://doi.org/10.3389/FPLS.2021.795274/TEXT

Huang W, Xiong T, Zhao Y, Heng J, Han G, Wang P, Zhao Z, Shi M, Li J, Wang J, Wu Y, Liu F, Xi JJ, Wang Y, Zhang QC (2024) Computational prediction and experimental validation identify functionally conserved lncRNAs from zebrafish to human. Nat Genet 56(1):124–135. https://doi.org/10.1038/s41588-023-01620-7

Huo C, Zhang B, Wang R (2021) Research progress on plant noncoding RNAs in response to low-temperature stress. Plant Signal Behav 17(1):2004035. https://doi.org/10.1080/15592324.2021.2004035

Hüttenhofer A, Vogel J (2006) Experimental approaches to identify non-coding RNAs. Nucleic Acids Res 34(2):635–646. https://doi.org/10.1093/NAR/GKJ469

Imaduwage IUH, Hewadikaram M (2024) Predicted roles of long non-coding RNAs in abiotic stress tolerance responses of plants. Mol Hortic 4(1):1–16. https://doi.org/10.1186/S43897-024-00094-3

Jacobs TB, LaFayette PR, Schmitz RJ, Parrott WA (2015) Targeted genome modifications in soybean with CRISPR/Cas9. BMC Biotechnol 15(1):16. https://doi.org/10.1186/S12896-015-0131-2

Jampala P, Garhewal A, Lodha M (2021) Functions of long non-coding RNA in Arabidopsis thaliana. Plant Signal Behav 16(9). https://doi.org/10.1080/15592324.2021.1925440

Jannesar M, Seyedi SM, Moazzam Jazi M, Niknam V, Ebrahimzadeh H, Botanga C (2020) A genome-wide identification, characterization and functional analysis of salt-related long non-coding RNAs in non-model plant Pistacia vera L. using transcriptome high throughput sequencing. Sci Rep 10(1):1–23. https://doi.org/10.1038/s41598-020-62108-6

Jiang H, Gao W, Jiang BL, Liu X, Jiang YT, Zhang LT, Zhang Y, Yan SN, Cao JJ, Lu J, Ma CX, Chang C, Zhang HP (2023) Identification and validation of coding and non-coding RNAs involved in high-temperature-mediated seed dormancy in common wheat. Front Plant Sci 14:1107277. https://doi.org/10.3389/FPLS.2023.1107277/BIBTEX

Jin X, Wang Z, Li X, Ai Q, Wong DCJ, Zhang F, Yang J, Zhang N, Si H (2023) Current perspectives of lncRNAs in abiotic and biotic stress tolerance in plants. Front Plant Sci 14:1334620. https://doi.org/10.3389/FPLS.2023.1334620/BIBTEX

Johnsson P, Lipovich L, Grandér D, Morris KV (2013) Evolutionary conservation of long non-coding RNAs; sequence, structure, function. Biochim Biophys Acta 1840(3):1063. https://doi.org/10.1016/J.BBAGEN.2013.10.035

Kaikkonen MU, Lam MTY, Glass CK (2011) Non-coding RNAs as regulators of gene expression and epigenetics. Cardiovasc Res 90(3):430. https://doi.org/10.1093/CVR/CVR097

Kęska K, Szcześniak MW, Adamus A, Czernicka M (2021) Waterlogging-stress-responsive LncRNAs, their regulatory relationships with miRNAs and target genes in cucumber (Cucumis sativus L.). Int J Mol Sci 22(15):8197. https://doi.org/10.3390/IJMS22158197/S1

Khalid R, Naveed H, Khalid Z (2023) Computational prediction of disease related lncRNAs using machine learning. Sci Rep 13(1):1–7. https://doi.org/10.1038/s41598-023-27680-7

Khraiwesh B, Zhu JK, Zhu J (2011) Role of miRNAs and siRNAs in biotic and abiotic stress responses of plants. Biochim Biophys Acta 1819(2):137. https://doi.org/10.1016/J.BBAGRM.2011.05.001

Kircher M, Ludwig KU (2022) Systematic assays and resources for the functional annotation of non-coding variants. Med Genet 34(4):275–286. https://doi.org/10.1515/MEDGEN-2022-2161/ASSET/GRAPHIC/J_MEDGEN-2022-2161_FIG_003.JPG

Kunz M, Wolf B, Fuchs M, Christoph J, Xiao K, Thum T, Atlan D, Prokosch HU, Dandekar T (2020) A comprehensive method protocol for annotation and integrated functional understanding of lncRNAs. Brief Bioinform 21(4):1391–1396. https://doi.org/10.1093/BIB/BBZ066

Lal MK, Kumar Tiwari R, Altaf MA, Kumar A, Kumar R (2023) Editorial: abiotic and biotic stress in horticultural crops: insight into recent advances in the underlying tolerance mechanism. https://doi.org/10.3389/fpls.2023.1212982

Lämke J, Bäurle I (2017) Epigenetic and chromatin-based mechanisms in environmental stress adaptation and stress memory in plants. Genome Biol 18(1):1–11. https://doi.org/10.1186/S13059-017-1263-6

Le NT, Harukawa Y, Miura S, Boer D, Kawabe A, Saze H (2020) Epigenetic regulation of spurious transcription initiation in Arabidopsis. Nat Commun 11(1):1–15. https://doi.org/10.1038/s41467-020-16951-w

Li J, Liu C (2019) Coding or noncoding, the converging concepts of RNAs. Front Genet 10(May):456150. https://doi.org/10.3389/FGENE.2019.00496/PDF

Li N, Wang Z, Wang B, Wang J, Xu R, Yang T, Huang S, Wang H, Yu Q (2022) Identification and characterization of long non-coding RNA in tomato roots under salt stress. Front Plant Sci 13:834027. https://doi.org/10.3389/FPLS.2022.834027/PDF

Liu D, Mewalal R, Hu R, Tuskan GA, Yang X (2017) New technologies accelerate the exploration of non-coding RNAs in horticultural plants. Hortic Res 4(1):1–8. https://doi.org/10.1038/hortres.2017.31

Lott SC, Voigt K, Lambrecht SJ, Hess WR, Steglich C (2020) A framework for the computational prediction and analysis of non-coding RNAs in microbial environmental populations and their experimental validation. ISME J 14(8):1955–1965. https://doi.org/10.1038/s41396-020-0658-7

Ma L, Bajic VB, Zhang Z (2013) On the classification of long non-coding RNAs. RNA Biol 10(6):924–933. https://doi.org/10.4161/RNA.24604

Ma X, Zhao F, Zhou B (2022) The characters of non-coding RNAs and their biological roles in plant development and abiotic stress response. Int J Mol Sci 23(8):4124. https://doi.org/10.3390/IJMS23084124

Magar ND, Shah P, Barbadikar KM, Bosamia TC, Madhav MS, Mangrauthia SK, Pandey MK, Sharma S, Shanker AK, Neeraja CN, Sundaram RM (2024) Long non-coding RNA-mediated epigenetic response for abiotic stress tolerance in plants. Plant Physiol Biochem 206:108165. https://doi.org/10.1016/J.PLAPHY.2023.108165

Mathews DH, Turner DH, Zuker M (2007) RNA secondary structure prediction. Curr Protoc Nucleic Acid Chem CHAPTER 11:Unit. https://doi.org/10.1002/0471142700.NC1102S28

Mattick JS, Makunin IV (2006) Non-coding RNA. Hum Mol Genet 15(suppl 1):R17–R29. https://doi.org/10.1093/HMG/DDL046

Mattick JS, Amaral PP, Carninci P, Carpenter S, Chang HY, Chen LL, Chen R, Dean C, Dinger ME, Fitzgerald KA, Gingeras TR, Guttman M, Hirose T, Huarte M, Johnson R, Kanduri C, Kapranov P, Lawrence JB, Lee JT et al (2023) Long non-coding RNAs: definitions, functions, challenges and recommendations. Nat Rev Mol Cell Biol 24(6):430–447. https://doi.org/10.1038/s41580-022-00566-8

Megha S, Basu U, Rahman MH, Kav NNV (2015) The role of long non-coding RNAs in abiotic stress tolerance in plants. In: Elucidation of Abiotic stress signaling in plants: functional genomics perspectives, vol 2. Springer, pp 93–106. https://doi.org/10.1007/978-1-4939-2540-7_4

Micheel J, Safrastyan A, Wollny D (2021) Advances in non-coding RNA sequencing. Non-Coding RNA 7(4):70. https://doi.org/10.3390/NCRNA7040070

Miguel V, Lamas S, Espinosa-Diez C (2020) Role of non-coding-RNAs in response to environmental stressors and consequences on human health. Redox Biol 37:101580. https://doi.org/10.1016/J.REDOX.2020.101580

Mirdar Mansuri R, Azizi AH, Sadri AH, Shobbar ZS (2022) Long non-coding RNAs as the regulatory hubs in rice response to salt stress. Sci Rep 12(1):1–14. https://doi.org/10.1038/s41598-022-26133-x

Morova T, Ding Y, Huang CCF, Sar F, Schwarz T, Giambartolomei C, Baca SC, Grishin D, Hach F, Gusev A, Freedman ML, Pasaniuc B, Lack NA (2023) Optimized high-throughput screening of non-coding variants identified from genome-wide association studies. Nucleic Acids Res 51(3):e18. https://doi.org/10.1093/NAR/GKAC1198

Naeem M, Majeed S, Hoque MZ, Ahmad I (2020) Latest developed strategies to minimize the off-target effects in CRISPR-Cas-mediated genome editing. Cells 9(7):1608. https://doi.org/10.3390/CELLS9071608

Nickerson JA, Momen-Heravi F (2024) Long non-coding RNAs: roles in cellular stress responses and epigenetic mechanisms regulating chromatin. Nucleus 15(1):2350180. https://doi.org/10.1080/19491034.2024.2350180

Ning B, Yu D, Yu AM (2019) Advances and challenges in studying noncoding RNA regulation of drug metabolism and development of RNA therapeutics. Biochem Pharmacol 169:113638. https://doi.org/10.1016/J.BCP.2019.113638

Noviello TMR, Ceccarelli F, Ceccarelli M, Cerulo L (2020) Deep learning predicts short non-coding RNA functions from only raw sequence data. PLoS Comput Biol 16(11):e1008415. https://doi.org/10.1371/JOURNAL.PCBI.1008415

Orro A, Trombetti GA (2023) High-accuracy ncRNA function prediction via deep learning using global and local sequence information. Biomedicine 11(6):1631. https://doi.org/10.3390/BIOMEDICINES11061631

Palos K, Nelson Dittrich AC, Yu L, Brock JR, Railey CE, Wu HYL, Sokolowska E, Skirycz A, Hsu PY, Gregory BD, Lyons E, Beilstein MA, Nelson ADL (2022) Identification and functional annotation of long intergenic non-coding RNAs in Brassicaceae. Plant Cell 34(9):3233–3260. https://doi.org/10.1093/PLCELL/KOAC166

Pandey P, Irulappan V, Bagavathiannan MV, Senthil-Kumar M (2017) Impact of combined abiotic and biotic stresses on plant growth and avenues for crop improvement by exploiting physio-morphological traits. Front Plant Sci 8:237767. https://doi.org/10.3389/FPLS.2017.00537/PDF

Patra GK, Gupta D, Rout GR, Panda SK (2023) Role of long non coding RNA in plants under abiotic and biotic stresses. Plant Physiol Biochem 194:96–110. https://doi.org/10.1016/J.PLAPHY.2022.10.030

Peña-Martínez EG, Rodríguez-Martínez JA (2024) Decoding non-coding variants: recent approaches to studying their role in gene regulation and human diseases. Front Biosci (Schol Ed) 16(1):4. https://doi.org/10.31083/J.FBS1601004/1945-0524-16-1-004/FIG3.PNG

Peschansky VJ, Wahlestedt C (2014) Non-coding RNAs as direct and indirect modulators of epigenetic regulation. Epigenetics 9(1):3–12. https://doi.org/10.4161/EPI.27473

Pesole G, Picardi E, Moro A, Tenenbaum SA, Vandevenne M, Delmarcelle M, Galleni M (2019) RNA regulatory networks as a control of stochasticity in biological systems. Front Genet 10(May):403. https://doi.org/10.3389/FGENE.2019.00403

Ponting CP, Oliver PL, Reik W (2009) Evolution and functions of long noncoding RNAs. Cell 136(4):629–641. https://doi.org/10.1016/J.CELL.2009.02.006

Ramakrishnan M, Zhang Z, Mullasseri S, Kalendar R, Ahmad Z, Sharma A, Liu G, Zhou M, Wei Q (2022) Epigenetic stress memory: a new approach to study cold and heat stress responses in plants. Front Plant Sci 13:1075279. https://doi.org/10.3389/FPLS.2022.1075279/PDF

Ravasi T, Suzuki H, Pang KC, Katayama S, Furuno M, Okunishi R, Fukuda S, Ru K, Frith MC, Gongora MM, Grimmond SM, Hume DA, Hayashizaki Y, Mattick JS (2006) Experimental validation of the regulated expression of large numbers of non-coding RNAs from the mouse genome. Genome Res 16(1):11. https://doi.org/10.1101/GR.4200206

Roulé T, Crespi M, Blein T (2022) Regulatory long non-coding RNAs in root growth and development. Biochem Soc Trans 50(1):403–412. https://doi.org/10.1042/BST20210743

Roychowdhury R, Das SP, Gupta A, Parihar P, Chandrasekhar K, Sarker U, Kumar A, Ramrao DP, Sudhakar C (2023) Multi-omics pipeline and omics-integration approach to decipher plant's abiotic stress tolerance responses. Genes 14(6):1281. https://doi.org/10.3390/GENES14061281

Sánchez Y, Huarte M (2013) Long non-coding RNAs: challenges for diagnosis and therapies. Nucleic Acid Ther 23(1):15. https://doi.org/10.1089/NAT.2012.0414

Sato K, Akiyama M, Sakakibara Y (2021) RNA secondary structure prediction using deep learning with thermodynamic integration. Nat Commun 12(1):1–9. https://doi.org/10.1038/s41467-021-21194-4

Schipper M, Posthuma D (2022) Demystifying non-coding GWAS variants: an overview of computational tools and methods. Hum Mol Genet 31(R1):R73–R83. https://doi.org/10.1093/HMG/DDAC198

Shanmugam A, Nagarajan A, Pramanayagam S (2017) Non-coding DNA-a brief review. J Appl Biol Biotechnol 5(05):42–47. https://doi.org/10.7324/JABB.2017.50507

Simna SP, Han Z (2022) Prospects of non-coding elements in genomic DNA based gene therapy. Curr Gene Ther 22(2):89. https://doi.org/10.2174/1566523221666210419090357

Singh A, Jain D, Pandey J, Yadav M, Bansal KC, Singh IK (2023) Deciphering the role of miRNA in reprogramming plant responses to drought stress. Crit Rev Biotechnol 43(4):613–627. https://doi.org/10.1080/07388551.2022.2047880

Song Y, Zhang D (2017) The role of long noncoding RNAs in plant stress tolerance. Methods Mol Biol 1631:41–68. https://doi.org/10.1007/978-1-4939-7136-7_3

Statello L, Guo CJ, Chen LL, Huarte M (2020) Gene regulation by long non-coding RNAs and its biological functions. Nat Rev Mol Cell Biol 22(2):96–118. https://doi.org/10.1038/s41580-020-00315-9

Sweeney BA, Petrov AI, Ribas CE, Finn RD, Bateman A, Szymanski M, Karlowski WM, Seemann SE, Gorodkin J, Cannone JJ, Gutell RR, Kay S, Marygold S, Dos Santos G, Frankish A, Mudge JM, Barshir R, Fishilevich S, Chan PP et al (2021) RNAcentral 2021: secondary structure integration, improved sequence search and new member databases. Nucleic Acids Res 49(D1):D212–D220. https://doi.org/10.1093/NAR/GKAA921

Szcześniak MW, Kubiak MR, Wanowska E, Makalowska I (2021) Comparative genomics in the search for conserved long noncoding RNAs. Essays Biochem 65(4):741. https://doi.org/10.1042/EBC20200069

Tan J, Li X, Zhang L, Du Z (2022) Recent advances in machine learning methods for predicting LncRNA and disease associations. Front Cell Infect Microbiol 12:1071972. https://doi.org/10.3389/FCIMB.2022.1071972/BIBTEX

Tao X, Li S, Chen G, Wang J, Xu S (2023) Approaches for modes of action study of long non-coding RNAs: from single verification to genome-wide determination. Int J Mol Sci 24(6):5562. https://doi.org/10.3390/IJMS24065562

Thiebaut F, Hemerly AS, Ferreira PCG (2019) A role for epigenetic regulation in the adaptation and stress responses of non-model plants. Front Plant Sci 10:413504. https://doi.org/10.3389/FPLS.2019.00246/BIBTEX

Tibbs Cortes L, Zhang Z, Yu J (2021) Status and prospects of genome-wide association studies in plants. Plant Genome 14(1):e20077. https://doi.org/10.1002/TPG2.20077

Tiwari S, Lata C (2021) Recent advancements in long noncoding RNA-mediated stress responses in rice. In: Long noncoding RNAs in plants: roles in development and stress, pp 63–74. https://doi.org/10.1016/B978-0-12-821452-7.00005-2

Traubenik S, Charon C, Blein T (2024) From environmental responses to adaptation: the roles of plant lncRNAs. Plant Physiol 195(1):232–244. https://doi.org/10.1093/PLPHYS/KIAE034

Uffelmann E, Huang QQ, Munung NS, de Vries J, Okada Y, Martin AR, Martin HC, Lappalainen T, Posthuma D (2021) Genome-wide association studies. Nat Rev Methods Primers 1(1):1–21. https://doi.org/10.1038/s43586-021-00056-9

Uffelmann E, Posthuma D, Peyrot WJ (2023) Genome-wide association studies of polygenic risk score-derived phenotypes may lead to inflated false positive rates. Sci Rep 13(1):1–4. https://doi.org/10.1038/s41598-023-29428-9

Urquiaga MCDO, Thiebaut F, Hemerly AS, Ferreira PCG (2021) From trash to luxury: the potential role of plant LncRNA in DNA methylation during abiotic stress. Front Plant Sci 11:603246. https://doi.org/10.3389/FPLS.2020.603246

Vitsios D, Dhindsa RS, Middleton L, Gussow AB, Petrovski S (2021) Prioritizing non-coding regions based on human genomic constraint and sequence context with deep learning. Nat Commun 12(1):1–14. https://doi.org/10.1038/s41467-021-21790-4

Waititu JK, Zhang C, Liu J, Wang H (2020) Plant non-coding RNAs: origin, biogenesis, mode of action and their roles in abiotic stress. Int J Mol Sci 21(21):8401. https://doi.org/10.3390/IJMS21218401

Wang J, Meng X, Dobrovolskaya OB, Orlov YL, Chen M (2017) Non-coding RNAs and their roles in stress response in plants. Genomics Proteomics Bioinformatics 15(5):301. https://doi.org/10.1016/J.GPB.2017.01.007

Wang X, Fan H, Wang B, Yuan F (2023) Research progress on the roles of lncRNAs in plant development and stress responses. Front Plant Sci 14:1138901. https://doi.org/10.3389/FPLS.2023.1138901/BIBTEX

Wei JW, Huang K, Yang C, Kang CS (2017) Non-coding RNAs as regulators in epigenetics (Review). Oncol Rep 37(1):3–9. https://doi.org/10.3892/OR.2016.5236

Winkle M, El-Daly SM, Fabbri M, Calin GA (2021) Noncoding RNA therapeutics – challenges and potential solutions. Nat Rev Drug Discov 20(8):629–651. https://doi.org/10.1038/s41573-021-00219-z

Wu L, Liu S, Qi H, Cai H, Xu M (2020) Research progress on plant long non-coding RNA. Plants 9(4):408. https://doi.org/10.3390/PLANTS9040408

Wu YZ, Su YH, Kuo CY (2022) Stressing the regulatory role of long non-coding RNA in the cellular stress response during cancer progression and therapy. Biomedicine 10(5):1212. https://doi.org/10.3390/BIOMEDICINES10051212

Xie F, Wang Q, Sun R, Zhang B (2015) Deep sequencing reveals important roles of microRNAs in response to drought and salinity stress in cotton. J Exp Bot 66(3):789–804. https://doi.org/10.1093/JXB/ERU437

Xu J, Hou QM, Khare T, Verma SK, Kumar V (2019) Exploring miRNAs for developing climate-resilient crops: a perspective review. Sci Total Environ 653:91–104. https://doi.org/10.1016/J.SCITOTENV.2018.10.340

Xu D, Yuan W, Fan C, Liu B, Lu MZ, Zhang J (2022) Opportunities and challenges of predictive approaches for the non-coding RNA in plants. Front Plant Sci 13:890663. https://doi.org/10.3389/FPLS.2022.890663/FULL

Yadav A, Mathan J, Dubey AK, Singh A (2024) The emerging role of non-coding RNAs (ncRNAs) in plant growth, development, and stress response signaling. Non-Coding RNA 10(1):13. https://doi.org/10.3390/NCRNA10010013

Yang Y, Saand MA, Huang L, Abdelaal WB, Zhang J, Wu Y, Li J, Sirohi MH, Wang F (2021) Applications of multi-omics technologies for crop improvement. Front Plant Sci 12:563953. https://doi.org/10.3389/FPLS.2021.563953/PDF

Yang H, Cui Y, Feng Y, Hu Y, Liu L, Duan L (2023) Long non-coding RNAs of plants in response to abiotic stresses and their regulating roles in promoting environmental adaption. Cells 12(5):729. https://doi.org/10.3390/CELLS12050729/S1

Yu Y, Zhang Y, Chen X, Chen Y (2019) Plant noncoding RNAs: hidden players in development and stress responses. Annu Rev Cell Dev Biol 35:407–431. https://doi.org/10.1146/ANNUREV-CELLBIO-100818-125218/CITE/REFWORKS

Zhang Y, Tao Y, Liao Q (2018) Long noncoding RNA: a crosslink in biological regulatory network. Brief Bioinform 19(5):930–945. https://doi.org/10.1093/BIB/BBX042

Zhang Y, Zhou Y, Zhu W, Liu J, Cheng F (2022) Non-coding RNAs fine-tune the balance between plant growth and abiotic stress tolerance. Front Plant Sci 13:965745. https://doi.org/10.3389/FPLS.2022.965745/FULL

Zhang Y, Xu J, Li R, Ge Y, Li Y, Li R (2023) Plants' response to abiotic stress: mechanisms and strategies. Int J Mol Sci 24(13):10915. https://doi.org/10.3390/IJMS241310915

Zhao Y, Teng H, Yao F, Yap S, Sun Y, Ma L (2020) Challenges and Strategies in Ascribing Functions to Long Noncoding RNAs. Cancer 12(6):1458. https://doi.org/10.3390/CANCERS12061458

Zhao L, Wang J, Li Y, Song T, Wu Y, Fang S, Bu D, Li H, Sun L, Pei D, Zheng Y, Huang J, Xu M, Chen R, Zhao Y, He S (2021) NONCODEV6: an updated database dedicated to long non-coding RNA annotation in both animals and plants. Nucleic Acids Res 49(D1):D165–D171. https://doi.org/10.1093/NAR/GKAA1046

Zhao Z, Yang Y, Iqbal A, Wu Q, Zhou L (2024) Biological insights and recent advances in plant long non-coding RNA. Int J Mol Sci 25(22):11964. https://doi.org/10.3390/IJMS252211964

Zhong W, Liu W, Chen J, Sun Q, Hu M, Li Y (2022) Understanding the function of regulatory DNA interactions in the interpretation of non-coding GWAS variants. Front Cell Dev Biol 10:957292. https://doi.org/10.3389/FCELL.2022.957292/PDF

Zhou Q, Jiang Y, Cai C, Li W, Leow MKS, Yang Y, Liu J, Xu D, Sun L (2023) Multidimensional conservation analysis decodes the expression of conserved long noncoding RNAs. Life Sci Alliance 6(6):10.26508/LSA.202302002

Zhou B, Ji B, Shen C, Zhang X, Yu X, Huang P, Yu R, Zhang H, Dou X, Chen Q, Zeng Q, Wang X, Cao Z, Hu G, Xu S, Zhao H, Yang Y, Zhou Y, Wang J (2024) EVLncRNAs 3.0: an updated comprehensive database for manually curated functional long non-coding RNAs validated by low-throughput experiments. Nucleic Acids Res 52(D1):D98–D106. https://doi.org/10.1093/NAR/GKAD1057

Zhu JK (2016) Abiotic stress signaling and responses in plants. Cell 167(2):313. https://doi.org/10.1016/J.CELL.2016.08.029

The manufacturer's authorised representative in the EU is Springer Nature Customer Service Centre GmbH, Europaplatz 3, 69115 Heidelberg, Germany. If you have any concerns regarding our products, please contact ProductSafety@springernature.com

Printed and bound by CPI Group (UK) Ltd, Croydon, CR0 4YY
01/10/2026
02225913-0013